R&D MANAGEMENT AND
CORPORATE FINANCIAL POLICY

ETM WILEY SERIES IN ENGINEERING & TECHNOLOGY MANAGEMENT

Series Editor: Dundar F. Kocaoglu, Portland State University

Badiru / PROJECT MANAGEMENT IN MANUFACTURING AND HIGH TECHNOLOGY OPERATIONS, Second Edition

Badiru / INDUSTRY'S GUIDE TO ISO 9000

Baird / MANAGERIAL DECISIONS UNDER UNCERTAINTY: AN INTRODUCTION TO THE ANALYSIS OF DECISION MAKING

Edosomwan / INTEGRATING INNOVATION AND TECHNOLOGY MANAGEMENT

Eisner / ESSENTIALS OF PROJECT AND SYSTEMS ENGINEERING MANAGEMENT

Eschenbach / CASES IN ENGINEERING ECONOMY

Gerwin and Kolodny / MANAGEMENT OF ADVANCED MANUFACTURING TECHNOLOGY: STRATEGY, ORGANIZATION, AND INNOVATION

Gönen / ENGINEERING ECONOMY FOR ENGINEERING MANAGERS

Guerard, Bean, and Stone / R&D MANAGEMENT AND CORPORATE FINANCIAL POLICY

Jain and Triandis / MANAGEMENT OF RESEARCH AND DEVELOPMENT ORGANIZATIONS: MANAGING THE UNMANAGEABLE, Second Edition

Lang and Merino / THE SELECTION PROCESS FOR CAPITAL PROJECTS

Martin / MANAGING INNOVATION AND ENTREPRENEURSHIP IN TECHNOLOGY-BASED FIRMS

Martino / RESEARCH AND DEVELOPMENT PROJECT SELECTION

Messina / STATISTICAL QUALITY CONTROL FOR MANUFACTURING MANAGERS

Morton and Pentico / HEURISTIC SCHEDULING SYSTEMS: WITH APPLICATION TO PRODUCTION SYSTEMS AND PROJECT MANAGEMENT

Niwa / KNOWLEDGE BASED RISK MANAGEMENT IN ENGINEERING: A CASE STUDY IN HUMAN COMPUTER COOPERATIVE SYSTEMS

Porter et al. / FORECASTING AND MANAGEMENT OF TECHNOLOGY

Riggs / FINANCIAL AND COST ANALYSIS FOR ENGINEERING AND TECHNOLOGY MANAGEMENT

Rubenstein / MANAGING TECHNOLOGY IN THE DECENTRALIZED FIRM

Sankar / MANAGEMENT OF INNOVATION AND CHANGE

Streeter / PROFESSIONAL LIABILITY OF ARCHITECTS AND ENGINEERS

Thamhain / ENGINEERING MANAGEMENT: MANAGING EFFECTIVELY IN TECHNOLOGY-BASED ORGANIZATIONS

R&D MANAGEMENT AND CORPORATE FINANCIAL POLICY

JOHN B. GUERARD, JR.
Director of Quantitative Research
Vantage Global Advisors
New York, New York

ALDEN S. BEAN
William R. Keenan, Jr. Professor
of Management & Technology
Lehigh University
Bethlehem, Pennsylvania

A WILEY-INTERSCIENCE PUBLICATION
JOHN WILEY & SONS, INC.
NEW YORK   CHICHESTER   WEINHEIM   BRISBANE   SINGAPORE   TORONTO

To Julie, Richard, Catherine, and Stephanie for their love and support.
—John Guerard

This book is dedicated with love to Diane, Scott, David, Andrew, Matthew and Timothy—but especially to Diane, my wife of 37 years, for her enduring support and patience.
—Alden Bean

This book is printed on acid-free paper. ∞

Copyright © 1998 by John Wiley & Sons, Inc. All rights reserved.

Published simultaneously in Canada.

No part of this publication may be reproduced, stored in a retrieval system or transmitted in any form or by any means, electronic, mechanical, photocopying, recording, scanning or otherwise, except as permitted under Sections 107 or 108 of the 1976 United States Copyright Act, without either the prior written permission of the Publisher, or authorization through payment of the appropriate per-copy fee to the Copyright Clearance Center, 222 Rosewood Drive, Danvers, MA 01923, (508) 750-8400, fax (508) 750-4744. Requests to the Publisher for permission should be addressed to the Permissions Department, John Wiley & Sons, Inc., 605 Third Avenue, New York, NY 10158-0012, (212) 850-6011, fax (212) 850-6008, E-Mail: PERMREQ @ WILEY.COM.

This publication is designed to provide accurate and authoritative information in regard to the subject matter covered. It is sold with the understanding that the publisher is not engaged in rendering professional services. If professional advice or other expert assistance is required, the services of a competent professional person should be sought.

Library of Congress Cataloging-in-Publication Data:

Guerard, John.
    R & D management and corporate financial policy / John B. Guerard, Alden S. Bean.
        p.   cm. – (Wiley series in engineering and technology management)
    Includes index.
    ISBN 0-471-61837-3 (cloth : alk. Paper)
    1. Research, Industrial–Finance–Management. 2. Corporations–Finance. I. Bean, Alden S. II. Title. III. Series: Wiley series in engineering and technology management.
HD30.4.G83   1997
658.5'7—dc21   97-12831

Printed in the United States of America.

10 9 8 7 6 5 4 3 2 1

CONTENTS

Preface		vii
1	R&D Management and Corporate Financial Policy	1
2	R&D and Technological Innovation in U.S. Industry: Are Firms Really Getting More for Less?	5
3	Meet Your Competition: The IRI/CIMS R&D Survey for FY 1995	19
4	The Interdependencies Among Corporate Financial Policies: Empirical Evidence Concerning the United States, Japan, and Major European Economies	37
5	Comparing Census/NSF R&D Data with Compustat R&D Data	95
6	More on the Interdependencies Among Corporate Financial Policies: The Case of Effective Debt	123
7	Historical Data and Analysts' Forecasts in the Creation of Efficient Portfolios	139

8	Estimation of Efficient Market-Neutral Japanese and U.S. Portfolios	205
9	The (Not So Special) Case of Social Investing	227
10	Summary and Conclusions	249

Index 253

PREFACE

The purpose of this book is to analyze the determinants of corporate research and development (R&D) expenditures in the United States during the 1975–1995 period, and of the other major industrialized, G7, countries (United Kingdom, France, Germany, Canada, Italy, and Japan) during the 1982–1995 period. We find that there are several similarities among the determinants of corporate R&D expenditures in the United States and the other G7 nations, which we note in Chapter 4. In Chapter 5 we find that stock prices are associated with R&D in the United States, and in Chapter 9 we show that R&D expenditures may be positively regarded in socially responsible investment and enhance stock selection capability.

This book reflects approximately 12 years of research on the R&D issue. Our research began with a study of the interactions between the R&D, capital investment, dividend, and new debt financing decisions of major industrial corporations. We found significant interdependencies, such that one must use a simultaneous equations model to adequately analyze a firm's financial decision-making process. Even the presence of federal financing of R&D was found to be insufficient to completely eliminate a firm's potentially binding budget constraint. The initial research used a 303-firm universe subset of Compustat data for the 1975–1982 period provided by Air Products and Chemicals, Inc. Lee Gaumer and James Sykes supported our initial research at Air Products. A corporate planning model was developed and estimated by the authors. We found significant correlations between stock prices and our targeted variables. In preparing the final draft of this book, we expanded our period of study in the United States to include the 3,000 largest U.S. firms for the 1978–1995 period and extended our modeling into Japan and Europe, finding interesting results. The expanded U.S. analysis supported our earlier results, and the European and Japanese results were generally supportive of the U.S. estimations, although we continue to refine and expand our G7 analysis. We have updated our analysis of federal aggregate activity through 1995, but, regrettably, could not update Chapter 5 on federal support of R&D at the firm (microeconomic) level.

The expansion of our analysis has resulted in five research publications, which are noted in Chapters 4, 7, 8, and 9. We would like to acknowledge the collaboration with several of our coauthors—Bernell Stone, John Blin, Steve Bender, and Mustafa Gultekin—on papers that led to the final forms of Chapters 4, 7, and 8. Their support and friendship is much appreciated. Greg Adams at Brigham Young University ran the estimated simultaneous equations of Chapters 4 and 6, and the IBES and value data regressions of Chapter 7 (which Guerard also ran at Vantage Global Advisors). We are delighted that John Wiley & Sons was so patient with us, as we believe our finished product is substantially better than that we initially proposed in the late 1980s. An earlier version of Chapter 9 won the first annual Moskowitz Award for the best research paper on socially responsible investing. We appreciate the support of I/B/E/S and WorldScope for providing us with academic databases to perform our research. We especially wish to acknowledge Ed Keon, Florence Eng, and Robert Lemmond for their assistance. Although all databases have their limitations, we believe that consistent and superior models can be developed with the domestic and international I/B/E/S and WorldScope databases.

1
R&D MANAGEMENT AND CORPORATE FINANCIAL POLICY

The purpose of this book is to analyze the determinants of corporate research and development (R&D) expenditures in the United States during the 1975–1995 period and of the other major industrialized countries (United Kingdom, France, Germany, Canada, Italy, and Japan) during the 1982–1995 period. Our research began with a study of the interactions between the R&D, capital investment, dividend, and new debt financing decisions of major industrial corporations. We found significant interdependencies, such that one must use a simultaneous equations model to adequately analyze a firm's financial decision-making process. Even the presence of federal financing of R&D was insufficient to completely eliminate the potentially binding budget constraints on firms. The initial research used a 303-firm universe of Compustat data for the 1975–1982 period. A corporate planning model was developed and estimated by the authors. We found significant correlations between stock prices and our targeted variables. In preparing the final draft of this book, we expanded our period of study for the 3,000 largest firms in the United States to include the period from 1978–1995 and extended our modeling into Japan and Europe, finding interesting results. The expanded U.S. analysis supported our earlier results, and the European and Japanese results were generally supportive of the U.S. estimations, although we continue to refine and expand our G7 analysis.

Among our goals was to develop an econometric model to analyze the interdependencies of decisions in regard to research and development, investment, dividends, and new debt financing. The strategic decision makers of a firm seek to allocate resources in accordance with a set of seemingly incompatible objectives. Management attempts to manage dividends, capital expenditures, and R&D activities while minimizing reliance on external funding to generate future profits.

Each firm has a "pool" of resources, composed of net income, depreciation, and new debt issues, and this pool is reduced by dividend payments, investment in capital projects, and expenditures for research and development activities. Miller and Modigliani (1961) put forth the *perfect markets hypothesis,* in regard to financial decisions, which holds that dividends are not influenced (limited) by investment decisions. There are no interdependencies between financial decisions in a perfect markets environment, except that new debt is issued to finance R&D, dividends, and investment.

The *imperfect markets hypothesis* concerning financial decisions holds that financial decisions are interdependent and that simultaneous equations must be used to efficiently estimate the equations. The interdependence hypothesis reflects the simultaneous equation financial decision modeling work of Dhrymes and Kurz (1967), Mueller (1967), Damon and Schramm (1972), McCabe (1979), Peterson and Benesh (1983), Jalilv and Harris (1984), Switzer (1984), and Guerard and McCabe (1992). Higgins (1972), Fama (1974), and McDonald, Jacquillat, and Nussenbaum (1975) found little evidence of significant interdependencies between financial decisions.

In Chapter 2 we survey current corporate R&D funding practices and, in Chapter 3, discuss how firms may meet the R&D challenge of competitors.

The estimation of simultaneous equations for financial decision making is the primary modeling effort of Chapters 4, 5, and 6. In Chapter 4, we estimate a set of simultaneous equations for the largest 3,000 securities in the United States during the 1978–1995 period and compare these results with those using the 303-firm sample described in Guerard and McCabe (1992). We also use an international database and extend our work to Europe and Japan. In Chapter 5, we review the federal financing impact on financial decisions during the 1975–1982 period. Recent restructuring has greatly changed the way many corporate officers think of new debt issuance, and we reexamine the problem of effective debt management in Chapter 6.

Security valuation and portfolio construction is a major issue and is developed in Chapters 7, 8, and 9. Chapter 7 presents our valuation analysis, using historical fundamental data from Compustat and earnings forecasts from I/B/E/S. We find statistically significant stock selection models in the United States, Europe, and Japan, using both historical and earnings forecasting data that violate the efficient markets hypothesis, which holds that securities are equilibriumly priced. Chapter 8 extends the basic portfolio strategies discussed in Chapter 7 to include market-neutral portfolios, and we find a much greater use of earnings forecasts in the United States. Socially responsible investing is examined in Chapter 9, and we find no difference between socially screened and socially unscreened portfolios. One can be socially responsible and produce efficient portfolios. Recent research by Guerard (1997) has found that R&D is significantly associated with a social strength (product) criteria. It may be possible for management to increase its R&D activities, be recognized as a "better" firm in the socially responsible investment community, and see its stock price rise. A brief summary and set of conclusions are presented in Chapter 10.

REFERENCES

Ben-Zion, U. "The R&D and Investment Decision and Its Relationship to the Firm's Market Value: Some Preliminary Results." In *R&D, Patents, and Productivity,* edited by Z. Griliches. Chicago: University of Chicago Press, 1984.

Damon, W. W., and R. Schramm. "A Simultaneous Decision Model for Production, Marketing and Finance." *Management Science* 18 (1972):161–172.

Dhrymes, P. J. *Econometrics: Statistical Foundations and Applications.* New York: Springer-Verlag, 1974.

Dhrymes, P. J., and M. Kurz. "Investment, Dividends, and External Finance Behavior of Firms." In *Determinants of Investment Behavior,* New York: edited by Robert Ferber. Columbia University Press, 1967.

Dhrymes, P. J., and M. Kurz. "On the Dividend Policy of Electric Utilities." *Review of Economics and Statistics* 46 (1964):76–81.

Fama, E. F. "The Empirical Relationship Between the Dividend and Investment Decisions of Firms." *American Economic Review* 63 (1974):304–318.

Guerard, J. B., Jr. "Additional Evidence on the Cost of Being Socially Responsible in Investing." *Journal of Investing* (in press).

Guerard, J. B., and G. M. McCabe. "The Integration of Research and Development Management into the Firm Decision Process." In *Management of R&D and Engineering,* edited by D. F. Kocaoglu. Amsterdam: North-Holland, 1992.

Hambrick, D. C., I. C. MacMillan and R. R. Barbosa. "Changes in Product R&D Budgets." *Management Science* 29 (1983):757–769.

Higgins, R. C. "The Corporate Dividend-Saving Decision." *Journal of Financial and Quantitative Analysis* 7 (1972):1527–1541.

Jalilvand, A., and R. S. Harris. "Corporate Behavior in Adjusting to Capital Structure and Dividend Targets: An Econometric Study." *Journal of Finance* 39 (1984):127–145.

Lintner, J. "Distributions of Incomes of Corporations Among Dividends, Retained Earnings and Taxes." *American Economic Review* 46 (1979):119–135.

Mansfield, E. "Size of Firm, Market Structure and Innovation." *Journal of Political Economy* 71 (1963):556–576.

McCabe, G. M. "The Empirical Relationship Between Investment and Financing: A New Look." *Journal of Financial and Quantitative Analysis* 14 (1979):119–135.

McDonald, J. G., B. Jacquillat, and M. Nussenbaum. "Dividend, Investment, and Financial Decisions: Empirical Evidence on French Firms." *Journal of Financial and Quantitative Analysis* 10 (1975):741–755.

Meyer, J. R., and E. Kuh. *The Investment Decision.* Cambridge: Harvard University Press, 1957.

Miller, M., and F. Modigliani, "Dividend Policy, Growth, and the Valuation of Shares." *Journal of Business* 34 (1961):411–433.

Mueller, D. C. "The Firm Decision Process: An Econometric Investigation." *Quarterly Journal of Economics* 81 (1967):58–87.

Peterson, P., and G. Benesh. "A Reexamination of the Empirical Relationship Between Investment and Financial Decisions." *Journal of Financial and Quantitative Analysis* 18 (1983):439–454.

Scherer, F. M. "Firm Size, Market Structure, Opportunity, and the Output of Patented Inventions." *American Economic Review* 55 (1965):1104–1113.

Switzer, L. "The Determinants of Industrial R&D: A Funds Flow Simultaneous Equation Approach." *Review of Economics and Statistics* 66 (1984):163–168.

2
R&D AND TECHNOLOGICAL INNOVATION IN U.S. INDUSTRY: ARE FIRMS REALLY GETTING MORE FOR LESS?

INTRODUCTION

In constant (1987) dollar terms, U.S. national expenditures for R&D have been declining since the early 1990s. The trend began with a decline in federal funding in 1989, followed by reduced industrial spending starting in 1991. Both trends continued through 1995, according to the National Science Foundation.[1] Since R&D expenditures are such an important element of the technological innovation process, their decline, particularly in industry, may indicate that executives in U.S. corporations are less optimistic than they have been about future returns to R&D and innovation. Alternately, industrial executives may believe that they have found ways of achieving desired levels of innovative outcomes more efficiently, requiring lower levels of R&D inputs.

This chapter explores this issue by reviewing the current state of R&D and innovation in U.S. industry and considers findings from several surveys of industrial R&D activities that may shed light on the question of declining returns versus increasing efficiency. We begin by reviewing National Science Foundation (NSF) data on trends in the funding and performance of R&D.

This chapter is based on a keynote address by Alden S. Bean, given at the Nineteenth Symposium on the Management of Technology in São Paulo, Brazil, October 22, 1996.

NATIONAL R&D FUNDING SOURCES AND PERFORMANCE PATTERNS IN THE UNITED STATES

As the cold war ended, U.S. citizens and public officials began to talk about a "peace dividend," a payback of some sort for the enormous cost of the defense and military infrastructure needed to wage the war. Along with cost savings realized through reductions in armed forces personnel and the closing and consolidation of military bases, came cutbacks in both energy and defense R&D expenditures. Department of Defense (DoD) expenditures in 1995 accounted for 50% of all federal R&D expenditures, down from 65% in 1986. As indicated in Figure 2.1, real (constant dollar) federal expenditures for all R&D reached a peak in 1987 and have declined steadily; industry spending on R&D peaked in 1991 and has remained constant since then. Other organizations, such as universities, increased their own internally funded R&D expenditures modestly during this period, but not enough to offset the loss of federal R&D funds.

Funding Patterns

The composition of the R&D activities is of interest. Specific industrial innovations that have economic impacts almost always involve some very focused design and development work, even though their operating principles may be based on fundamental knowledge generated and refined by basic and applied research of a more general nature. Thus, it is instructive to see whether the reductions in R&D funding described earlier have been evenly distributed across three categories of R&D work (shown in Figure 2.1).

Figure 2.2 shows that basic research funding in the United States increased steadily in terms of constant (1987) dollars between 1972 and 1993, when it reached a peak of $23,464 million. The amount declined to $23,220 million in 1994, and the NSF estimated that it would continue to decline in 1995. Applied research (see Figure 2.3) peaked in 1991 at almost $33 billion and has drawn back to about $31 billion since then. As indicated by Figure 2.4, development, which is most directly associated with industrial innovation, reached a peak of $83,453 million in 1990, then declined to $80,024 million by 1994 and to an estimated $78,470 million in 1995.

Thus, according to NSF data, constant dollar declines have occurred in funding for all three categories of R&D, although the timing of the peaks differ. Revised 1995 data released by the NSF in October, 1996,[2] suggest that the actual 1995 expenditures were higher than the original estimates, yielding a 4% inflation-adjusted overall increase in 1995 over 1994. This report also indicates that real 1996 expenditures were expected to be about

(Text continues on page 11.)

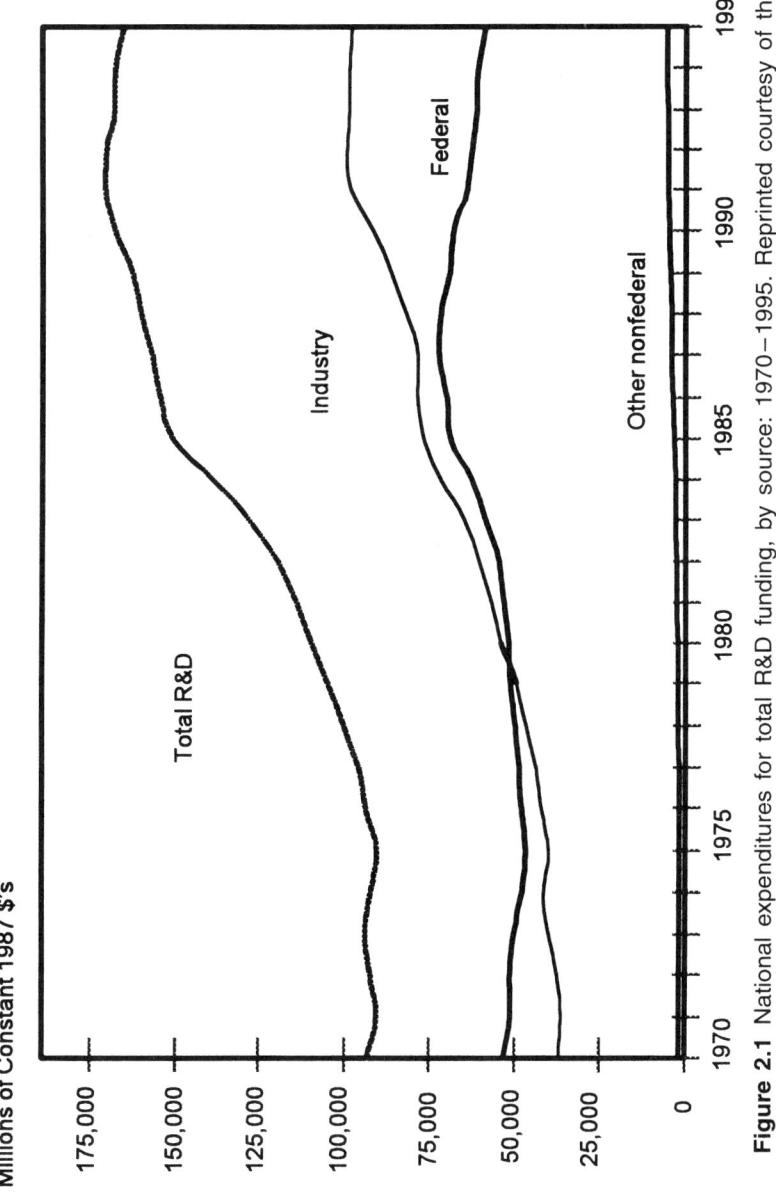

Figure 2.1 National expenditures for total R&D funding, by source: 1970–1995. Reprinted courtesy of the National Science Foundation.

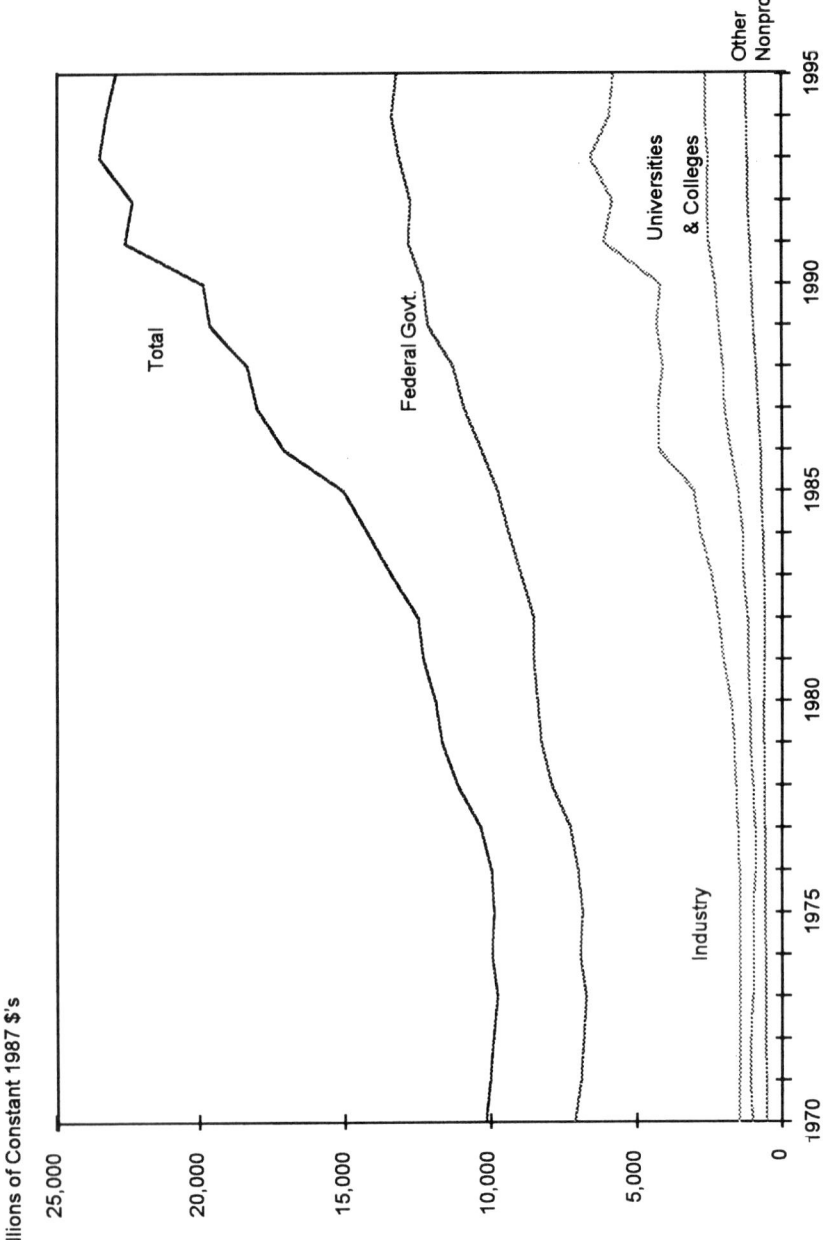

Figure 2.2 National R&D expenditures for basic research, by source of funds: 1970–1995. Reprinted courtesy of the National Science Foundation.

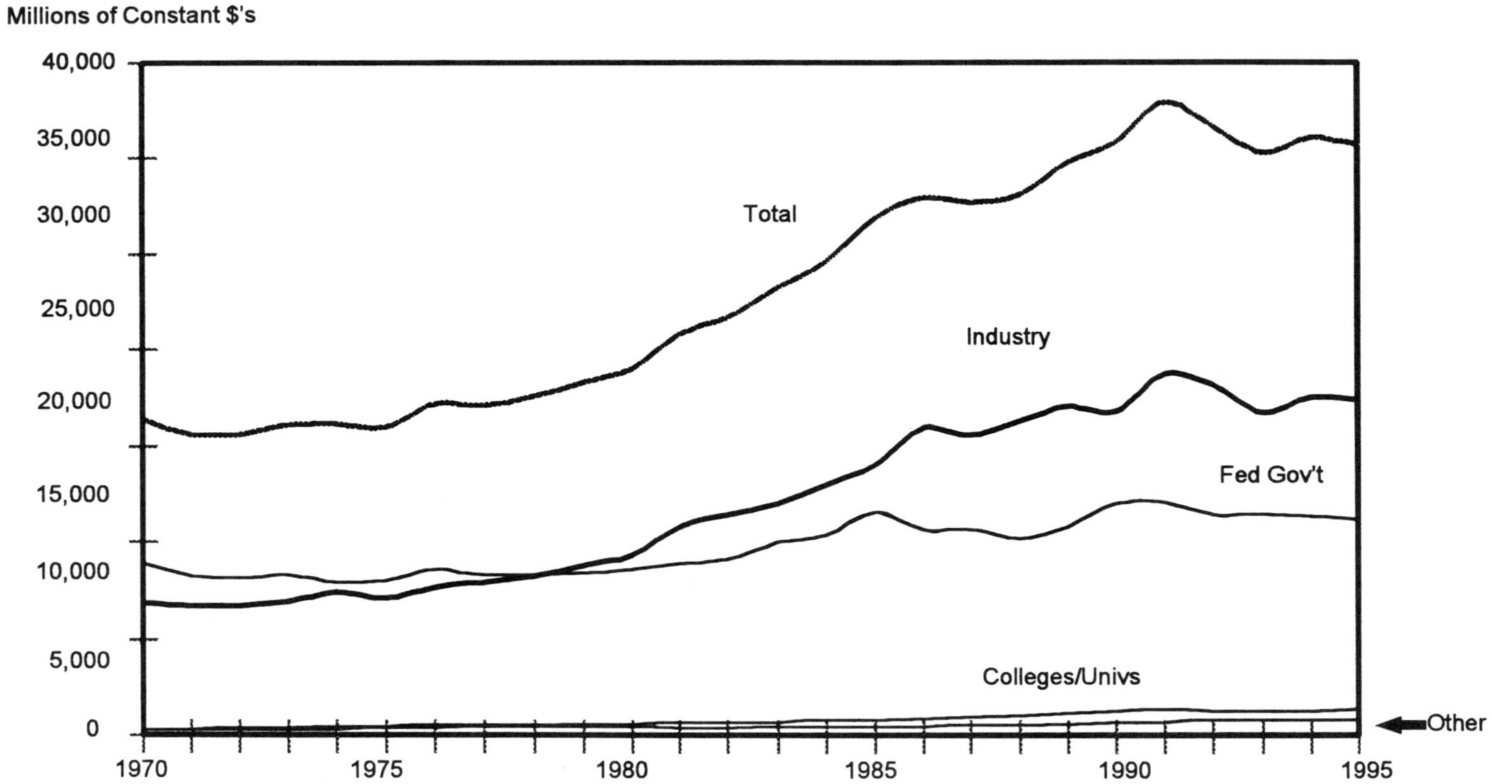

Figure 2.3 National expenditures for applied research, by source of funds: 1970–1995. Reprinted courtesy of the National Science Foundation.

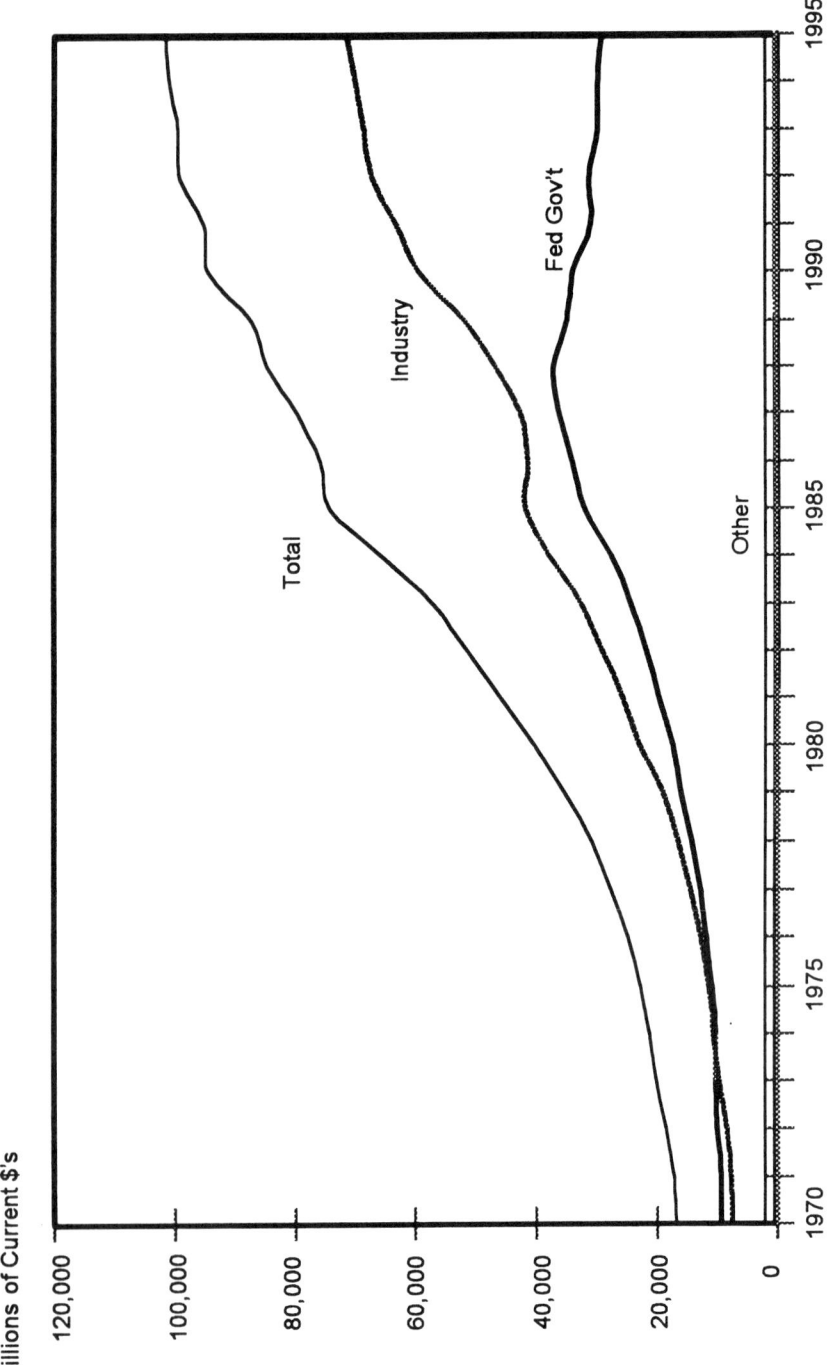

Figure 2.4 National expenditures for development, by source of funds: 1970–1995. Reprinted courtesy of the National Science Foundation.

1% higher than in 1995, whereas real gross domestic product (GDP) growth were expected to be about 2.7%. If these 1996 projections are borne out, 1996 would show a substantial decline in overall R&D funding as a percentage of GDP. Regarding composition effects, the new NSF report projects that basic research funding will decline .9%, applied research will increase by 1.0%, and development will increase by 1.5% in 1996, all in real terms, relative to 1995.

The Performer Base

These funding patterns, while clearly indicating that U.S. national R&D funds have reached a plateau and may be in a slight decline, mask changes in the objectives of the R&D and the makeup of the performer base. Figures 2.5, 2.6, 2.7, and 2.8[3] can be used to examine the performer base question. These NSF data indicate that industry's importance as an overall R&D performer has been unchanged since 1991. New NSF data for 1995[4] suggest that industry performance increased more than NSF originally expected, and that the nation will be dependent on industry and industry-operated federal laboratories for 73% of its R&D in 1996. This proportion was 34% in 1960 and has increased constantly over the 36-year period (see Figure 2.9). Examination of Figure 2.6 shows that universities and colleges continue to be the most important performers of basic research and that industry has been declining in importance as a basic research performer since 1991. The NSF's revised data[5] further bear this out, as mentioned earlier. Figure 2.7 reveals a similar pattern for industrial performance of applied research. Thus, as indicated by Figure 2.8, industry's forte as an R&D performer continues to be developed, where small increases have occurred each year since the early 1990s. NSF's newly released data[6] suggests that a relatively large increase in real development expenditures occurred in 1995 and that a 1.5% increase is expected in 1996.

Sectoral Differences

Underlying the changes in R&D funding sources, R&D performers, and the composition of R&D work described earlier, are two additional changes in the institutional structure of R&D. The first is the differential effect of declining federal funds on specific industries, and the second is a substantial rise in the importance of nonmanufacturing industries as R&D performers.

First, federal funding for industrially performed R&D reached a peak of $30,752 million in 1986, then declined each year through 1993 in both current and constant dollars.[7] The largest impact has been in the transportation

(Text continues on page 17.)

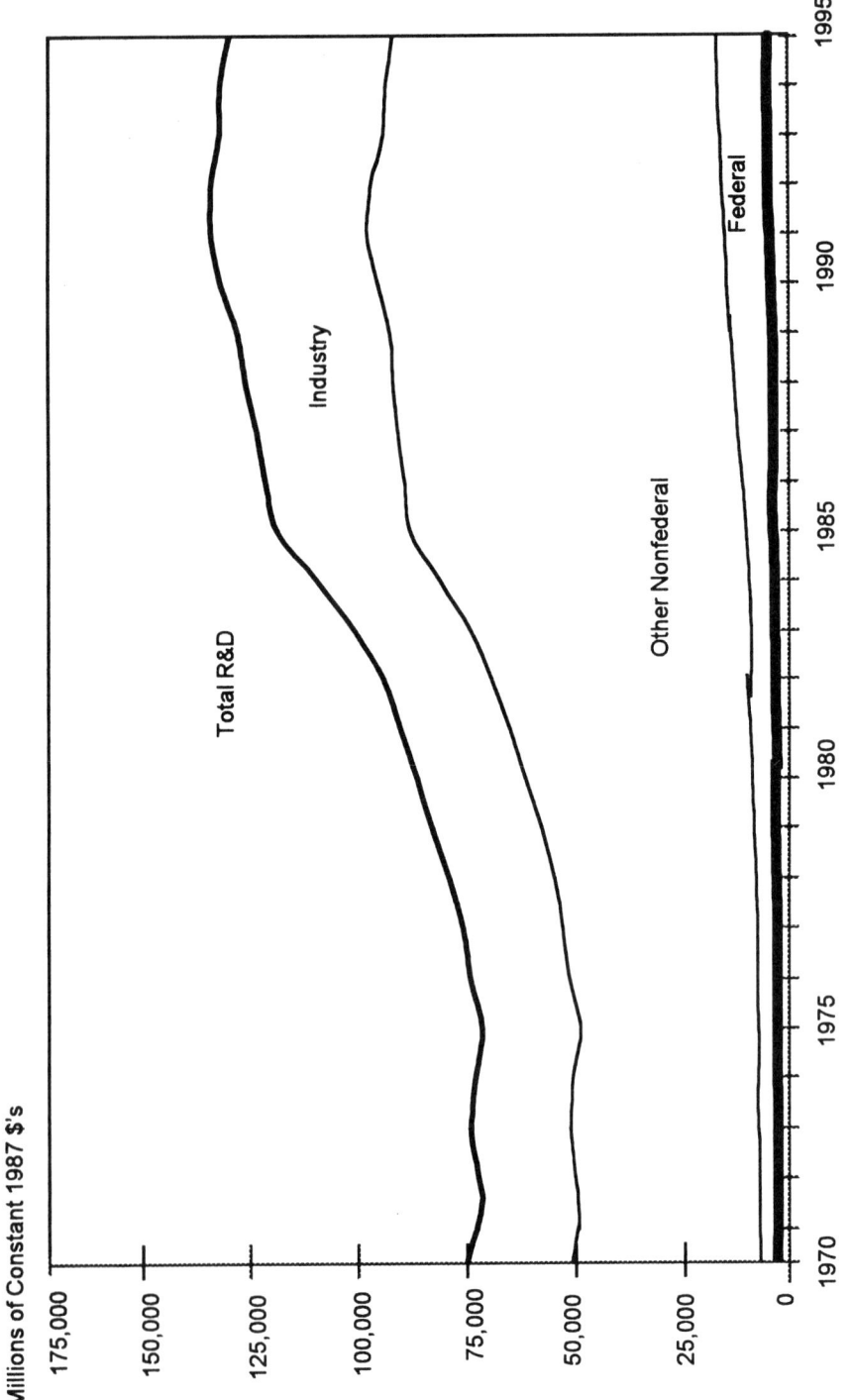

Figure 2.5 National expenditures for total R&D funding, by performer: 1970–1995. Reprinted courtesy of the National Science Foundation.

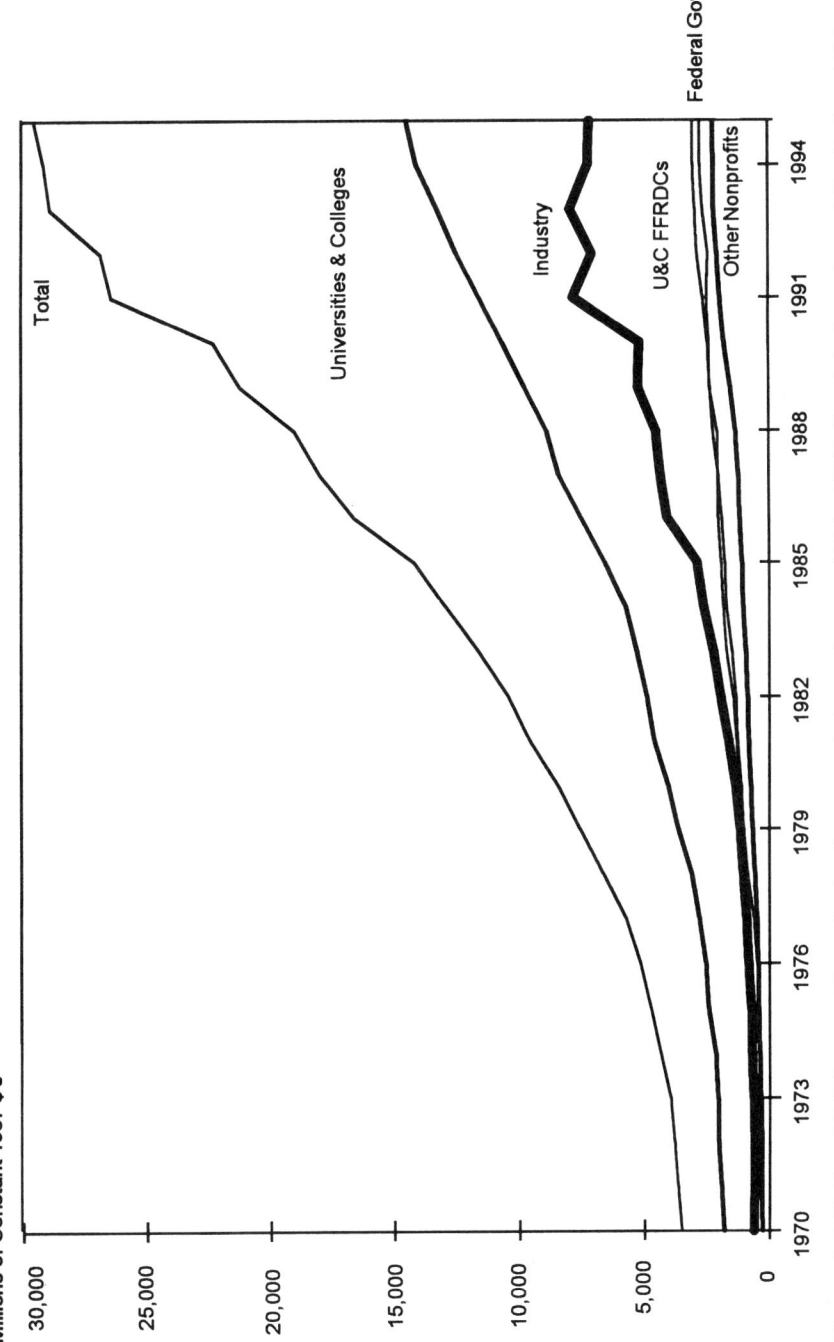

Figure 2.6 National R&D expenditures for basic research, by performer: 1970–1995. Reprinted courtesy of the National Science Foundation.

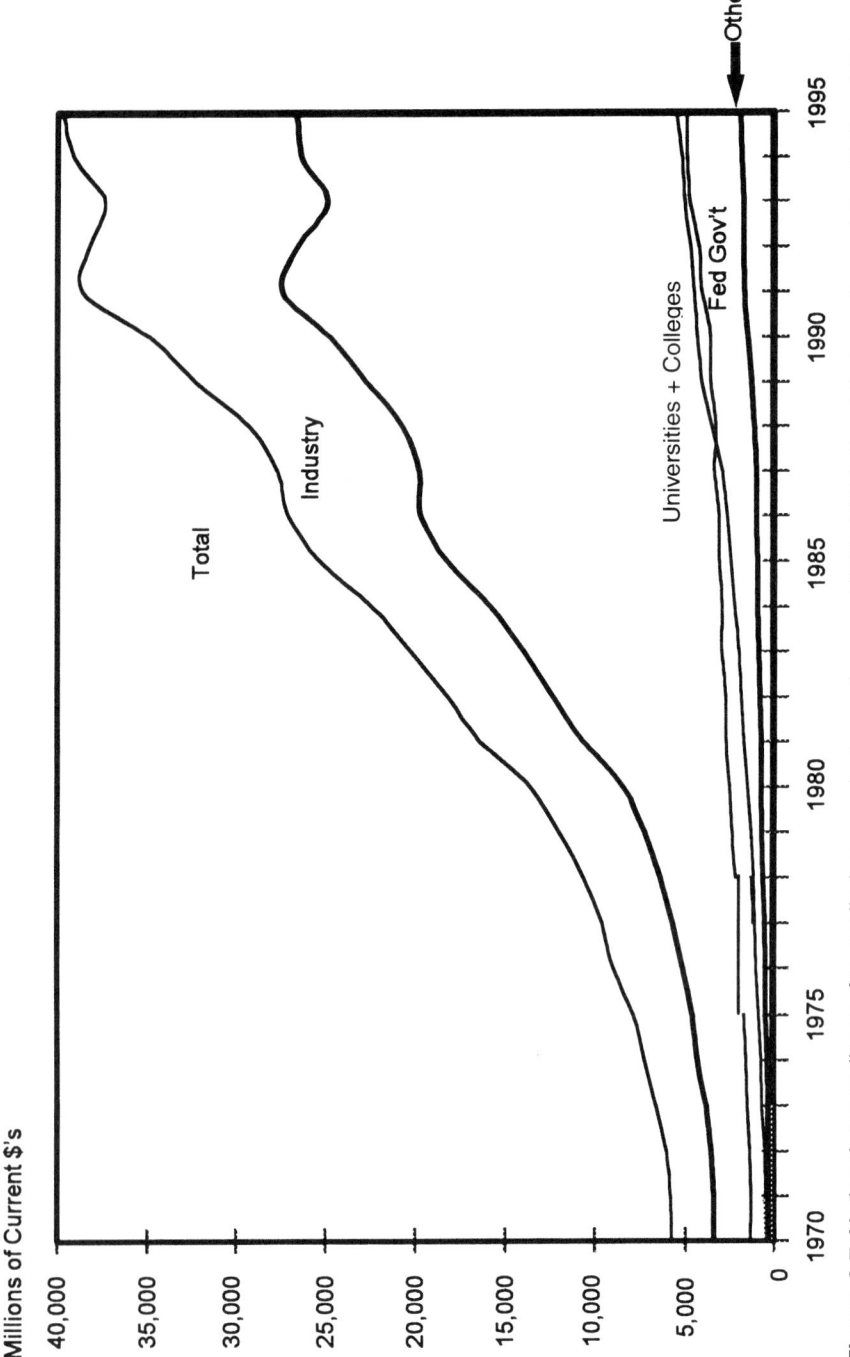

Figure 2.7 National expenditures for applied research, by performer: 1970–1995. Reprinted courtesy of the National Science Foundation.

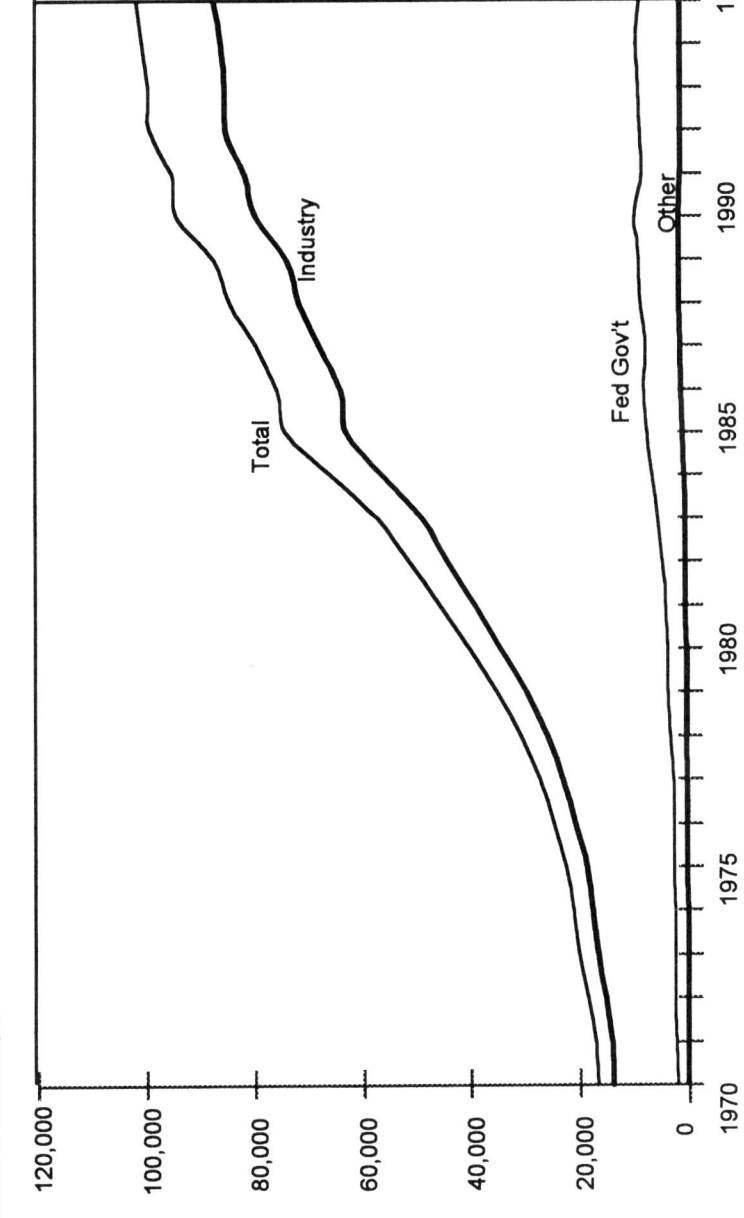

Figure 2.8 National expenditures for development, by performer: 1970–1995. Reprinted courtesy of the National Science Foundation.

16 R&D AND TECHNOLOGICAL INNOVATION IN U.S. INDUSTRY

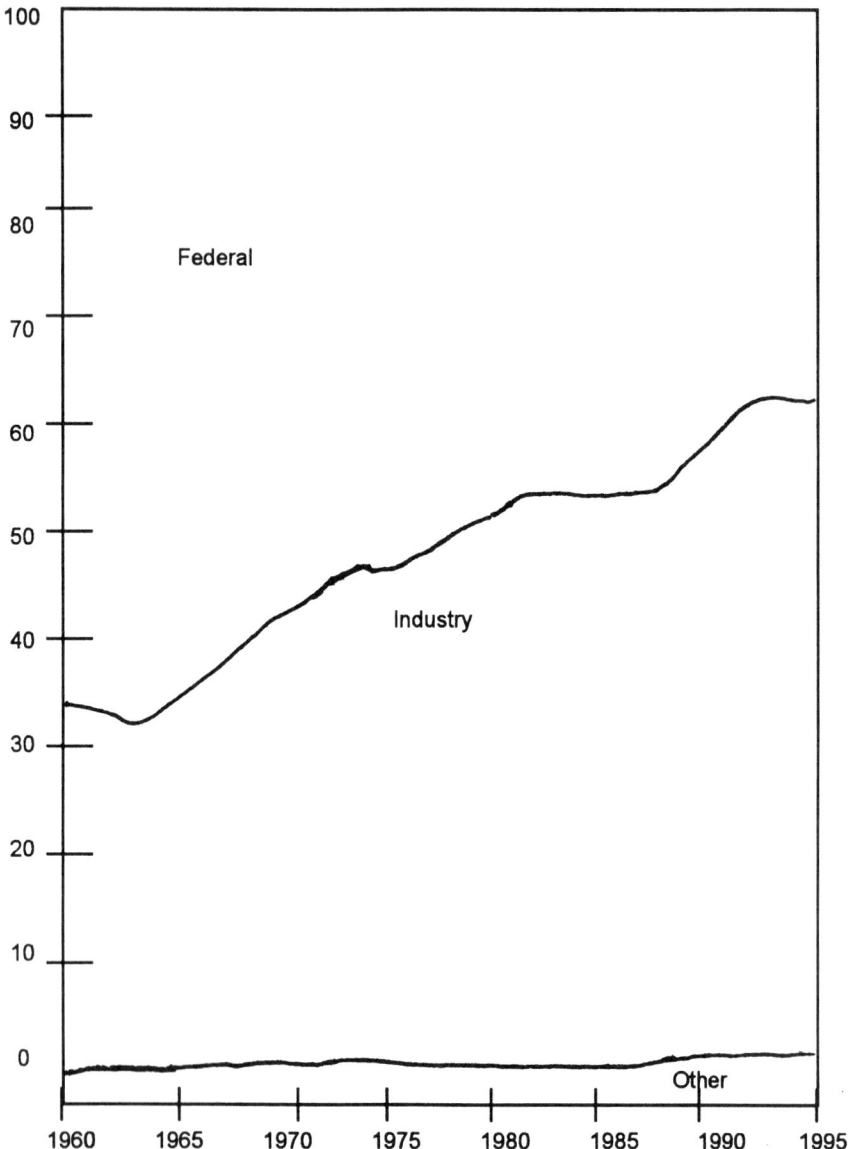

Figure 2.9 National R&D expenditures, by source of funds. Reprinted courtesy of the National Science Foundation.

and equipment sector, which includes aerospace and missiles, where federal R&D funds declined, in *current* dollar terms, from $20,865 million in 1987 to $10,438 million in 1993. The second most heavily affected industry has been the electrical equipment sector, which includes communications equipment and electronic components. From a high of $5,399 million in 1987, federal funding declined to $1,402 million in current dollars in 1993. Thus, these sectors have borne a disproportionate share of the "peace dividend" effect as it relates to R&D. NSF data show, furthermore, that the companies whose R&D programs were affected by these cutbacks were not able to replace the lost federal dollars with R&D funds from other sources.[8] Thus, they were net losers as national R&D performers.

Of the remaining industrial sectors with substantial R&D programs, only the professional and scientific instruments sector has realized a gain in both federal and non-federal R&D funding since 1987. This sector includes various scientific, surgical, optical, and mechanical measuring instruments used in many applications, including health care diagnostics and treatment.[9] Only the chemical and allied products sector, which has received very little federal R&D funding over the years, showed a pattern of growth in R&D performance, increasing from $8,843 million to $15,381 million (in current dollars) between 1983 and 1992.

CONCLUSIONS

If the reader infers from the preceding discussion that industrially performed R&D should have declined *more* than the aggregate data suggest it did, this would clearly be correct if it were not for the nonmanufacturing industries. In constant dollar terms, R&D performed in manufacturing firms peaked at $84,311 million in 1987 and declined to $70,538 million in 1993. As for the nonmanufacturing sector, its rise as an R&D performer began in 1975 and has continued. From a low of $1,494 million in 1975, it reached $25,279 million constant 1987 dollars, in 1995.[10] Thus, R&D in nonmanufacturing industries now accounts for 26% of the nation's total R&D; 71% of the nation's total R&D is now performed in industry or industry-operated federally owned laboratories, and 66% of the nation's R&D is funded by industry, the largest percentage in the history of NSF's 44-year-old data series.[11]

One may therefore conclude that, as a nation, the United States is much more dependent for its future scientific and technical knowledge base, and its near-term prospects for growth-enhancing innovation, on the R&D decisions made in U.S. industrial firms. Thus, it is more than a matter of academic curiosity to understand how these decisions are made, how they compare with similar decisions made in other national contexts, and whether industry's recent reductions in R&D investment reflect diminishing economic returns to innovation.

NOTES

1. *Science and Technology Indicators,* NSB-96-21 (Washington, D.C.: National Science Foundation, 1996, Appendix Table 4-4.
2. Steven Payson, "Science Resource Studies Division Data Brief," National Science Foundation, Washington, D.C. (October 25, 1996).
3. NSB-96-21, *Science and Technology Indicators,* Appendix Table 4-6.
4. Payson, "Science Resource Studies Division Data Brief."
5. Ibid.
6. Ibid.
7. NSB-96-21, *Science and Technology Indicators,* Appendix Table 4-7.
8. Ibid.
9. Ibid.
10. NSB-96-21, *Science and Technology Indicators,* Appendix Table 4-11.
11. Ibid.

3
MEET YOUR COMPETITION: THE IRI/CIMS R&D SURVEY FOR FY 1995

As evidenced by the results of the Industrial Research Institute/Center for Innovation Management Studies (IRI/CIMS) Annual R&D Survey for fiscal year (FY) 1995, the fourth compilation of year-to-year information on U.S. industry R&D spending and performance, downsizing is taking a toll on R&D's contribution to the business performance of firms.

For the last four years, the new sales ratio of a representative sample of Industrial Research Institute (IRI) companies has been decreasing—by nearly 20%. The new sales ratio is the percentage of total annual sales that can be attributed to new and improved products commercialized during the past five years.

This metric has been used by many companies to measure and monitor their new product innovation performance (Brenner 1994; Germeraard 1994). It is frequently cited by academics and others as a measure of R&D effectiveness (see Cooper and Kleinschmidt 1996). The downward trend in the new sales ratio should be a concern to R&D executives and managers, and to corporate executives as well. It suggests that companies were less

---

The authors, on behalf of the IRI Database Subcommittee, wish to acknowledge the efforts of the IRI industry focus leaders who encouraged member participation in the survey, and to express appreciation to all the IRI member companies who took time to diligently complete the survey forms. We also wish to acknowledge the financial support of the National Science Foundation. Alden S. Bean would like to acknowledge *Research Technology Management* and Roger L. Whiteley and M. Jean Russo of the Center for Innovation Management Studies, Lehigh University.

successful in revitalizing their product lines in 1995 than they were in 1992, and that the effectiveness of R&D in support of the future growth of the firm is in decline; this should come as no surprise.

The investment in R&D by these same companies as measured by R&D intensity—the ratio of R&D spending to total sales—has also been in decline over these four years. In previous articles, the authors have demonstrated the strong correlations between the new sales ratio and R&D intensity and the share of R&D spending allocated to product development. In a recent article in *Research·Technology Management* (*RTM*), Robert Cooper (1996), based on an independent study of R&D management practices, states that R&D spending devoted to new products is the single strongest driver of the impact of new products on both sales and profits.

The decline in R&D intensity over the four years is also reflected in the number of personnel committed to R&D efforts. Since 1992 the average number of R&D personnel in a paired sample of IRI companies has declined by more than 10%. The decline in the new sales ratio can be linked justifiably to the decline in the commitment to R&D by corporate management during the recent flurry of downsizing and reengineering. The question is, "Will the decline in commitment of resources to new product development continue?" If so, what impact will this have on the long-term competitiveness of U.S. industrial firms?

Based on IRI's 1996 Annual R&D Trends Forecast, R&D intensity is expected to increase in 1996 over 1995 (see Germeraad, 1996). IRI's 1997 forecast indicates a further increase in 1997. If these forecasts are borne out, it will be interesting to see what impact this will have on the new sales ratio in future years.

SOURCE OF THE DATA

This information derives from four years of data collected by the IRI/CIMS Annual R&D Survey established by the Database Subcommittee of IRI's Research-on-Research Committee. The intent was to build a unique database that would allow R&D executives and general management to benchmark and track information on R&D spending and performance of firms and of business segments in specific industries represented by those firms. The resultant database is a valuable source of data for use by scholars in studying the impact of R&D management decisions and practices on the financial performance of firms within specific industries.

The survey has now been carried out for four years, covering IRI company data for fiscal years 1992, 1993, 1994, and 1995. The survey is done in cooperation with the Center for Innovation Management Studies (CIMS) at Lehigh University. CIMS serves as a repository for the data in order to protect confidentiality and promote objectivity. The authors are all members of the CIMS staff. The cost of administering the survey is currently being underwritten by the National Science Foundation (NSF) under a three-year grant to CIMS, and by the IRI.

Tables of statistics (average, standard deviation, median, and count) for each variable for firms and laboratories grouped by size, and for segments by industry, are available for IRI members and may be obtained by others by special request to IRI. While use of these data is subject to IRI restrictions regarding confidentiality, academic investigators seeking research data should contact Dr. Alden S. Bean, Director, Center for Innovation Management Studies, Rauch Business Center, Lehigh University, Bethlehem, PA 18015. E-mail address: ASB2@Lehigh.EDU.

The survey questionnaire requests financial data about the source and allocation of R&D funds at three levels within the responding company: at the firm or corporate level, at the business segment level for one or more business segments, and at the laboratory level. The survey also requests information about the number and use of R&D personnel and measures of R&D performance at the firm and business segment levels. The survey questionnaire used this year is similar to those used in previous years. (See Figure 3.1.) In order to assure credible time series data, the primary questions and definitions relating to the sources and allocation of funds and personnel and performance metrics have remained constant over the four-year period.

WHO RESPONDED

In the latest survey, covering fiscal year 1995, responses were received from 100 IRI companies, about 40% of the IRI membership. Usable data were received from 88 firms, 163 business segments, and 70 laboratories. Almost all of the respondents provided quantitative data about sources of R&D funds and the allocation of R&D expenditures and personnel. A few companies did not report firm-level data, but did provide data at the business segment or laboratory level.

The total sales revenue and total R&D expenditures of the 88 IRI companies that reported firm-level data for the 1995 fiscal year were $732 billion and $20.6 billion, respectively.[1] The National Science Foundation estimates that the total R&D spending by U.S. industrial firms in 1995 was $98.5 billion (*Industrial Research and Development Facts* 1996). Thus, the companies reporting data in the 1995 IRI/CIMS Annual R&D Survey account for about 20% of the total

	Laboratory Type	Primary SIC Code
Firm Segment or Laboratory Name _____	L	S

	$Millions
1. Firm, Segment, or Laboratory Profile	
a. Total Net Sales	F,S
b. Total R&D Expenditures	F,S,L

	$Millions
2. Sources of R&D Funding	
a. Company Financed—Corporate Sponsored	F,S,L
b. Company Financed—Business Unit/Project Sponsored	F,S,L
c. Federal Government	F,S,L
d. Other Outside Contract	F,S,L
e. Total R&D Funds = (2a+2b+2c+2d)	F,S,L

	$Millions
3. R&D and Technical Service Expenditures by Activities	
a. Basic Research	F,S,L
b. Applied Research	F,S,L
c. Product Development	F,S,L
d. Process Development	F,S,L
e. Technical Service	F,S,L

	$MIllions
4. Total R&D Expenditures by Expense Accounts	
a. Support Services (See instructions)	F,S,L
b. Technical R&D = (1b − 4a)	F,S,L

	Number
5. Distribution of R&D Personnel	
a. Total Personnel	F,S,L
b. Support Services Personnel	F,S,L
c. Technical R&D Personnel = (5a − 5b)	F,S,L
d. Technical R&D Personnel—PhDs and MDs	F,S,L
e. Technical R&D Personnel—Exempt (Including PhDs and MDs)	F,S,L

	$Millions
6. Innovation Performance	
a. What were your annual sales in 1995 attributed to new or improved *products and services* commercialized during the period 1990–1994?	F,S
b. What were your annual cost savings in 1995 attributed to new or improved *processes* commercialized during the period 1990–1994?	F,S

	Number
7. Patent Performance	
a. How many U.S. patents were granted to your firm in 1995?	F
b. How many non-U.S. patents were granted to your firm in 1995?	F

	($Millions)
8. Special Issues	
a. What were your 1995 R&D expenditures required to meet compliance with health, safety, and environmental regulations *within your own company*?	S
b. What were your 1995 R&D expenditures to *provide your customers with products* in compliance with health, safety, and environmental regulations?	S
c. What were your 1995 R&D expenditures to support outside contracts for R&D at colleges, universities, research institutes, and consortia?	F
d. What were your 1995 R&D capital expenditures?	F
e. What was your R&D annual depreciation expense in 1995?	F

F,S,L = Information requested at the Firm Level, Segment Level, or Laboratory Level

Figure 3.1 Information requested in the 1994 IRI/CIMS Annual R&D Survey. All data requested were for FY 1995. Separate forms were designed for each level.

industry spending for R&D as reported by NSF. Based on these figures, the survey sample represents a significant portion of the internal R&D resources of industrial firms in the United States. The survey results provide detailed information about the utilization of these R&D resources and their impact on the innovation performance of U.S. industrial firms and industry segments.

COMPILATION OF RESULTS

The average of reported values and calculated ratios for key variables for all firms, segments, and laboratories that reported data for the 1995 fiscal year were calculated and are shown in Table 3.1. The results presented in the tables and graphs are arithmetic averages or medians of either reported values or calculated ratios of a particular set of firms, segments, or laboratories. In some cases, the averages were determined after "extreme values" were identified and deleted following statistical convention.

The firm-level data were also aggregated into four quartiles, based on the net sales revenues of the firms, and quartile averages determined. A similar aggregation was also made for laboratory data based on the type of laboratory (corporate or segment), and size of laboratory based on the total number of laboratory personnel.

In requesting business-segment data, the survey asked respondents to identify the four-digit SIC Code that most nearly matched the principal product lines for that business segment. The Standard Industrial Classification (SIC) system used in the United States is a hierarchical system, which, at the broadest level, divides the U.S. economy into 10 divisions, ranging from agriculture, mining, and manufacturing to public administration. Each division is subdivided into two-digit Major Groups, three-digit Industry Groups, and four-digit Industries. In requesting business-segment data, the survey asked respondents to identify the four-digit SIC code which most nearly matched the product lines for that business segment. The segment-level data were then aggregated on the basis of these SIC codes into three-digit Industry Groups or two-digit Major Groups, resulting in the 18 broad industry categories listed in Figures 3.2 to 3.6.

The data from the 163 business segments were subsequently aggregated into industry categories based on the SIC codes provided, and the average values and ratios were calculated for each industry. Only averages based on business-segment data from five or more companies could be reported. This limitation was imposed by the need to assure confidentiality of data from any one firm. Following this protocol, business-segment results could be published for only 14 industry categories based on the 1995 survey data alone. To provide results for additional industry categories, we also calculated industry averages and related statistics based on combined survey data for all four years—1992 to 1995.

TABLE 3.1 IRI Annual R&D Survey Results for Fiscal Year 1995

	Firms	Segments	Laboratories
Average Values			
Sales Revenue—$ million	8320.9	3368.2	
Total R&D Funds—$ million	234.1	103.2	61.7
Total R&D Personnel	1272.5	471.5	386.3
Average Ratios			
RESEARCH INTENSITY—%	3.1	3.6	
SOURCE OF R&D FUNDS—%			
Corporate	26.8	14.3	38.5
Business Units	68.0	80.1	56.2
Federal Government	3.9	4.4	3.5
Other Outside Contracts	1.3	1.4	1.8
Use of R&D and Technical Service Funds—%			
Basic Research	3.0	2.8	4.6
Applied Research	18.3	17.1	31.4
Product Development	41.0	39.0	28.4
Process Development	20.5	22.9	18.3
Technical Service	17.3	18.3	17.3
Allocation of R&D Expense—%			
Support Services	22.4	23.0	27.0
Technical Projects	77.6	77.0	73.0
R&D Personnel			
Support Services—% of total R&D personnel	17.2	15.9	19.7
Ph.D.s and M.D.s—% of technical R&D personnel	21.0	21.9	30.5
Exempt—% of technical R&D personnel	72.6	74.7	73.9
R&D Expense/R&D Employees—$(000)			
Based on Total R&D Personnel	132.1	139.2	140.4
Based on Technical R&D Personnel	163.3	170.5	177.6
Achievements			
New/Improved Products—% of total sales	21.3	21.1	
Process Cost Savings—% of R&D expenditures	51.3	77.8	
U.S. Patents per 100 R&D Employees	6.3		
Other Ratios—% of Total R&D Expenditures			
To Meet Regulations—internal processes		4.9	
To Meet Regulations—customer products		4.7	
For External Research	2.2		
For Depreciation	8.8		

Firm Data: 1995 versus 1994

The average ratios reported in Table 3.1 for the 1995 fiscal year may be compared with the survey results of earlier years, published in the January-February issues of *RTM* in 1994, 1995, and 1996 (Wolff 1994, 1995; Whitely, Bean, and Russo 1996). The average ratios do not show much change from one year to the next. But year-to-year comparisons of the firm-level average ratios from paired sets of companies indicate rather marked declines in both allocation of resources (R&D intensity and number of R&D personnel) and R&D outcomes (patents and the new sales ratio) over the last few years.

Based on a comparison of firm-level 1994 data with that of 1993 for the same 50 companies, there was a net decrease of 5% in R&D intensity between 1993 and 1994 and a drop of nearly 10% in total R&D personnel. There was also a corresponding drop in the new sales ratio of more than 10% and a modest decline in the number of patents granted.

A similar analysis was made in 1996 comparing 1995 results with those of 1994 for 67 companies that reported data in both years. The results are very similar to those of earlier years, as can be seen in Figure 3.2. Further evidence of the declining trend in resources and outcomes can be seen in Table 3.2, which compares the firm-level average values and average ratios for 25 IRI companies that have reported survey data over the four years. Although the percentage changes noted in Figure 3.2 are quite pronounced, we did not find all the year-to-year differences to be statistically significant.

While some year-to-year differences may be questioned, the downward trend of these metrics over the four years is real. In our analysis of earlier survey data, strong correlations were found between the new sales ratio and both R&D intensity and the percentage of R&D funds that were allocated to product development (see Whitely, Bean, and Russo 1996). The same correlations were found in analyzing 1995 data.

Given the downward yearly trend in both new sales ratio and R&D intensity, and the consistent correlation between the two variables based on individual firm data each year, one may justifiably conclude that the decline in new product sales is linked to a cutback in the resources committed to R&D activities by corporate management.

Corporate versus Segment Labs

Corporate laboratories are defined as those serving the technology interests of the corporation as a whole, while a segment laboratory primarily serves the interest of a single business segment. Table 3.3 shows a striking contrast between the two types of laboratories in both the sources and use of R&D funds. We saw this same contrast in previous years.

26 MEET YOUR COMPETITION: THE IRI/CIMS R&D SURVEY FOR FY 1995

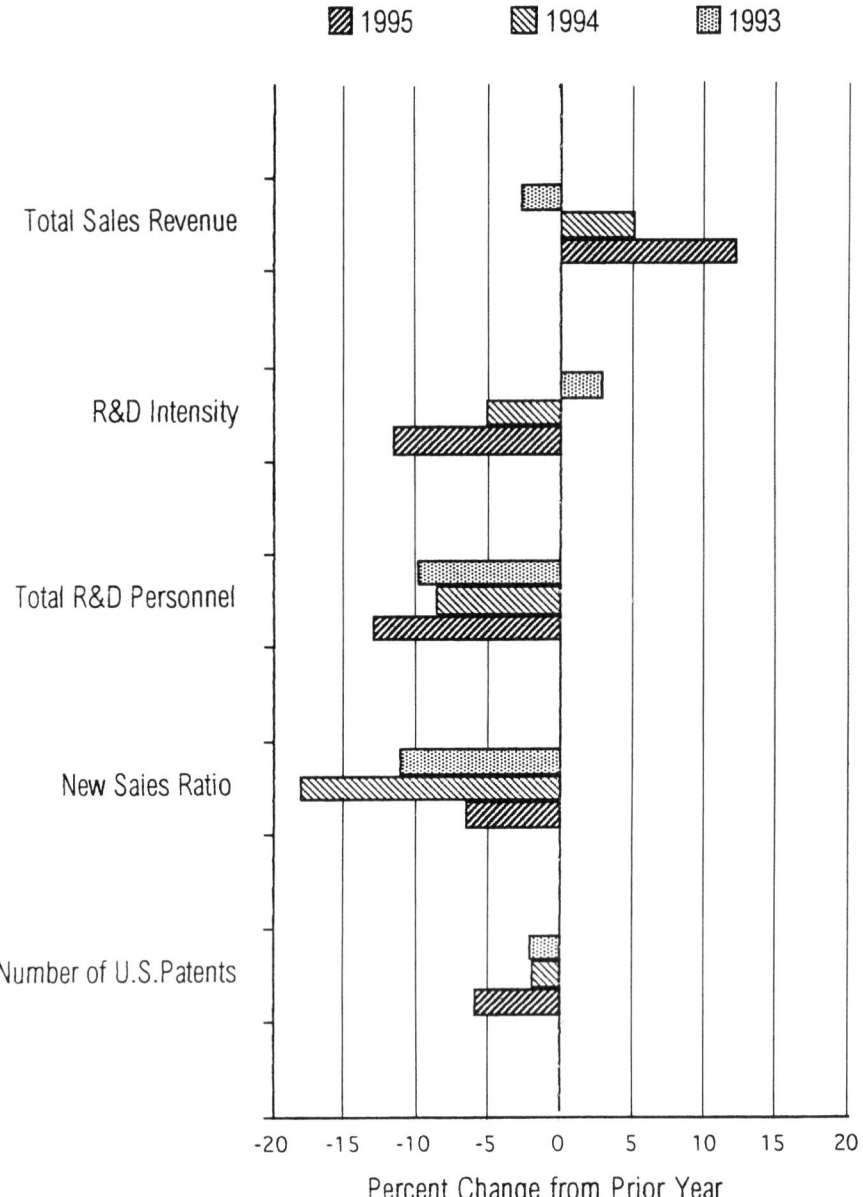

Figure 3.2 Changes in the average values reported in 1993, 1994, and 1995 from those reported by the same companies in the prior year. In 1995, the values are based on 65 common companies. In 1994 and 1993, the values are based on 40 and 30 common companies, respectively.

TABLE 3.2 Comparison of Firm-Level Data for Fiscal Years 1992 to 1995

	Based on the Same 25 Firms			
	1992	1993	1994	1995
Average Values				
Sales Revenue—$ million	8278.5	8505.2	8837.4	9926.5
Total R&D Funds—$ million	205.1	195.0	192.8	181.5
Number of R&D Personnel	1517.6	1618.6	1522.1	1420.2
Number of U.S. Patents Granted	117.2	117.8	107.0	101.1
Average Ratios				
RESEARCH INTENSITY—%	2.71	2.70	2.56	2.41
SOURCE OF R&D FUNDS—%				
Corporate	17.3	13.7	15.9	14.4
Business Units	77.6	82.2	80.1	81.7
Federal Government	3.6	2.9	3.4	3.3
Other Outside Contracts	1.5	1.1	0.5	0.5
Use of R&D and Technical Service Funds—%				
Basic Research	3.2	3.5	3.2	3.1
Applied Research	15.2	14.6	16.4	17.3
Product Development	38.6	39.7	39.2	38.2
Process Development	22.9	23.4	21.8	23.3
Technical Service	20.1	18.7	19.4	18.1
Achievements				
New/Improved Products—% of total sales	19.4	17.1	15.7	16.1

The percentage of spending allocated for basic and applied research in the corporate laboratories is nearly double that in the segment laboratories. Segment laboratories allocated almost no funds to basic research, reflecting the short-term view of R&D usually assigned by business unit management that is providing the bulk of the R&D funds for the segment laboratories.

The differences between corporate and segment laboratories are also seen in the results of the recently completed IRI/CIMS R&D Management Practices Survey. This survey is administered periodically and complements the Annual R&D Survey by requesting more detailed information about specific management practices related to planning and budgeting, and to program goals and achievements. Comparable IRI surveys were done in 1988 and 1996.

The results indicate that the contrast between corporate and segment laboratories is more pronounced today than in 1988.[2] Other data from the two

TABLE 3.3 Comparison of Corporate vs. Segment Labs—FY 1995

Average Ratios	Corporate	Segment
Number in Sample	58	12
Source of R&D Funds—%		
Corporate	43.9	12.7
Business Units	49.7	87.2
Federal Government	4.2	0.1
Other Outside Contracts	2.2	0.0
Use of R&D and Technical Service Funds by Activity—%		
Basic Research	5.3	1.2
Applied Research	33.8	20.3
Product Development	26.8	35.7
Process Development	17.7	21.1
Technical Service	16.4	21.6
Allocation of R&D Expense—%		
Support Services	27.7	23.3
Technical Projects	72.3	76.7
R&D Personnel Costs—$(000)		
R&D Expense per R&D Employee	143.6	124.2

management practices surveys suggest that the management of segment laboratories has become increasingly divorced from direct corporate influence in setting budgets and planning over that same period.

Business Segment Data

Although we reported on 16 industries in 1995, we were unable to report on two of those industries for the 1995 fiscal year because not enough companies reported data for the relevant SIC categories.

To provide more complete reporting of segment-level results, we first combined the business segment data for all four years before aggregating the combined data by SIC code. We felt justified in this approach because we found that the levels of average R&D intensity and the differences from industry-to-industry were similar to those reported in the three previous years. This approach allows us to report the average values and ratios of business-segment data for 18 distinct industry categories. The averages and median values reported in the subsequent figures are based on the combined data of all four years, 1992 to 1995.

COMPILATION OF RESULTS   29

▨ Internal R&D Funds   ▧ Total R&D Funds

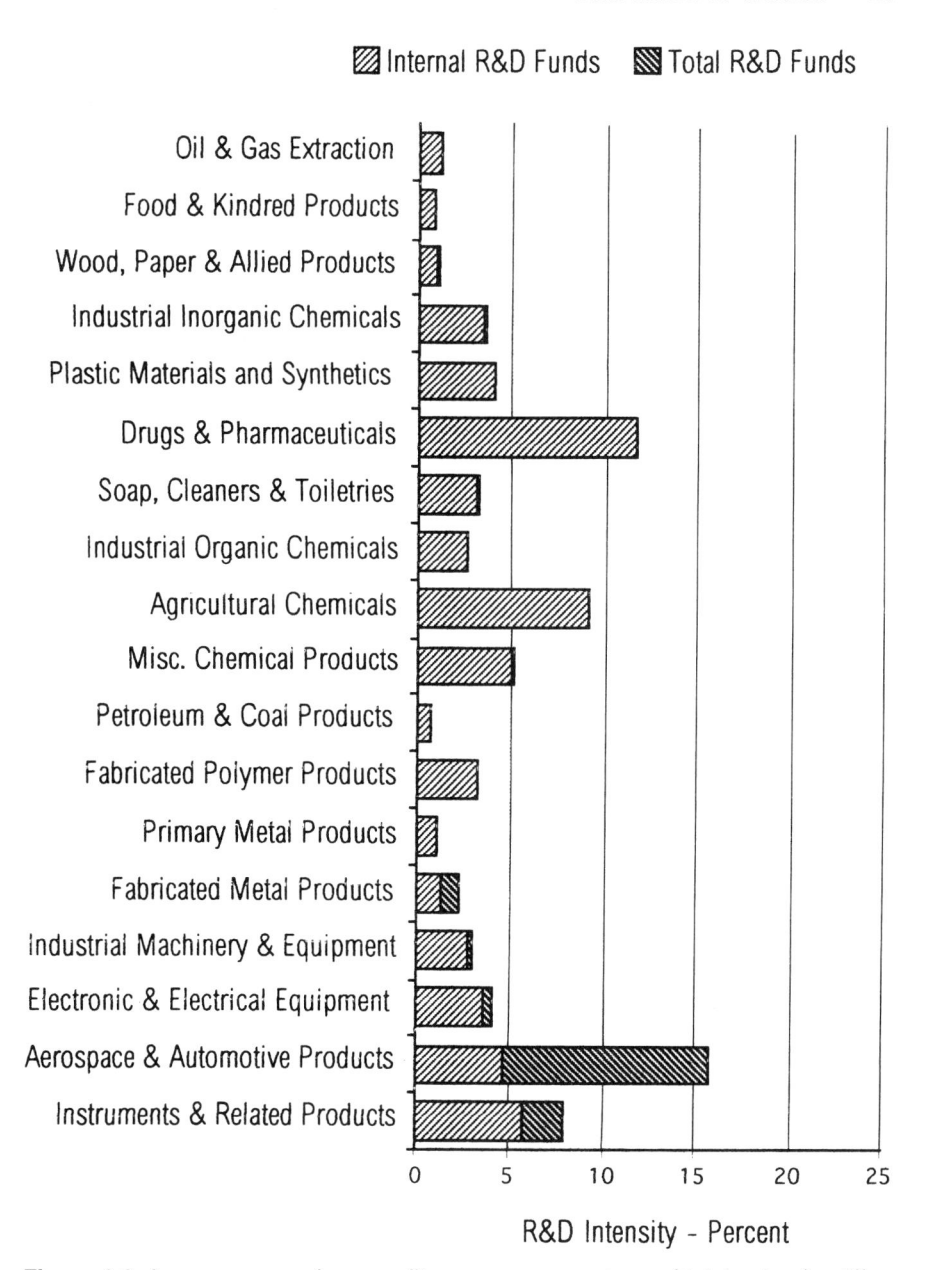

Figure 3.3 Average research expenditures as a percentage of total sales for different industries based on internal R & D funding and total R & D funding. Total R & D funds include internal funds plus funds from federal sources and other outside contracts. Averages are based on the combined data for fiscal years 1992 to 1995.

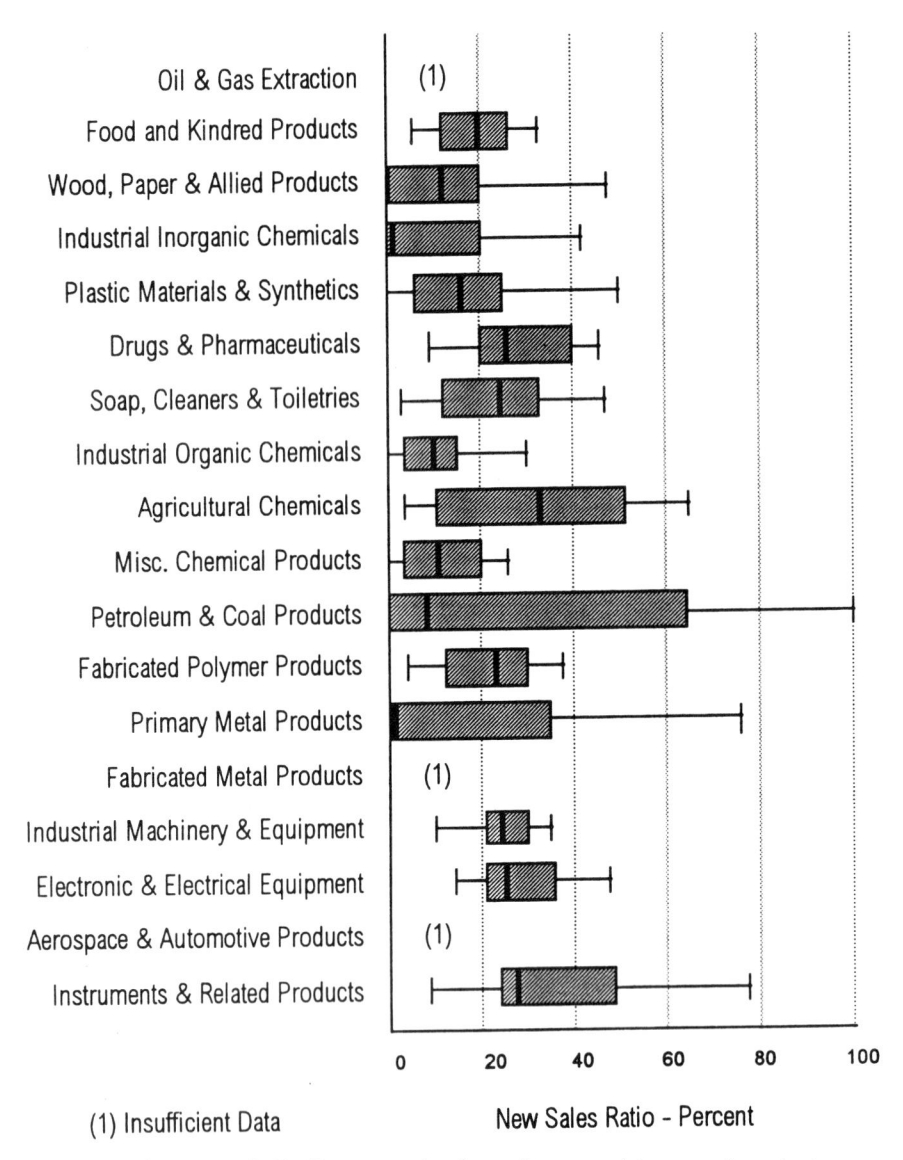

30 MEET YOUR COMPETITION: THE IRI/CIMS R&D SURVEY FOR FY 1995

Oil & Gas Extraction	(1)
Food and Kindred Products	
Wood, Paper & Allied Products	
Industrial Inorganic Chemicals	
Plastic Materials & Synthetics	
Drugs & Pharmaceuticals	
Soap, Cleaners & Toiletries	
Industrial Organic Chemicals	
Agricultural Chemicals	
Misc. Chemical Products	
Petroleum & Coal Products	
Fabricated Polymer Products	
Primary Metal Products	
Fabricated Metal Products	(1)
Industrial Machinery & Equipment	
Electronic & Electrical Equipment	
Aerospace & Automotive Products	(1)
Instruments & Related Products	

(1) Insufficient Data

New Sales Ratio - Percent

Figure 3.3 (*continued*) Median annual sales of new and improved products commercialized in the prior five years as a percentage of total annual sales. Box and whisker charts are based on all data for fiscal years 1992 to 1995.

The average R&D intensity of the IRI company business segments in the 18 industry categories are shown in Figure 3.3a. The averages are shown for two levels of R&D funding—internal funding and total funding. Internal R&D funds are those from internal corporate- and business-segment sources. Total R&D funds include internal funds as well as funds from federal government and other outside contracts.

As can be seen in Figure 3.3a, the use of contract funds to supplement internal sources of funds for R&D is very industry specific. Most IRI business segments in our sample (74%) rely solely on internal sources for their R&D funds. In Figure 3.3b the R&D intensity data are displayed in the form of "box-and-whisker" charts. Only the R&D intensities based on internal funds are shown.

The dark bar within the "box" is the median value for the data, rather than the average. Half of the observations or cases lie above the median value, half below. The ends of the box represent upper and lower bounds of the middle 50% of the cases. The ends of the "whiskers" define the limits within which 95% of the cases on either side of the median would lie and beyond which any observations are considered to be "outliers" or "extreme values." We believe this type of diagram provides more useful information for benchmarking than the simple bar chart of averages used in reporting research intensity in previous years.

Comparison of Figures 3.3a and 3.3b shows a good correspondence between the average and median values of R&D intensity for most industries. The box-and-whisker charts provide a more complete picture than simple averages of the differences in the R&D intensity of IRI company business segments from industry to industry.

The variation in the new sales ratio of business segments within and across the 18 industry categories is shown in Figure 3.4. This metric has been cited often as a benchmark for innovation performance, a figure of 30% being representative of good innovation performance.

The distributions portrayed in Figure 3.4 clearly indicate that no single target is likely to apply across all industries—or across all business segments within a single company. For example, the data indicate that a new sales ratio of 20% would represent outstanding innovation performance for an industrial organic chemicals business, but very poor performance for an agricultural chemicals business. Figures 3.5, 3.6, and 3.7 illustrate the sources of R&D funding and the allocation of those funds over the last four years.

NEED FOR MORE OUTCOME DATA

The company response rate to the IRI/CIMS Annual R&D Survey has steadily increased over the past four years to about 40% of the total IRI membership. This is a good rate of response, but the response has not been uniform. Only about half of the IRI respondents provide data on R&D performance as measured by the sales of new and improved products, and only about a third provide data on cost savings resulting from new and improved processes.

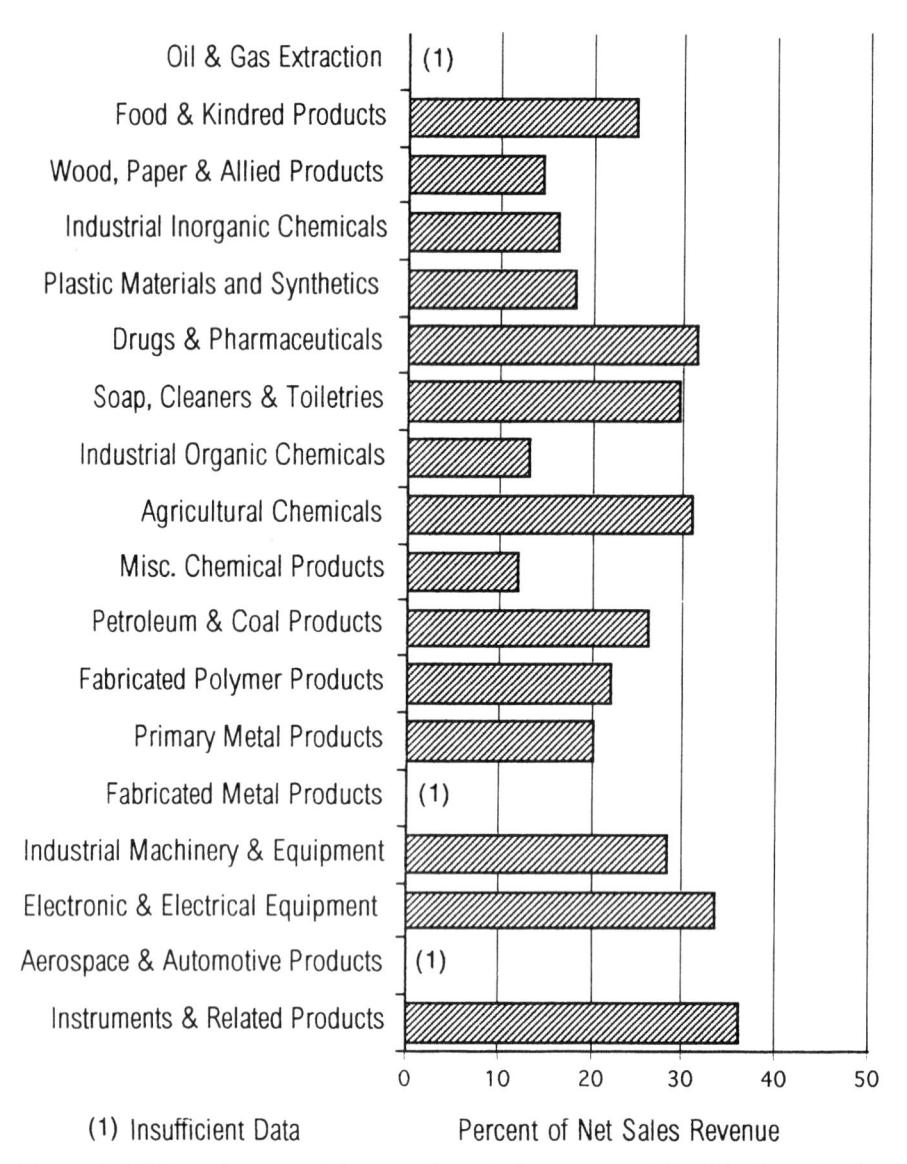

Industry	Percent
Oil & Gas Extraction	(1)
Food & Kindred Products	~25
Wood, Paper & Allied Products	~15
Industrial Inorganic Chemicals	~16
Plastic Materials and Synthetics	~19
Drugs & Pharmaceuticals	~33
Soap, Cleaners & Toiletries	~30
Industrial Organic Chemicals	~13
Agricultural Chemicals	~32
Misc. Chemical Products	~13
Petroleum & Coal Products	~26
Fabricated Polymer Products	~23
Primary Metal Products	~20
Fabricated Metal Products	(1)
Industrial Machinery & Equipment	~28
Electronic & Electrical Equipment	~34
Aerospace & Automotive Products	(1)
Instruments & Related Products	~36

(1) Insufficient Data   Percent of Net Sales Revenue

Figure 3.4 Sales of new and improved products, commercialized in the prior five years as a percentage of net sales revenue. Averages are based on all data for fiscal years 1992 to 1995.

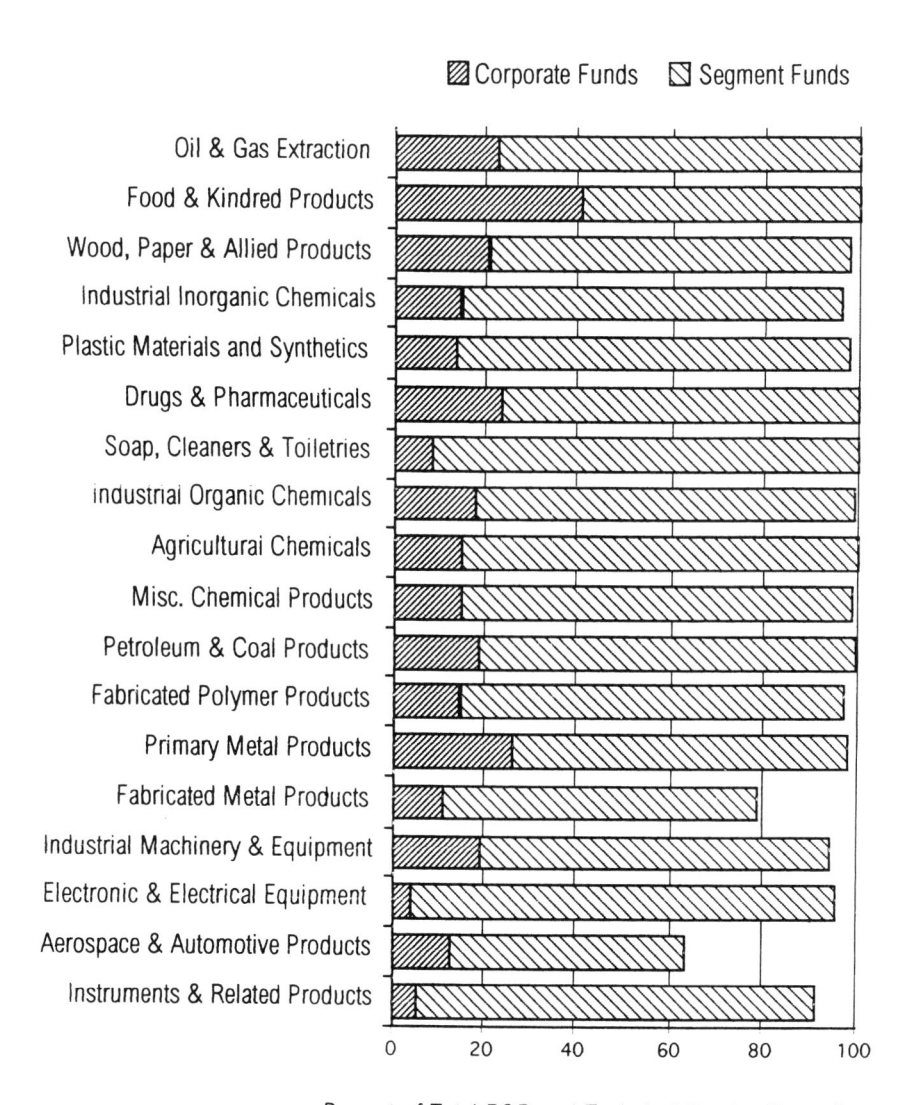

NEED FOR MORE OUTCOME DATA   33

Percent of Total R&D and Technical Service Expenditures

Figure 3.5 Percentage of R & D funds from corporate and business segment sources for different industries. Difference between the total internal sources and 100% represents the percentage of R & D funds from external sources. These are largely federal funds. Averages are based on data for fiscal years 1992 to 1995.

Figure 3.6 Allocation of R & D funds for basic research and applied research for different industries. Averages are based on all data for fiscal years 1992 to 1995.

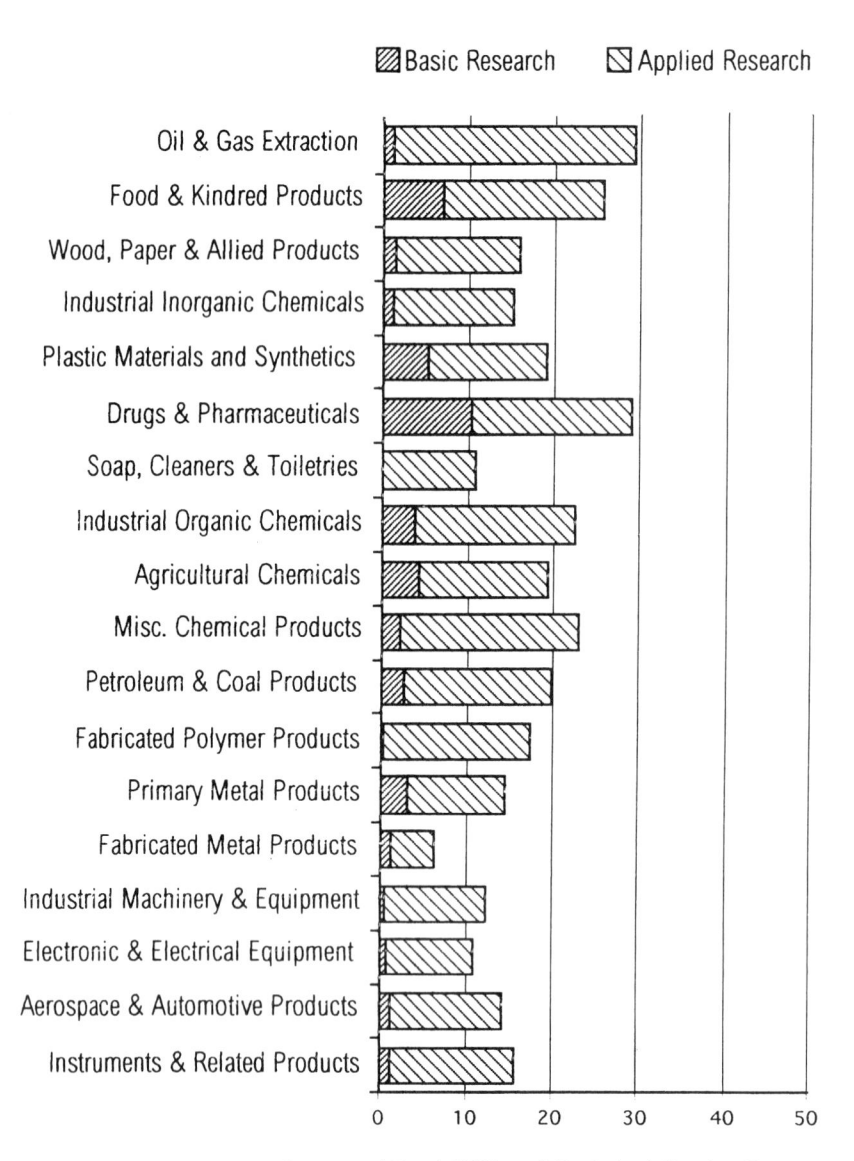

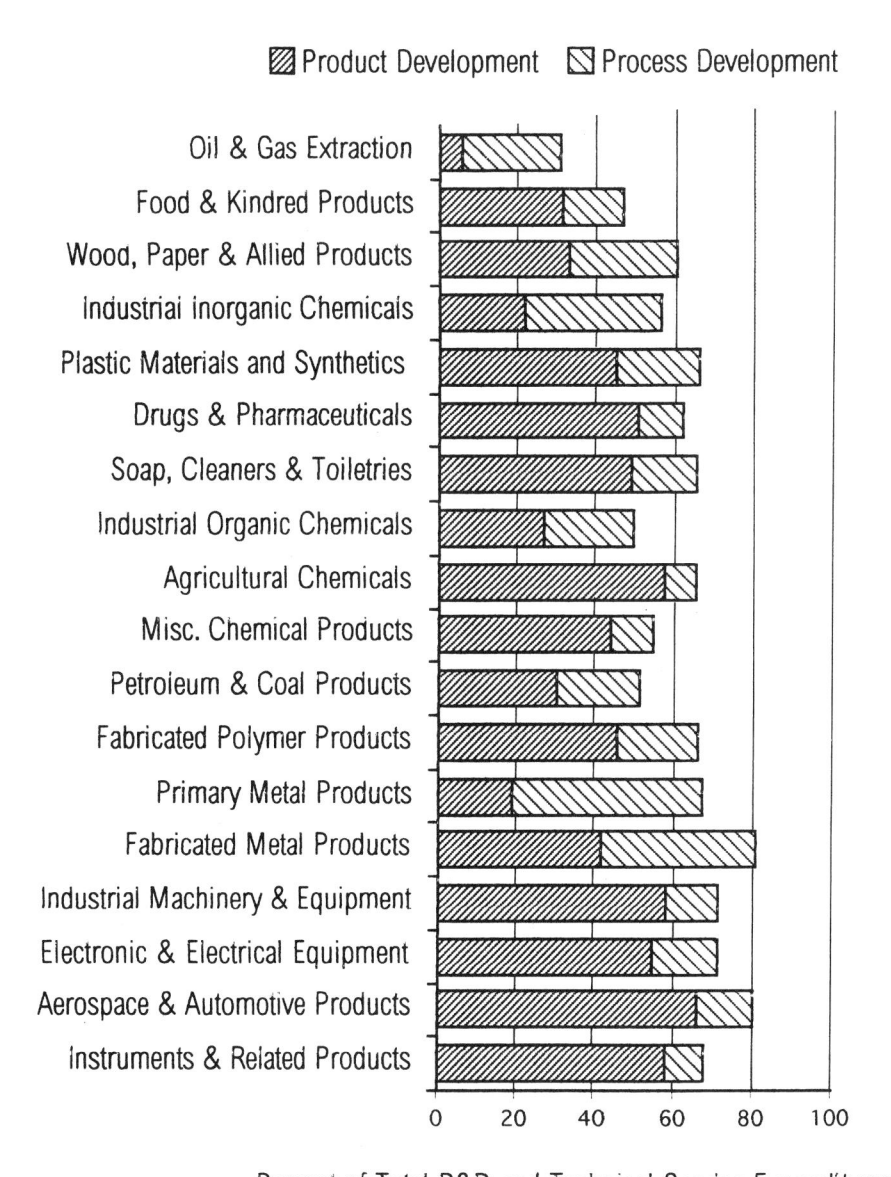

NEED FOR MORE OUTCOME DATA   35

▨ Product Development   ⬚ Process Development

Oil & Gas Extraction	
Food & Kindred Products	
Wood, Paper & Allied Products	
Industrial Inorganic Chemicals	
Plastic Materials and Synthetics	
Drugs & Pharmaceuticals	
Soap, Cleaners & Toiletries	
Industrial Organic Chemicals	
Agricultural Chemicals	
Misc. Chemical Products	
Petroleum & Coal Products	
Fabricated Polymer Products	
Primary Metal Products	
Fabricated Metal Products	
Industrial Machinery & Equipment	
Electronic & Electrical Equipment	
Aerospace & Automotive Products	
Instruments & Related Products	

Percent of Total R&D and Technical Service Expenditures

Figure 3.7 Allocation of R & D funds for product development and process development for different industries. Averages are based on all data for fiscal years 1992 to

If we seek good benchmarks of R&D performance or wish to better understand the factors that determine the effectiveness and productivity of R&D, we need better and more complete information on outcomes, particularly at the business-segment level.

NOTES

1. Preliminary evidence for 1996 by Whiteley, Bean, and Russo (1997) finds 78 IRI Companies reporting $972 billion sales and $20.6 billion R&D expenditures, maintaining a 2.6 % R&D intensity ratio.
2. The corporate data for 1996 shows a larger percentage of R&D going to product development and process development (see Whiteley, Bean, and Russo, 1997).

REFERENCES

Brenner, Merrill S. "Tracking New Products: A Practitioner's Guide." *Research · Technology Management* (November-December 1994):36–40.

Cooper, Robert G., and Elko J. Kleinschmidt. "Winning Business in Product Development: The Critical Success Factors." *Research · Technology Management* (July-August 1996):18–29.

Germeraad, Paul. "Industrial Research Institute's Annual R&D Trends Forecast for 1996." *Research · Technology Management* (January-February 1996):15–17.

Industrial Research and Development Facts. IRI publication, July 1996.

Tipping, James W., Eugene, Zeffren, and Alan R. Fusfeld, "Assessing the Value of Your Technology." *Research · Technology Management,* (September-October 1995):22–39.

Whiteley, Roger L., Alden S. Bean, and M. Jean Russo. "Meet Your Competition: the 1997 IRI/CIMS Annual R&D Survey." *Research · Technology Management* (in press).

Whiteley, Roger L., Alden S. Bean, and M. Jean Russo. "Meet Your Competition: Results from the 1994 IRI/CIMS Annual R&D Survey." *Research · Technology Management* (January-February 1996):18–25.

Wolff, Michael F. "Meet Your Competition: Data from the IRI R&D Survey." *Research · Technology Management* (January-February 1995):17–25.

Wolff, Michael F. "Meet Your Competition: Data from the IRI R&D Survey." *Research · Technology Management* (January-February 1994):18–24.

4
THE INTERDEPENDENCIES AMONG CORPORATE FINANCIAL POLICIES: EMPIRICAL EVIDENCE CONCERNING THE UNITED STATES, JAPAN, AND MAJOR EUROPEAN ECONOMIES

The hypothesis that firms simultaneously determine their research and development, investment, dividend, and new debt policies is generally substantiated in the financial economcs literature. The determinants of research and development, dividend, investment, and financing decisions of 3,000 firms were estimated econometrically during the 1978–1993 period and compared with a 303-firm sample previously studied by the authors. We also compared our U.S. results, with an international database, to Japanese and European firms.

This chapter estimates an econometric model to analyze the interdependencies between decisions in regard to research and development (R&D), investment, dividends, and new debt financing. Management attempts to manage dividends, capital expenditures, and R&D activities while minimizing reliance on external funding to generate future profits and maximize the stock price.

Each firm has a "pool" of resources, composed of net income, depreciation, and new debt issues, and this pool is reduced by dividend payments, investment in capital projects, and expenditures for research and development activities. The *perfect markets hypothesis,* in regard to financial decisions, holds that dividends are not influenced (limited) by investment decisions. There are no interdependencies of financial decisions in a perfect markets environment, except that new debt is issued to finance R&D, dividends, and investment. The *imperfect markets hypothesis* concerning financial decisions holds that financial decisions are interdependent and that simultaneous equations must be used to estimate the equations econometrically. Empirical evidence supporting the *interdependence hypothesis* is found in the simultaneous equation financial decision modeling work of Dhrymes and Kurz (1967), Mueller (1967), Damon and Schramm (1972), Dhrymes (1974), McCabe (1979), Peterson and Benesh (1983), Jalilvand and Harris (1984), Switzer (1984), Guerard, Bean, and Andrews (1987), Guerard and Stone (1987), and Guerard and McCabe (1992).

The goal of this study is to test empirically the *independence of financial decisions hypothesis,* using the Guerard and McCabe (1992) sample of 303 predominately manufacturing firms during the 1978–1995 period, and compare its results to a 3,000-firm Compustat database for the corresponding and extended (1978–1995) period. We also examine the independence of United States, Japanese, and G5 countries' (United Kingdom, France, Germany, Italy, and Canada) financial decisions using the WorldScope database for the 1982–1995 period. We find stronger evidence of the interdependence of financial decisions using the larger 3,000-firm database than using the Guerard and McCabe 303-firm database for the 1978–1995 period. There is stronger evidence for the interdependence of the U.S. financial decisions than for the Japanese and G5 financial decisions.

INTRODUCTION AND REVIEW OF THE PERFECT MARKETS HYPOTHESIS LITERATURE

Miller and Modigliani (1961) formulated the perfect markets hypothesis, in which the dividend decision is independent of the investment decision, by determining that the valuation process of the firm is independent of dividend policy and firm value is dependent on investment opportunities to produce earnings, dividends, or cash flow. In addition, the R&D and investment decisions are independent of its financing decision. The firm's dividend policy is generally maintained until a permanent change in operations (earnings) has occurred. New capital issues raise funds with which research and development, dividends, and investments are undertaken. It is assumed that increases in dividends and investments lead to new capital issues. The only interdependence of financial decisions that Miller and Modigliani found to be consistent with the perfect markets hypothesis is that new debt could be issued to finance capital expenditures.

Dhrymes and Kurz (1967) modeled the interdependence of the dividend, investment, and new debt decisions of 181 industrial and commercial firms during the 1947–1968 period and found (1) strong interdependence between the investment and dividend decisions; new debt issues result from increased investments and dividends but do not directly affect them; (2) that the interdependence among the two-stage least squares residuals compel the use of full information (three-stage least squares regression methods); and (3) that the accelerator theory in which the change in sales positively affected investment, as well as the profit theory, is necessary to explain investment.

The Dhrymes and Kurz study generated much interest in testing the perfect markets hypothesis. Mueller (1967) found significant interdependencies between investment and new debt and R&D and dividends (this study uses a subset of the Guerard and McCabe sample). Higgins (1972), Fama (1974), and McDonald, Jacquillat, and Nussenbaum (1975) found no statistically significant evidence of the interdependence of financial decisions, whereas Peterson and Benesh (1983), Switzer (1984), Jalilvand and Harris (1984), Guerard, Bean, and Andrews (1987), Guerard and Stone (1987), and Guerard and McCabe (1992) found significant interdependence among financial decisions. Thus, the evidence on the perfect markets hypothesis is mixed.

THE MODEL

The model used employs investment, dividends, and new capital financing equations to describe the budget constraint facing the manager of a manufacturing firm. The manager may use the available funds to undertake capitalized research and development activities (R&D) or new investment (CE), to pay dividends (DIV), or to increase net working capital (LIQ). The sources of funds are represented by net income (NI), depreciation (DEP), and new debt issues (EF):

$$RD + CE + DIV + LIQ = NI + DEP + EF$$

The financial variables are divided by total assets to reduce heteroscedatisicity (the standardized variables are denoted by a preceding "S," for example, standardized dividends are SDIV). The use of annual current (not a distributed lag formulation) variables has been criticized by McCabe (1979); however, use of current variables is supported by Dhrymes and Kurz (1967), Mueller (1967), and McDonald, Jacquillat, and Nussenbaum (1975), and Peterson and Benesh (1983).

The research and development equation employed in our model reflects the work of Mansfield (1963), Mueller (1967), Grabowski (1968), Grabowski and Mueller (1972), Switzer (1984), and Guerard and McCabe (1990). Research and development expenditures are modeled in terms of investments, dividends, and new capital issues, to reflect the imperfect markets hypothesis. We use the previous one-year research and development expenditures ($R\&D_{t-1}$) to serve as a surrogate for previous patents and previous research and development activities, as did Switzer (1984). Net income and depreciation should serve to increase research and development expenditures because the firm has the resources to pursue the activities that a smaller, less profitable firm could not afford.

The investment equation used the rate of profit theory in investment; Tinbergen (1939), Meyer and Kuh (1957) and Dhrymes and Kurz (1967) employed investment equations in which net income positively affects investment. The accelerator position on investment holds that an increase in sales (LSAL), increases capital investment. Depreciation is normally included in the investment analysis, because depreciation describes the deterioration of capital in the productive process. Investment, in an imperfect market, should increase as new capital is issued; funds for capital expenditures increase with borrowing. Given that dividends and investments are alternative uses of funds, as dividends increase in the imperfect market one would expect investment to fall. An increase in net working capital (LIQ), should serve to decrease investment, according to the *alternative use of funds* argument.

Dividends should increase in an imperfect market as new debt is issued because more funds are obtained; dividends should decrease as investment increases, as dividends are an alternative use of funds. Dividends are a positive function of net income because income increases the firm's amount of available funds and retained earnings. An increase in net working capital should serve to decrease dividends in an imperfect market. Dividends should be a positive function of last year's dividends (LDIV), because management is reluctant to cut dividends (Lintner 1956; Fama and Babiak 1968; Switzer 1984). Miller and Modigliani (1961) argue that dividends do not affect the financing of profitable investment, because external funds can be raised in a perfect market.

New debt issues are defined as changes in long-term debt; previous studies used only new debt financing because manufacturing firms traditionally finance only 2% to 3% of investment with equity issues (Higgins 1972). New debt issued should be positively correlated with dividends, R&D, and investment, because increases in the uses of funds must be financed with corresponding increases in the sources of funds. An increase in net income and depreciation, the firm's primary components of cash flow, should reduce the new debt issued in an imperfect market, because cash flow and net debt issues are alternative sources of funds. As the cost of debt (KD), and total debt to equity (DE) rise, new debt financing should fall because of the additional expense and risk.

The following is a summary of the hypothesized equation system:

$$SCE = f(\overset{-}{SDIV}, \overset{-}{SRD}, \overset{+}{SEF}, \overset{-}{SLIQ}, \overset{+}{SNI}, \overset{+}{SLSAL}) \quad (4.1)$$

$$SDIV = f(\overset{-}{SCE}, \overset{-}{SRD}, \overset{-}{SLIQ}, \overset{+}{SEF}, \overset{+}{SLDIV}, \overset{+}{SNI}) \quad (4.2)$$

$$SEF = f(\overset{+}{SCE}, \overset{+}{SRD}, \overset{+}{SDIV}, \overset{-}{SNI}, \overset{-}{SDEP}, \overset{-}{SKD}, \overset{-}{SDE}) \quad (4.3)$$

$$SRD = f(\overset{+}{SLRD}, \overset{-}{SCE}, \overset{-}{SDIV}, \overset{+}{SEF}, \overset{+}{SNI}) \quad (4.4)$$

THE DATA

The original Guerard and McCabe (1992) sample of 303 predominately manufacturing firms for the 1975–1982 period is used in this study for the 1978–1995 period, and the results are compared with those derived from using the largest 3,000 firms in June of each year as the relevant universe. The financial variables are drawn from the Compustat files. We also use annual financial data for all firms in the WorldScope database during the 1982–1995 period in the United States, the United Kingdom, Japan, France, Germany, Canada, and Italy.

SIMULTANEOUS EQUATION ESTIMATION RESULTS

Ordinary least squares (OLS) analysis is used to initially estimate equations (4.1) through (4.4). The simultaneous equation results reported in the study are produced with the use of two-stage (2SLS) and three-stage (3SLS) least squares analysis. Although Dhrymes and Kurz (1967) found that the insignificantly negative association between capital expenditures and dividends in the two-stage least squares regression estimation became a significantly negative association in the three-stage least squares estimations, we found no statistically significant differences between the limited-information (two-stage least squares) and full-information (three-stage least squares) procedures. The two-stage least squares regression equation residuals are not highly correlated, providing the statistical basis for the insignificant coefficient differences of the two- and three-stage least squares estimations. The highest correlations are found in the annual regression residuals between the

(Text continued on page 50)

TABLE 4.1 Estimated System Equations 303-Firm Sample

OLS   SCE

	1978 (t)		1979		1980		1981	
SDIV	−0.1634	(−0.73)	−0.0146	(−0.07)	−0.4293	(−1.73)	−0.1949	(−0.85)	−0.2697
SEF	0.1828	(3.05)	0.2376	(4.25)	0.2327	(4.59)	0.3093	(6.24)	0.1343
SRD	0.4211	(3.29)	0.5594	(4.64)	0.5651	(4.77)	0.4582	(4.03)	0.5344
SNI	0.3116	(3.89)	0.3637	(3.73)	0.5597	(5.77)	0.4422	(5.51)	0.4530
SLIQ	−0.1596	(−6.70)	−0.2013	(−8.74)	−0.2085	(−8.15)	−0.1881	(−7.41)	−0.2104
SLSAL	0.0001	(0.00)	0.0176	(1.10)	−0.0196	(−1.63)	0.0164	(1.28)	−0.0069
F−Value	12.27		17.18		20.01		19.06		11.93
adjusted r²	0.267		0.343		0.380		0.367		0.258

	1987		1988		1989		1990	
SDIV	−0.1300	(−0.96)	−0.1392	(−1.21)	−0.3805	(−2.76)	−0.0990	(−0.68)	−0.0716
SEF	0.0010	(0.04)	0.0663	(2.17)	0.0648	(1.71)	0.0634	(1.85)	0.0625
SRD	0.2841	(3.46)	0.2632	(3.58)	0.2559	(3.07)	0.2831	(3.13)	0.3844
SNI	0.0531	(0.86)	−0.0144	(−0.31)	0.1662	(2.34)	0.0951	(1.20)	0.0447
SLIQ	−0.0594	(−3.49)	−0.0380	(−2.28)	−0.0423	(−2.27)	−0.0542	(−2.73)	−0.0634
SLSAL	0.0147	(0.93)	0.0106	(0.83)	−0.0309	(−2.29)	0.0108	(0.58)	0.0130
F−Value	3.76		4.78		5.39		3.27		4.66
adjusted r²	0.087		0.116		0.134		0.074		0.113

OLS   SDIV

	1978		1979		1980		1981	
SLDIV	1.1025	(34.01)	0.9838	(26.71)	0.8581	(26.78)	1.0980	(52.00)	1.0311
SCE	−0.0095	(−1.10)	0.0115	(1.01)	−0.0419	(−4.96)	0.0085	(1.42)	0.0003
SEF	0.0035	(0.50)	−0.0132	(−1.48)	−0.0031	(−0.50)	0.0030	(0.70)	0.0017
SRD	0.0116	(0.76)	−0.0252	(−1.30)	0.0105	(0.71)	−0.0084	(−0.92)	−0.0034
SNI	0.0214	(2.03)	0.0356	(2.16)	0.0399	(3.18)	0.0341	(4.27)	0.0296
SLNI	0.0512	(3.18)	0.0355	(2.54)	0.0662	(3.70)	0.0020	(0.17)	−0.0019
SLIQ	−0.0040	(−1.24)	−0.0014	(−0.35)	−0.0100	(−2.79)	−0.0019	(−0.80)	−0.0040
F−Value	269.63		178.52		223.83		621.17		988.76
adjusted r²	0.910		0.870		0.893		0.959		0.973

	1987		1988		1989		1990	
SLDIV	1.0492	(29.09)	0.8635	(20.17)	0.9952	(43.63)	0.3440	(7.09)	0.9682
SCE	0.0110	(0.62)	−0.0038	(−0.14)	−0.0185	(−1.53)	−0.0517	(−1.42)	−0.0057
SEF	0.0718	(11.28)	−0.0249	(−2.69)	0.0010	(0.17)	0.0091	(0.58)	−0.0000
SRD	0.0325	(1.72)	−0.0251	(−0.97)	−0.0230	(−1.73)	−0.0248	(−0.59)	0.0144
SNI	0.0608	(4.80)	0.0565	(3.75)	0.0497	(4.52)	0.2138	(5.49)	0.0258
SLNI	−0.0028	(−0.28)	0.0448	(2.16)	0.0055	(0.69)	−0.0611	(−1.45)	0.0337
SLIQ	−0.0040	(−0.98)	0.0071	(1.17)	0.0085	(2.90)	0.0182	(1.94)	0.0014
F−Value	188.76		103.97		475.02		37.53		429.04
adjusted r²	0.883		0.807		0.951		0.599		0.945

OLS   SEF

	1978		1979		1980		1981	
SDIV	−0.7377	(−2.80)	−0.7882	(−3.06)	−0.6494	(−1.89)	−0.7264	(−2.25)	−1.6267
SCE	0.3591	(3.99)	0.4527	(5.10)	0.5522	(5.78)	0.6765	(6.52)	0.4872
SRD	−0.1330	(−0.89)	−0.1077	(−0.73)	−0.0329	(−0.21)	0.2426	(1.59)	−0.3545
KD	0.7826	(3.13)	0.4948	(2.04)	0.9563	(3.66)	0.5363	(2.16)	0.9375
SDEP	−0.4236	(−1.69)	−0.9285	(−4.00)	−0.9827	(−3.46)	−0.9806	(−3.32)	−0.7641
SNI	0.2042	(2.05)	−0.0246	(−0.20)	0.0088	(0.07)	−0.2984	(−2.54)	0.1129
DE	−0.0845	(−2.05)	−0.0467	(−1.09)	−0.0013	(−0.03)	−0.0216	(0.44)	0.0735
F−Value	9.36		10.41		8.85		10.74		5.86
adjusted r²	0.239		0.262		0.228		0.267		0.152

1982		1983		1984		1985		1986	
(−0.99)	−0.2105	(−0.91)	−0.2400	(−1.66)	−0.4400	(−2.25)	−0.0334	(−0.19)	
(3.41)	0.0124	(−0.36)	0.0689	(1.88)	0.2013	(3.52)	0.0280	(0.80)	
(4.50)	0.4087	(4.10)	0.4747	(5.58)	0.6556	(6.36)	0.3701	(4.38)	
(4.15)	0.3440	(3.83)	0.2160	(3.39)	0.2553	(2.96)	0.0170	(0.33)	
(−7.37)	−0.1820	(−7.71)	−0.1397	(−8.43)	−0.1263	(−5.74)	−0.0821	(−4.34)	
(−0.37)	−0.0061	(−0.43)	0.0190	(−1.67)	0.0250	(1.15)	−0.0164	(−0.83)	
	11.91		22.15		18.44		5.55		
	0.262		0.407		0.361		0.132		

1991		1992		1993		1994		1995	
(−0.50)	−0.0113	(−0.08)	−0.0289	(−0.24)	−0.1314	(−1.92)	−0.4438	(−1.84)	
(1.71)	0.0308	(1.23)	0.0185	(0.87)	−0.0008	(−0.03)	0.0302	(1.11)	
(4.24)	0.2694	(2.97)	0.4477	(5.30)	0.3267	(3.52)	0.3655	(2.74)	
(0.57)	0.0330	(0.53)	0.0171	(0.36)	0.1166	(2.58)	0.1913	(1.89)	
(−3.14)	−0.0518	(−2.52)	−0.0501	(−2.47)	−0.0595	(−2.59)	−0.0688	(−2.40)	
(0.57)	−0.0021	(−0.08)	−0.0038	(−0.18)	−0.0279	(−1.17)	−0.0091	(−0.81)	
	2.55		5.09		4.04		3.02		
	0.051		0.125		0.096		0.067		

1982		1983		1984		1985		1986	
(63.64)	0.8585	(26.18)	0.9662	(56.83)	0.7831	(23.20)	1.0185	(73.11)	
(0.08)	0.0069	(0.66)	−0.0178	(−2.29)	−0.0110	(−0.77)	−0.0005	(−0.09)	
(0.76)	−0.0271	(−5.54)	−0.0011	(−0.29)	−0.0054	(−0.47)	−0.0017	(−0.67)	
(−0.49)	−0.0078	(−0.55)	0.0032	(0.34)	−0.0470	(−2.18)	−0.0071	(−1.08)	
(4.63)	0.0414	(3.62)	0.0448	(6.89)	0.0750	(3.81)	0.0104	(2.63)	
(−0.33)	0.0074	(0.59)	0.0265	(3.24)	0.0297	(1.61)	0.0117	(1.60)	
(−2.25)	−0.0061	(−1.57)	−0.0083	(−3.94)	−0.0043	(−0.95)	0.0011	(0.74)	
	178.87		979.57		150.49		1340.05		
	0.871		0.974		0.850		0.981		

1991		1992		1993		1994		1995	
(32.12)	0.7946	(20.06)	0.8766	(23.24)	0.5902	(4.07)	0.1031	(4.18)	
(−0.43)	−0.0165	(−0.76)	0.0153	(0.70)	−0.1540	(−1.97)	−0.0366	(−1.55)	
(−0.01)	−0.0043	(−0.61)	0.0049	(0.83)	−0.0197	(−0.89)	−0.0000	(−0.00)	
(0.89)	0.0314	(1.20)	−0.0137	(−0.54)	−0.0396	(−0.41)	0.0205	(0.49)	
(1.80)	0.0210	(1.12)	0.0197	(1.40)	0.1251	(2.65)	0.1380	(4.69)	
(2.19)	0.0758	(2.89)	0.0531	(3.40)	0.2021	(3.65)	0.0141	(1.12)	
(0.39)	−0.0148	(−2.47)	0.0117	(2.04)	0.0004	(0.02)	−0.0233	(−2.65)	
	117.69		143.09		16.39		9.14		
	0.824		0.853		0.387		0.252		

1982		1983		1984		1985		1986	
(−3.43)	1.5742	(3.19)	−0.5307	(−1.84)	−0.6170	(−2.59)	−0.2996	(−0.82)	
(3.68)	0.1025	(0.16)	0.4955	(3.32)	0.2500	(2.73)	0.0821	(0.46)	
(−1.59)	0.0313	(0.14)	−0.1244	(−0.69)	−0.2464	(−1.76)	0.3015	(1.48)	
(2.93)	0.7319	(2.09)	0.1460	(0.54)	0.7988	(3.90)	0.8943	(2.58)	
(−2.06)	−0.2259	(−0.54)	−0.7491	(−2.52)	0.2077	(0.94)	−0.0138	(−0.04)	
(0.56)	0.1880	(0.99)	−0.0368	(−0.28)	0.1534	(1.45)	0.2066	(1.67)	
(0.99)	−0.2123	(−3.08)	0.0203	(0.39)	0.0299	(0.75)	−0.1702	(−2.89)	
	4.04		2.80		5.93		2.97		
	0.104		0.064		0.157		0.072		

(Continued on following page)

44 THE INTERDEPENDENCIES AMONG CORPORATE FINANCIAL POLICIES

TABLE 4.1 (Continued)

	1987		1988		1989		1990	
SDIV	1.4504	(4.34)	−0.2524	(−0.84)	0.5685	(2.06)	0.1032	(0.32)	−0.0232
SCE	−0.0051	(−0.03)	0.4155	(1.89)	0.1922	(0.99)	0.5002	(2.41)	0.2565
SRD	0.2090	(0.88)	0.3090	(1.41)	0.0175	(0.09)	−0.0306	(−0.14)	−0.1515
KD	0.5441	(1.43)	0.0696	(0.17)	0.7893	(2.08)	0.3099	(0.71)	0.6227
SDEP	−0.5592	(−1.68)	−0.1380	(−0.35)	0.3425	(0.89)	−0.4333	(−1.06)	−0.0140
SNI	−0.3687	(−2.35)	0.6331	(5.91)	−0.1220	(−0.80)	0.1269	(0.71)	0.0243
TE	−0.2095	(−3.44)	−0.3483	(−6.33)	−0.0715	(−1.38)	−0.2208	(−3.94)	−0.1701
F−Value	5.91		10.98		2.54		4.64		4.20
adjusted r²	0.165		0.289		0.060		0.130		0.121

OLS SRD

	1978		1979		1980		1981	
SLRD	1.1771	(66.00)	1.2245	(69.22)	1.2793	(41.02)	1.2366	(77.80)	1.1650
SCE	0.0260	(3.13)	0.0279	(3.74)	0.0356	(2.83)	−0.0050	(−0.64)	0.0042
SDIV	−0.0601	(−2.19)	−0.0942	(−3.76)	−0.1713	(−3.53)	−0.1115	(−4.01)	−0.1239
SEF	−0.0017	(−0.22)	0.0000	(0.00)	0.0004	(0.03)	0.0130	(2.00)	−0.0032
SNI	0.0245	(2.65)	0.0304	(2.90)	0.0530	(3.01)	0.0455	(4.95)	0.0435
F−Value	1047.26		1242.89		399.96		1292.41		1742.31
adjusted r²	0.966		0.971		0.915		0.972		0.979

	1987		1988		1989		1990	
SLRD	1.0211	(71.18)	1.1465	(68.20)	0.9887	(70.45)	1.0335	(69.36)	0.9782
SCE	0.0152	(1.12)	0.0158	(1.01)	−0.0011	(−0.09)	0.0056	(0.45)	0.0178
SDIV	−0.0524	(−2.15)	−0.0358	(−1.57)	−0.0410	(−1.76)	−0.0620	(−2.64)	−0.0541
SEF	−0.0084	(−1.61)	−0.0237	(−4.28)	−0.0016	(−0.24)	0.0006	(0.11)	0.0010
SNI	0.0329	(3.29)	0.0081	(0.93)	0.0531	(4.66)	0.0628	(5.03)	0.0595
F−Value	1088.78		1009.68		1108.99		1034.54		1459.47
adjusted r²	0.969		0.967		0.970		0.968		0.977

2SLS SCE

	1978		1979		1980		1981	
SDIV	0.1233	(0.43)	−0.0638	(−0.23)	0.4386	(0.88)	0.3222	(0.89)	0.7541
SEF	0.4549	(3.06)	0.3741	(2.53)	1.0365	(3.15)	0.8492	(4.49)	0.8605
SRD	0.4569	(3.17)	0.5467	(4.22)	0.6647	(3.35)	0.4122	(2.69)	0.8382
SNI	0.2805	(3.20)	0.4272	(4.14)	0.4659	(3.00)	0.4977	(4.74)	0.3294
SLIQ	−0.1363	(−5.03)	−0.2003	(−8.53)	−0.1777	(−4.29)	−0.1779	(−5.36)	−0.2177
SLSAL	−0.0176	(−0.87)	0.0139	(0.85)	−0.0216	(−1.15)	0.0245	(1.46)	−0.0422
F−Value	10.83		14.62		7.94		11.61		5.69
adjusted r²	0.241		0.305		0.183		0.254		0.130

	1987		1988		1989		1990	
SDIV	−0.2315	(−1.42)	−0.0308	(−0.20)	−0.7157	(−1.90)	0.2360	(0.83)	−0.0492
SEF	−0.1439	(−1.47)	0.1623	(2.30)	1.1303	(2.36)	0.0465	(0.62)	0.0898
SRD	0.2282	(2.36)	0.2175	(2.75)	0.2315	(1.13)	0.3090	(3.19)	0.3761
SNI	0.0437	(0.63)	−0.0835	(−1.23)	0.3884	(1.95)	−0.0185	(−0.16)	0.0405
SLIQ	−0.0597	(−3.22)	−0.0275	(−1.50)	0.0033	(0.07)	−0.0615	(−2.91)	−0.0611
SLSAL	0.0138	(0.80)	0.0294	(1.62)	−0.0003	(−0.01)	0.0159	(0.83)	0.0137
F−Value	3.52		4.39		1.76		2.69		4.01
adjusted r²	0.080		0.106		0.026		0.056		0.095

SIMULTANEOUS EQUATION ESTIMATION RESULTS   45

1991		1992		1993		1994		1995	
(−0.08)	−0.2017	(−0.49)	−0.2593	(−0.56)	−0.1454	(−0.61)	−0.0024	(−0.00)	
(1.39)	0.5356	(1.92)	0.3919	(1.12)	0.1991	(0.60)	0.5254	(1.92)	
(−0.74)	0.1766	(0.56)	−0.3496	(−0.98)	0.2445	(0.70)	−0.0631	(−0.16)	
(1.62)	0.5827	(1.14)	0.6631	(0.90)	3.3061	(4.14)	0.2581	(0.33)	
(−0.04)	−0.5622	(−0.98)	0.0553	(0.09)	−1.7959	(−2.65)	−0.8683	(−1.45)	
(0.15)	0.1037	(0.56)	0.1849	(0.91)	−0.1241	(−0.67)	−0.2529	(−0.82)	
(−3.30)	−0.2045	(−2.71)	−0.1392	(−1.59)	−0.0111	(−0.13)	−0.1194	(−1.54)	
	2.33		1.38		4.71		1.37		
	0.051		0.015		0.132		0.015		

1982		1983		1984		1985		1986	
(90.22)	1.1364	(67.71)	1.1245	(64.59)	1.1730	(68.35)	1.0803	(102.65)	
(0.65)	0.0263	(2.60)	0.0652	(5.67)	0.0546	(6.08)	0.0213	(2.53)	
(−4.79)	−0.1206	(−3.45)	−0.0465	(−1.77)	−0.0694	(−2.69)	−0.0356	(−1.78)	
(−0.80)	−0.0019	(−0.36)	−0.0017	(−0.25)	0.0021	(0.27)	0.0028	(0.66)	
(4.53)	0.0460	(3.74)	0.0153	(1.44)	0.0179	(1.67)	0.0099	(1.72)	
	974.67		1026.86		1368.13		2261.05		
	0.964		0.965		0.974		0.984		

1991		1992		1993		1994		1995	
(80.62)	1.0369	(67.54)	1.0485	(92.34)	1.0103	(76.51)	0.8704	(34.87)	
(1.76)	0.0320	(2.58)	0.0207	(2.19)	0.0276	(2.49)	0.0229	(1.43)	
(−2.88)	−0.0073	(−0.33)	0.0152	(1.00)	−0.0335	(−3.34)	−0.0025	(−0.05)	
(0.21)	0.0003	(0.07)	0.0004	(0.17)	0.0031	(1.00)	0.0025	(0.44)	
(5.90)	0.0160	(1.79)	0.0081	(1.41)	0.0294	(4.74)	0.0718	(3.58)	
	957.97		2001.05		1255.92		317.53		
	0.965		0.983		0.974		0.904		

1982		1983		1984		1985		1986	
(1.31)	−0.3004	(−0.90)	−0.2850	(−1.68)	−0.5699	(−2.27)	0.1147	(0.54)	
(3.67)	−0.0387	(−0.37)	−0.0152	(−0.09)	0.4449	(3.44)	0.3306	(3.03)	
(3.71))	0.4122	(3.99)	0.3927	(4.18)	0.6262	(5.62)	0.3630	(3.57)	
(1.73)	0.3717	(4.03)	0.2218	(3.39)	0.2937	(2.99)	0.0140	(0.22)	
(−4.49)	−0.1874	(−7.05)	−0.1383	(−7.88)	−0.1155	(−4.88)	−0.0757	(−3.33)	
(−1.26)	−0.0095	(−0.65)	0.0250	(1.89)	0.0208	(0.89)	−0.0264	(−1.11)	
	12.04		19.30		18.00		5.29		
	0.265		0.372		0.355		0.126		

1991		1992		1993		1994		1995	
(−0.32)	−0.0440	(−0.27)	−0.0343	(−0.25)	−0.2758	(−2.12)	−0.0524	(−0.09)	
(1.16)	0.0309	(0.19)	0.0197	(0.76)	−0.0089	(−0.18)	0.0519	(0.77)	
(3.99)	0.2448	(2.64)	0.4393	(5.16)	0.2984	(2.99)	0.3540	(2.50)	
(0.50)	0.0412	(0.62)	0.0181	(0.37)	0.1561	(2.59)	0.1397	(1.01)	
(−2.96)	−0.0498	(−1.34)	−0.0492	(−2.39)	−0.0520	(−2.14)	−0.0617	(−2.06)	
(0.59)	−0.0032	(−0.12)	−0.0037	(−0.18)	−0.0351	(−1.21)	−0.0085	(−0.74)	
	2.02		4.83		3.71		2.05		
	0.034		0.118		0.087		0.036		

(Continued on following page)

TABLE 4.1 (Continued)

2SLS SDIV

	1978		1979		1980		1981	
SLDIV	1.1050	(29.34)	0.9805	(21.90)	0.8416	(17.08)	1.0852	(45.06)	1.0491
SCE	−0.0148	(−1.19)	−0.0118	(−0.75)	−0.0319	(−2.62)	0.0066	(0.65)	−0.0112
SEF	0.0071	(0.47)	−0.0048	(−0.20)	−0.0224	(−0.67)	−0.0081	(−0.51)	0.0177
SRD	0.0148	(0.90)	−0.0121	(−0.57)	0.0070	(0.37)	−0.0017	(−0.17)	0.0095
SNI	0.0221	(2.04)	0.0442	(2.45)	0.0343	(2.45)	0.0346	(4.18)	0.0295
SLNI	0.0522	(3.21)	0.0380	(2.60)	0.0694	(3.19)	−0.0008	(−0.05)	0.0027
SLIQ	−0.0048	(−1.33)	−0.0060	(−1.30)	−0.0090	(−2.23)	−0.0027	(−0.92)	−0.0067
F−Value	268.86		174.05		209.61		589.32		769.22
adjusted r^2	0.910		0.867		0.887		0.957		0.966

	1987		1988		1989		1990	
SLDIV	1.0459	(27.58)	0.8571	(18.64)	1.0251	(25.61)	0.3391	(6.70)	0.9670
SCE	0.0118	(0.41)	0.0168	(0.45)	0.0005	(0.02)	0.0021	(0.05)	0.0034
SEF	0.0557	(2.46)	−0.0674	(−4.27)	−0.0505	(−1.00)	−0.0282	(−0.86)	−0.0071
SRD	0.0331	(1.55)	−0.0291	(−1.02)	−0.0254	(−1.50)	−0.0397	(−0.88)	0.0089
SNI	0.0591	(4.49)	0.0800	(4.61)	0.0382	(2.21)	0.2281	(5.66)	0.0245
SLNI	−0.0052	(−0.49)	0.0372	(1.68)	0.0004	(0.03)	−0.0743	(−1.69)	0.0346
SLIQ	−0.0044	(−1.02)	0.0028	(0.42)	0.0072	(1.89)	0.0177	(1.73)	0.0015
F−Value	165.05		93.99		326.02		35.76		425.05
adjusted r^2	0.868		0.791		0.930		0.587		0.945

2SLS SEF

	1978		1979		1980		1981	
SDIV	−0.8687	(−3.07)	−0.6927	(−2.38)	−0.5672	(−1.46)	−0.8734	(−2.56)	−1.6995
SCE	0.4222	(3.12)	0.2847	(2.23)	0.3288	(2.45)	0.3886	(2.40)	0.4923
SRD	−0.1463	(−0.94)	−0.0904	(−0.59)	−0.0324	(−0.19)	0.1675	(1.06)	−0.3354
KD	0.7717	(3.08)	0.4223	(1.71)	0.9690	(3.64)	0.6257	(2.46)	0.8456
SDEP	−0.5153	(−1.75)	−0.6788	(−2.51)	−0.5852	(−1.74)	−0.4411	(−1.18)	−0.7891
SNI	0.1998	(1.90)	−0.0055	(−0.04)	0.0724	(0.53)	−0.1791	(−1.40)	0.1133
DE	−0.0733	(−1.75)	−0.0704	(−1.58)	−0.0197	(−0.38)	0.0189	(0.37)	0.0717
F−Value	8.89		6.91		4.46		5.51		4.94
adjusted r^2	0.229		0.182		0.115		0.144		0.127

	1987		1988		1989		1990	
SDIV	−0.0181	(−0.04)	−0.3210	(−0.96)	0.5755	(2.00)	−0.7732	(−1.31)	−0.0912
SCE	−0.2886	(−0.81)	0.3905	(1.22)	0.1040	(0.30)	0.0964	(0.31)	0.0023
SRD	0.2279	(0.88)	0.3725	(1.67)	0.0093	(0.05)	−0.1170	(−0.51)	−0.1778
KD	0.6252	(1.55)	0.0558	(0.13)	0.7930	(2.08)	0.2232	(0.50)	0.6433
SDEP	−0.3642	(−0.95)	−0.1794	(−0.40)	0.4581	(0.87)	0.0642	(0.12)	0.3060
SNI	−0.0599	(−0.34)	0.6387	(5.87)	−0.1125	(−0.71)	0.4855	(1.96)	0.0822
DE	−0.1946	(−2.98)	−0.3520	(−6.32)	−0.0739	(−1.41)	−0.2146	(−3.58)	−0.1767
F−Value	3.01		10.85		2.41		3.83		4.10
adjusted r^2	0.075		0.286		0.055		0.104		0.111

2SLS SRD

	1978		1979		1980		1981	
SLRD	1.1752	(61.37)	1.2312	(64.83)	1.2971	(35.58)	1.2303	(73.23)	1.1766
SCE	0.0164	(1.45)	0.0192	(2.06)	0.0167	(0.96)	0.0097	(0.90)	−0.0127
SDIV	−0.0724	(−2.16)	−0.0889	(−2.73)	−0.1402	(−2.08)	−0.0763	(−2.14)	−0.0993
SEF	−0.0035	(−0.22)	0.0203	(1.11)	0.0521	(1.03)	0.0303	(1.46)	0.0136
SNI	0.0277	(2.90)	0.0358	(3.21)	0.0570	(2.99)	0.0428	(4.15)	0.0423
F−Value	1035.87		1183.42		352.96		1179.61		1563.35
adjusted r^2	0.965		0.970		0.904		0.969		0.976

1982		1983		1984		1985		1986	
(45.29)	0.9384	(16.70)	0.9457	(41.18)	0.7343	(18.83)	1.0182	(67.49)	
(−1.34)	−0.0106	(−0.63)	−0.0221	(−1.56)	−0.0987	(−3.76)	−0.0012	(−0.14)	
(1.63)	−0.0714	(−3.50)	−0.0383	(−2.13)	−0.0007	(−0.02)	−0.0018	(−0.26)	
(0.90)	−0.0015	(−0.08)	−0.0007	(−0.06)	0.0211	(0.74)	−0.0068	(−0.96)	
(3.98)	0.0419	(2.94)	0.0516	(5.94)	0.0893	(3.86)	0.0104	(2.58)	
(0.38)	0.0153	(0.98)	0.0220	(2.08)	0.0395	(1.71)	0.0119	(1.51)	
(−2.63)	−0.0159	(−2.71)	−0.0100	(−3.33)	−0.0161	(−2.94)	0.0010	(0.63)	
	119.84		631.24		125.50		1339.80		
	0.819		0.960		0.825		0.981		

1991		1992		1993		1994		1995	
(35.27)	0.8023	(16.96)	0.8769	(23.15)	0.6126	(4.08)	0.1096	(4.33)	
(0.21)	−0.0337	(−1.04)	0.0251	(0.90)	−0.1172	(−1.27)	0.0162	(0.50)	
(−0.52)	0.0516	(0.99)	0.0028	(0.40)	−0.0014	(−0.03)	−0.0006	(−0.03)	
(0.52)	0.0357	(1.13)	−0.0182	(−0.68)	0.0136	(0.13)	−0.0039	(−0.09)	
(1.67)	0.0244	(1.09)	0.0200	(1.42)	0.1304	(2.71)	0.1343	(4.38)	
(2.18)	0.0669	(2.04)	0.0525	(3.36)	0.1958	(3.51)	0.0110	(0.86)	
(0.42)	−0.0048	(−0.39)	0.0118	(2.00)	−0.0053	(−0.21)	−0.0197	(−2.18)	
	85.44		142.79		15.78		8.56		
	0.773		0.852		0.377		0.239		

1982		1983		1984		1985		1986	
(−3.52)	2.8613	(5.28)	−0.5499	(−1.83)	−0.7401	(−2.55)	−0.2134	(−0.56)	
(2.58)	0.1298	(0.56)	0.5013	(2.11)	0.1851	(1.22)	0.5169	(1.64)	
(−1.48)	0.0481	(0.20)	−0.1212	(−0.66)	−0.2608	(−1.82)	0.2715	(1.29)	
(2.62)	0.7040	(1.95)	0.0700	(0.25)	0.7678	(3.74)	0.8065	(2.27)	
(−1.87)	−0.1295	(−0.27)	−0.7625	(−2.10)	0.2866	(1.10)	−0.4295	(−1.04)	
(0.55)	0.0400	(0.20)	−0.0421	(−0.29)	0.1873	(1.60)	0.1195	(0.88)	
(0.96)	−0.2569	(−3.61)	0.0182	(0.35)	0.0331	(0.83)	−0.1596	(−2.64)	
	6.50		1.90		5.45		3.16		
	0.173		0.033		0.144		0.078		

1991		1992		1993		1994		1995	
(−0.30)	−0.0025	(−0.00)	−0.4614	(−0.89)	−0.4783	(−1.11)	−1.2274	(−0.68)	
(0.01)	0.2791	(0.65)	0.6384	(1.19)	0.3202	(0.71)	0.8902	(1.68)	
(−0.84)	0.1506	(0.46)	−0.3790	(−1.04)	0.3412	(0.95)	−0.1190	(−0.28)	
(1.66)	0.5959	(1.14)	0.5226	(0.70)	3.3587	(4.28)	0.1278	(0.16)	
(0.67)	−0.2410	(−0.34)	−0.2163	(−0.29)	−2.0412	(−2.59)	−1.3929	(−1.59)	
(0.48)	0.0991	(0.50)	0.2093	(1.01)	−0.0270	(−0.12)	−0.1456	(−0.39)	
(−3.37)	−0.2122	(−2.79)	−0.1417	(−1.61)	−0.0254	(−0.29)	−0.1190	(−1.47)	
	1.82		1.45		4.80		1.22		
	0.032		0.018		0.134		0.009		

1982		1983		1984		1985		1986	
(72.51)	1.1229	(57.26)	1.1297	(59.69)	1.1779	(61.07)	1.0767	(96.70)	
(−1.34)	0.0367	(2.65)	0.0335	(2.16)	0.0672	(4.82)	0.0334	(2.62)	
(−2.74)	−0.0221	(−0.40)	−0.0729	(−2.28)	0.0069	(0.21)	−0.0402	(−1.87)	
(0.87)	−0.0340	(−2.11)	−0.0085	(−0.32)	0.0214	(1.15)	−0.0087	(−0.74)	
(4.01)	0.0331	(2.26)	0.0239	(2.12)	−0.0005	(−0.04)	0.0104	(1.75)	
	796.06		968.72		1260.57		2147.86		
	0.956		0.963		0.971		0.984		

(Continued on following page)

TABLE 4.1 (Continued)

	1987		1988		1989		1990	
SLRD	1.0221	(65.64)	1.1535	(63.20)	0.9834	(52.57)	1.0346	(63.74)	0.9823
SCE	0.0037	(0.18)	−0.0129	(−0.59)	0.0231	(0.83)	0.0097	(0.64)	−0.0008
SDIV	−0.0486	(−1.77)	−0.0912	(−3.29)	−0.0083	(−0.22)	−0.1069	(−2.37)	−0.0646
SEF	−0.0125	(−0.75)	−0.0448	(−4.77)	−0.0695	(−1.46)	0.0088	(0.74)	−0.0002
SNI	0.0313	(2.99)	0.0240	(2.33)	0.0309	(1.46)	0.0788	(4.24)	0.0622
F−Value	1078.66		889.84		671.97		994.75		1426.74
adjusted r^2	0.969		0.963		0.952		0.967		0.976

3SLS   SCE

	1978		1979		1980		1981	
SDIV	0.4804	(1.72)	0.0638	(0.23)	0.8640	(1.92)	0.7065	(2.13)	1.3640
SEF	0.7140	(5.37)	0.4823	(3.31)	1.3228	(5.39)	1.1541	(8.35)	1.0388
SRD	0.4659	(3.27)	0.5560	(4.29)	0.6110	(3.13)	0.2435	(1.68)	0.7027
SNI	0.1885	(2.20)	0.4123	(4.01)	0.2891	(1.98)	0.4576	(4.45)	0.0501
SLIQ	−0.0941	(−3.85)	−0.1892	(−8.16)	−0.1025	(−3.40)	−0.1096	(−4.17)	−0.0830
SLSAL	−0.0086	(−0.49)	0.0172	(1.09)	0.0070	(0.58)	0.0268	(2.47)	−0.0107

	1987		1988		1989		1990	
SDIV	−0.2527	(−1.55)	−0.0264	(−0.17)	−0.6313	(−1.83)	0.2756	(0.97)	−0.0540
SEF	−0.1725	(−1.84)	0.1842	(2.67)	1.0803	(3.43)	0.0509	(0.68)	0.1134
SRD	0.2240	(2.35)	0.2248	(2.86)	0.2767	(1.41)	0.3147	(3.25)	0.3802
SNI	0.0471	(0.69)	−0.0892	(−1.34)	0.3484	(1.92)	−0.0385	(−0.33)	0.0468
SLIQ	−0.0617	(−3.63)	−0.0187	(−1.07)	−0.0002	(−0.01)	−0.0584	(−2.78)	−0.0588
SLSAL	0.0108	(0.70)	0.0118	(0.68)	0.0110	(0.74)	0.0261	(1.40)	0.0077

3SLS   SDIV

	1978		1979		1980		1981	
SLDIV	1.1024	(29.30)	0.9819	(22.01)	0.8278	(17.19)	1.0752	(45.02)	1.0728
SCE	−0.0183	(−1.48)	−0.0343	(−2.23)	−0.0128	(−1.09)	0.0156	(1.55)	−0.0265
SEF	0.0060	(0.40)	0.0058	(0.24)	−0.0481	(−1.47)	−0.0209	(−1.35)	0.0343
SRD	0.0159	(0.97)	−0.0003	(−0.01)	−0.0039	(−0.21)	−0.0026	(−0.26)	0.0197
SNI	0.0240	(2.21)	0.0547	(3.06)	0.0289	(2.10)	0.0303	(3.72)	0.0296
SLNI	0.0512	(3.16)	0.0361	(2.52)	0.0678	(3.22)	−0.0012	(−0.08)	0.0032
SLIQ	−0.0055	(−1.54)	−0.0098	(−2.15)	−0.0074	(−1.88)	−0.0021	(−0.72)	−0.0076

	1987		1988		1989		1990	
SLDIV	1.0366	(27.38)	0.8170	(18.27)	1.0516	(28.36)	0.3323	(6.71)	0.9624
SCE	−0.0136	(−0.47)	0.0586	(1.59)	0.0404	(1.76)	0.0946	(2.10)	0.0110
SEF	0.0508	(2.26)	−0.1036	(−6.99)	−0.0927	(−2.09)	−0.0837	(−2.69)	−0.0116
SRD	0.0377	(1.77)	−0.0350	(−1.25)	−0.0369	(−2.24)	−0.0758	(−1.70)	0.0051
SNI	0.0593	(4.50)	0.1014	(6.03)	0.0258	(1.58)	0.2428	(6.29)	0.0257
SLNI	−0.0029	(−0.28)	0.0394	(1.95)	−0.0004	(−0.05)	−0.0818	(−2.06)	0.0311
SLIQ	−0.0054	(−1.26)	−0.0018	(−0.29)	0.0074	(2.26)	0.0130	(1.40)	0.0014

3SLS   SEF

	1978		1979		1980		1981	
SDIV	−0.9375	(−3.34)	−0.7358	(−2.54)	−0.5091	(−1.42)	−0.8706	(−2.66)	−1.5488
SCE	0.6606	(5.33)	0.3913	(3.14)	0.5406	(4.76)	0.5136	(3.91)	0.7739
SRD	−0.2763	(−1.81)	−0.1649	(−1.08)	−0.2160	(−1.33)	−0.0636	(−0.43)	−0.5494
KD	0.6876	(3.17)	0.2469	(1.06)	0.8486	(5.11)	0.4174	(2.60)	0.5183
SDEP	−0.1577	(−0.61)	−0.5114	(−1.97)	0.1426	(0.60)	0.3909	(1.54)	0.0742
SNI	0.0819	(0.81)	−0.0541	(−0.43)	−0.0440	(−0.35)	−0.2081	(−1.83)	0.1172
DE	−0.0056	(−0.15)	−0.0508	(−1.20)	0.0236	(0.73)	0.0506	(1.55)	0.0523

1991		1992		1993		1994		1995	
(75.55)	1.0384	(64.71)	1.0521	(89.31)	1.0228	(69.78)	0.8683	(32.48)	
(−0.06)	0.0168	(1.05)	0.0138	(1.15)	0.0180	(1.27)	0.0138	(0.64)	
(−3.23)	−0.0099	(−0.38)	0.0195	(1.15)	−0.0208	(−1.06)	−0.1558	(−1.29)	
(−0.02)	−0.0033	(−0.23)	0.0022	(0.72)	0.0185	(3.12)	0.0090	(0.63)	
(5.94)	0.0163	(1.73)	0.0072	(1.23)	0.0292	(3.54)	0.0981	(3.69)	
	942.10		1988.86		1079.71		298.20		
	0.964		0.983		0.969		0.898		

1982		1983		1984		1985		1986	
(2.74)	−0.4221	(−1.27)	−0.2976	(−1.75)	−0.6136	(−2.57)	0.1911	(0.90)	
(6.40)	−0.0317	(−0.31)	0.0083	(0.05)	0.6409	(5.47)	0.5070	(5.40)	
(3.32)	0.4230	(4.10)	0.3975	(4.25)	0.6347	(5.73)	0.3144	(3.14)	
(0.29)	0.4086	(4.46)	0.2308	(3.55)	0.3015	(3.20)	−0.0159	(−0.26)	
(−2.54)	−0.1904	(−7.20)	−0.1354	(−7.78)	−0.0844	(−3.80)	−0.0295	(−1.53)	
(−0.63)	−0.0187	(−1.31)	0.0206	(1.60)	−0.0204	(−1.15)	−0.0149	(−0.82)	

1991		1992		1993		1994		1995	
(−0.35)	−0.0485	(−0.30)	−0.0137	(−0.10)	−0.3493	(−2.70)	0.1515	(0.25)	
(1.47)	0.0326	(0.21)	0.0549	(2.14)	−0.0387	(−0.81)	0.1183	(1.79)	
(4.03)	0.2463	(2.66)	0.4492	(5.28)	0.2776	(2.78)	0.3729	(2.64)	
(0.58)	0.0466	(0.70)	0.0135	(0.28)	0.1725	(2.88)	0.1234	(0.89)	
(−2.87)	−0.0512	(−1.38)	−0.0450	(−2.20)	−0.0505	(−2.10)	−0.0531	(−1.80)	
(0.34)	−0.0071	(−0.26)	−0.0034	(−0.17)	−0.0391	(−1.39)	−0.0094	(−0.84)	

1982		1983		1984		1985		1986	
(48.23)	0.9531	(17.63)	0.9152	(42.64)	0.6900	(18.40)	1.0187	(67.54)	
(−3.50)	−0.0410	(−2.57)	−0.0347	(−2.65)	−0.1643	(−7.33)	−0.0022	(−0.26)	
(3.46)	−0.0867	(−4.51)	−0.0740	(−4.63)	0.0200	(0.67)	−0.0014	(−0.20)	
(1.96)	0.0081	(0.45)	−0.0007	(−0.05)	0.0616	(2.13)	−0.0062	(−0.88)	
(4.15)	0.0527	(3.76)	0.0594	(7.20)	0.1121	(5.45)	0.0104	(2.60)	
(0.50)	0.0150	(1.07)	0.0206	(2.37)	0.0302	(1.69)	0.0117	(1.48)	
(−3.28)	−0.0233	(−4.37)	−0.0122	(−4.73)	−0.0208	(−4.35)	0.0009	(0.55)	

1991		1992		1993		1994		1995	
(35.41)	0.7959	(17.44)	0.8738	(23.09)	0.5348	(3.66)	0.1066	(4.31)	
(0.66)	−0.0622	(−1.94)	0.0342	(1.23)	−0.3456	(−3.84)	0.0691	(2.18)	
(−0.85)	0.0605	(1.29)	−0.0033	(−0.46)	−0.0451	(−1.10)	−0.0191	(−0.96)	
(0.29)	0.0471	(1.52)	−0.0237	(−0.89)	0.0688	(0.66)	−0.0530	(−1.20)	
(1.76)	0.0215	(1.01)	0.0203	(1.44)	0.1476	(3.08)	0.1338	(4.39)	
(1.97)	0.0788	(2.81)	0.0529	(3.40)	0.1968	(3.64)	0.0050	(0.41)	
(0.38)	−0.0085	(−0.80)	0.0115	(1.96)	−0.0205	(−0.83)	−0.0154	(−1.77)	

1982		1983		1984		1985		1986	
(−3.31)	2.6984	(5.03)	−0.6411	(−2.16)	−0.3344	(−1.18)	−0.3627	(−0.97)	
(5.36)	−0.0450	(−0.21)	0.2547	(1.17)	0.6136	(4.53)	1.1907	(4.68)	
(−2.69)	−0.0380	(−0.16)	−0.2405	(−1.34)	−0.4128	(−2.96)	−0.1986	(−1.01)	
(3.16)	0.4768	(1.57)	−0.2881	(−1.23)	0.4017	(2.31)	0.5818	(2.16)	
(0.30)	0.3007	(0.75)	−0.1374	(−0.43)	−0.0239	(−0.11)	−0.0487	(−0.15)	
(0.64)	0.1423	(0.73)	0.0411	(0.30)	0.0284	(0.25)	0.1316	(1.07)	
(1.30)	−0.2842	(−4.33)	−0.0009	(−0.02)	0.0285	(0.80)	−0.0531	(−1.01)	

(Continued on following page)

TABLE 4.1 (Continued)

	1987		1988		1989		1990	
SDIV	−0.2486	(−0.61)	−0.5136	(−1.55)	0.6220	(2.21)	−0.7699	(−1.30)	−0.0881
SCE	−1.2665	(−4.03)	1.1041	(3.75)	0.7605	(5.90)	0.1488	(0.49)	0.1133
SRD	0.4066	(1.60)	0.0763	(0.35)	−0.1732	(−1.04)	−0.0538	(−0.24)	−0.2462
KD	0.6609	(1.89)	−0.0098	(−0.03)	0.6965	(4.86)	0.5271	(1.29)	0.6323
SDEP	−0.3504	(−1.04)	−0.2090	(−0.55)	0.1309	(0.58)	−0.0217	(−0.05)	0.3846
SNI	−0.0195	(−0.11)	0.6312	(5.84)	−0.2510	(−1.78)	0.4941	(2.01)	0.0854
DE	−0.1854	(−2.94)	−0.2592	(−5.00)	0.0032	(0.10)	−0.2131	(−3.70)	−0.1661

3SLS	SRD							
	1978		1979		1980		1981	
SLRD	1.1738	(61.31)	1.2357	(65.20)	1.3138	(36.28)	1.2290	(73.19)	1.1879
SCE	0.0165	(1.46)	0.0195	(2.10)	−0.0089	(−0.52)	0.0081	(0.75)	−0.0291
SDIV	−0.0760	(−2.27)	−0.0752	(−2.32)	−0.1117	(−1.67)	−0.0677	(−1.91)	−0.0745
SEF	−0.0063	(−0.39)	0.0344	(1.91)	0.0966	(1.97)	0.0384	(1.89)	0.0312
SNI	0.0281	(2.94)	0.0369	(3.32)	0.0620	(3.26)	0.0440	(4.28)	0.0411

	1987		1988		1989		1990	
SLRD	1.0234	(65.75)	1.1480	(62.93)	0.9684	(53.06)	1.0306	(63.56)	0.9853
SCE	−0.0064	(−0.13)	−0.0027	(−0.12)	0.0791	(3.64)	0.0283	(1.88)	−0.0131
SDIV	−0.0503	(−1.83)	−0.1127	(−4.07)	0.0275	(0.77)	−0.1072	(−2.38)	−0.0672
SEF	−0.0146	(−0.88)	−0.0587	(−6.39)	−0.1257	(−3.15)	0.0093	(0.79)	−0.0007
SNI	0.0312	(2.98)	0.0332	(3.23)	0.0111	(0.57)	0.0786	(4.24)	0.0625

new debt and capital expenditures equations. We find little differences among the 2SLS and 3SLS results and report the OLS, 2SLS, and 3SLS squares regression results.

The estimated OLS, 2SLS, and 3SLS system equations are shown in Table 4.1 for the 303-firm Guerard and McCabe (1992) sample. The initial regression results for the 1975–1982 period are substantiated in the longer 1978–1995 period. Capital expenditures are associated with higher R&D and new debt financing, and R&D is associated with higher capital expenditures. New debt is issued to finance capital expenditures.

The estimated system equations for the largest 3,000 securities annually in the United States are shown in Table 4.2. Here we find stronger evidence of the interdependencies of financial decisions in the larger universe than in the original Guerard and McCabe (1992) study. Dividends are an alternative use of funds in R&D management, whereas investments are positively associated with increasing research activities and new debt financing at the 10% level of significance. In the estimated dividend equation in Table 4.2, dividends are negatively associated with capital expenditures and positively associated with R&D and new debt financing, as one would expect in using

1991		1992		1993		1994		1995	
(−0.29)	0.0382	(0.08)	−0.4567	(−0.89)	−0.6685	(−1.55)	−1.4610	(−0.81)	
(0.41)	0.3842	(0.97)	1.1510	(2.17)	−0.2937	(−0.68)	1.2608	(2.44)	
(−1.17)	0.1085	(0.34)	−0.5721	(−1.57)	0.3599	(1.02)	−0.3790	(−0.89)	
(1.64)	0.7542	(1.71)	0.3756	(0.51)	3.1074	(4.21)	−0.3578	(−0.46)	
(0.84)	−0.1463	(−0.25)	−0.2793	(−0.37)	−1.5695	(−2.11)	−0.8828	(−1.04)	
(0.50)	0.0917	(0.48)	0.2066	(1.00)	0.0389	(0.18)	−0.1217	(−0.33)	
(−3.18)	−0.1968	(−2.77)	−0.1365	(−1.57)	−0.0021	(−0.03)	−0.1170	(−1.48)	

1982		1983		1984		1985		1986	
(74.07)	1.1154	(57.12)	1.1341	(59.97)	1.1780	(61.27)	1.0706	(96.38)	
(−3.20)	0.0478	(3.46)	0.0160	(1.04)	0.0754	(5.44)	0.0528	(4.23)	
(−2.08)	0.0020	(0.04)	−0.0827	(−2.60)	0.0385	(1.18)	−0.0452	(−2.11)	
(2.07)	−0.0474	(−3.06)	−0.0069	(−0.27)	0.0296	(1.60)	−0.0212	(−1.85)	
(3.90)	0.0293	(2.01)	0.0278	(2.47)	−0.0082	(−0.68)	0.0112	(1.88)	

1991		1992		1993		1994		1995	
(75.83)	1.0406	(64.86)	1.0532	(89.41)	1.0295	(70.38)	0.8626	(32.35)	
(−1.03)	0.0059	(0.37)	0.0125	(1.04)	0.0154	(1.09)	0.0163	(0.75)	
(−3.36)	−0.0108	(−0.42)	0.0204	(1.20)	−0.0122	(−0.63)	−0.2006	(−1.67)	
(−0.06)	−0.0024	(−0.17)	0.0036	(1.17)	0.0300	(5.27)	0.0104	(0.73)	
(5.97)	0.0166	(1.76)	0.0071	(1.21)	0.0291	(3.54)	0.1062	(4.01)	

the imperfect markets hypothesis. New debt financing is associated with higher capital expenditures, dividends, and R&D variables. R&D is associated with higher new debt financing and dividend levels and negatively associated with capital expenditures. There are significant violations of the independence (perfect markets) hypothesis in the capital expenditures, dividend, new debt, and research activities equation estimations, particularly in the larger 3,000-firm analysis.

The statistically significantly positive coefficient of the external funds issued variable is convincing in the investment equation and complements the work of McCabe (1979), Peterson and Benesh (1983), and Guerard and McCabe (1992); Dhrymes and Kurz (1967) and Switzer (1984) did not always find a significantly positive relationship between new debt and investments. Mueller found an inverse relationship between investments and research in his earlier study (1967) and no relationship between the variables in his later study with Grabowski (1972). Switzer (1984) found no significant association between R&D and investment. Decreases in net liquidity are associated with rising investment and dividends. Net income and depreciation positively affect investment and negatively affect new debt financing.

(Text continued on page 72)

TABLE 4.2 Estimated System Equations 3,000-Firm Compustat Universe

OLS   SCE

	1978 (t)		1979		1980		1981	
SDIV	0.0018	(0.12)	−0.0683	(−3.18)	−0.1996	(−8.78)	−0.2424	(−11.27)	−0.2194
SEF	0.1081	(11.68)	0.2762	(20.78)	0.1642	(14.12)	0.2642	(24.55)	0.3007
SRD	0.0757	(1.37)	0.0192	(0.38)	0.1979	(3.54)	0.1537	(3.00)	0.1653
SNI	0.0102	(0.57)	0.0324	(4.15)	0.1885	(9.12)	0.2458	(12.38)	0.1853
SLIQ	−0.0689	(−10.80)	−0.0663	(−10.42)	−0.0835	(−12.40)	−0.1113	(−16.19)	−0.1227
SLSAL	0.0277	(5.19)	0.0193	(3.49)	0.0015	(0.30)	0.0074	(1.50)	0.0024
F−Value	52.96		108.38		72.27		156.82		180.67
adjusted r²	0.120		0.218		0.154		0.278		0.300

	1987		1988		1989		1990	
SDIV	−0.0759	(−4.33)	−0.0876	(−4.07)	−0.0319	(−2.21)	−0.1259	(−6.11)	−0.0623
SEF	0.1117	(12.27)	0.0360	(4.71)	0.0785	(8.37)	0.0405	(5.47)	0.0695
SRD	0.0469	(1.53)	0.0372	(1.22)	−0.0263	(−1.07)	0.0922	(3.25)	0.0438
SNI	0.0526	(3.80)	0.0445	(2.95)	0.0067	(0.49)	0.0935	(5.70)	0.0326
SLIQ	−0.0574	(−7.94)	−0.0455	(−6.67)	−0.0550	(−7.02)	−0.0615	(−8.88)	−0.0455
SLSAL	0.0180	(3.36)	0.0286	(4.83)	0.0310	(4.41)	0.0270	(4.90)	0.0180
F−Value	42.30		17.15		27.06		27.57		25.61
adjusted r²	0.098		0.041		0.064		0.065		0.061

OLS   SDIV

	1978		1979		1980		1981	
SLDIV	0.7891	(153.22)	1.2708	(232.40)	1.0969	(116.77)	0.6933	(81.11)	1.0022
SCE	0.0015	(0.30)	−0.0014	(−0.37)	−0.0207	(−3.24)	−0.0442	(−5.97)	−0.0161
SEF	−0.0144	(−5.41)	0.0079	(2.95)	0.0199	(5.37)	0.0178	(4.03)	0.0078
SRD	−0.0004	(−0.03)	0.0133	(1.41)	0.0300	(1.75)	0.0891	(4.78)	0.0316
SNI	0.0753	(18.91)	0.0012	(0.80)	0.0811	(13.47)	0.1778	(22.69)	0.1348
SLNI	−0.0171	(−3.77)	0.0171	(5.63)	0.0299	(6.39)	−0.0760	(−7.63)	−0.0619
SLIQ	−0.0008	(−0.47)	−0.0022	(−1.83)	−0.0088	(−4.11)	−0.0178	(−6.75)	−0.0115
F−Value	57833.77		10888.74		10778.86		8258.84		6735.49
adjusted r²	0.994		0.971		0.970		0.960		0.949

	1987		1988		1989		1990	
SLDIV	0.6987	(53.85)	0.7234	(101.39)	0.5235	(23.34)	0.3026	(28.84)	0.9158
SCE	−0.0429	(−2.75)	−0.0265	(−3.25)	−0.0939	(−3.74)	−0.0953	(−5.74)	−0.0516
SEF	0.0342	(4.97)	0.0078	(2.58)	0.0231	(2.05)	0.0248	(4.21)	0.0153
SRD	0.1757	(7.74)	0.0681	(5.52)	0.2748	(8.80)	0.2541	(11.27)	0.1322
SNI	0.1917	(18.03)	0.0581	(8.85)	0.0912	(4.39)	0.2628	(19.69)	0.1424
SLNI	−0.0670	(−5.11)	0.0081	(1.38)	0.1728	(7.58)	0.1833	(14.10)	−0.0158
SLIQ	−0.0242	(−4.44)	−0.0155	(−5.78)	−0.0519	(−5.55)	−0.0623	(−11.34)	−0.0188
F−Value	1432.66		3555.75		537.49		951.48		815.08
adjusted r²	0.816		0.916		0.619		0.745		0.714

OLS   SEF

	1978		1979		1980		1981	
SDIV	−0.1887	(−6.95)	−0.0950	(−3.16)	0.3263	(8.87)	0.4125	(12.33)	0.1764
SCE	0.6022	(13.35)	0.6473	(21.82)	0.5989	(15.40)	0.8998	(27.83)	0.8918
SRD	0.5512	(5.25)	0.3262	(4.76)	0.3339	(3.75)	0.7862	(9.89)	0.4275
KD	2.2105	(17.47)	0.0679	(1.53)	0.4296	(8.49)	0.8635	(11.17)	0.8354
SDEP	−0.4794	(−3.78)	−0.5046	(−5.80)	−0.5470	(−4.72)	−1.2234	(−11.35)	−1.3605
SNI	0.2755	(8.04)	0.0277	(2.28)	−0.3732	(−11.26)	−0.4721	(−15.13)	−0.2060
DE	−0.1678	(−10.85)	−0.1500	(−13.79)	−0.1416	(−9.97)	0.0215	(1.45)	0.0540
F−Value	124.49		121.97		106.87		190.35		171.99
adjusted r²	0.275		0.268		0.240		0.353		0.322

52

1982		1983		1984		1985		1986	
(−8.26)	−0.1184	(−7.30)	−0.0871	(−4.68)	−0.1559	(−7.39)	−0.1098	(−5.74)	
(27.45)	0.2245	(17.93)	0.1220	(13.64)	0.1296	(11.61)	0.0856	(8.20)	
(2.97)	0.2368	(6.15)	0.1649	(5.04)	0.1686	(5.78)	0.1872	(5.54)	
(8.40)	0.1274	(7.71)	0.0569	(3.97)	0.1271	(7.05)	0.0746	(4.93)	
(−15.86)	−0.0887	(−12.81)	−0.0819	(−13.68)	−0.0683	(−9.74)	−0.0902	(−11.76)	
(0.45)	−0.0058	(−1.19)	0.0234	(4.75)	0.0178	(4.59)	0.0159	(2.65)	
	90.66		66.68		49.10		40.70		
	0.196		0.145		0.110		0.093		

1991		1992		1993		1994		1995	
(−4.25)	−0.0710	(−4.48)	−0.0763	(−5.12)	−0.0984	(−4.88)	−0.1028	(−5.18)	
(8.48)	0.0298	(4.94)	0.0389	(5.72)	0.0388	(4.91)	0.0252	(3.31)	
(1.96)	0.0306	(1.40)	0.1082	(4.90)	0.1129	(4.36)	0.0685	(2.50)	
(2.92)	0.0404	(3.37)	0.0526	(4.33)	0.0773	(4.51)	0.0830	(4.86)	
(−6.98)	−0.0441	(−7.76)	−0.0475	(−7.58)	−0.0425	(−5.73)	−0.0413	(−5.24)	
(3.22)	0.0243	(4.45)	0.0277	(4.60)	0.0251	(3.66)	0.0228	(3.14)	
	17.99		21.14		14.84		11.32		
	0.044		0.054		0.040		0.030		

1982		1983		1984		1985		1986	
(97.32)	0.7825	(52.38)	0.7868	(136.24)	0.6930	(66.50)	0.9131	(105.77)	
(−2.52)	−0.0584	(−4.66)	−0.0180	(−2.95)	−0.0453	(−4.79)	−0.0167	(−1.86)	
(1.96)	0.0291	(3.75)	0.0029	(1.07)	0.0128	(2.44)	0.0091	(2.04)	
(1.77)	0.2600	(11.77)	0.0267	(2.77)	0.0996	(7.16)	0.0535	(3.61)	
(19.07)	0.3818	(50.34)	−0.0210	(−4.04)	0.1807	(20.90)	0.1118	(15.32)	
(−6.53)	−0.2929	(−20.74)	0.0901	(13.51)	−0.0373	(−3.39)	−0.0505	(−7.26)	
(−4.39)	−0.0320	(−7.52)	−0.0089	(−4.88)	−0.0188	(−5.73)	−0.0109	(−3.20)	
	7961.26		12962.69		5860.36		4916.24		
	0.962		0.975		0.946		0.936		

1991		1992		1993		1994		1995	
(46.90)	0.8953	(93.74)	0.7932	(91.30)	0.8164	(97.91)	0.8127	(112.83)	
(−2.53)	−0.0333	(−2.75)	−0.0346	(−2.82)	−0.0283	(−3.05)	−0.0195	(−2.28)	
(1.90)	0.0084	(2.45)	0.0127	(3.30)	0.0015	(0.46)	0.0065	(2.23)	
(5.86)	0.0745	(6.02)	0.1055	(8.46)	0.0666	(6.29)	0.0322	(2.94)	
(11.72)	0.0823	(10.62)	0.1057	(15.35)	0.0852	(11.03)	0.0490	(6.59)	
(−0.95)	0.0041	(0.49)	0.0263	(3.27)	−0.0140	(−1.79)	−0.0048	(−0.72)	
(−2.88)	−0.0216	(−6.73)	−0.0208	(−5.87)	−0.0128	(−4.18)	−0.0148	(−4.93)	
	3672.15		5074.63		6943.46		8732.96		
	0.921		0.943		0.961		0.968		

1982		1983		1984		1985		1986	
(4.41)	0.1034	(4.92)	0.2295	(5.79)	0.2171	(5.79)	0.1255	(3.48)	
(31.51)	0.7115	(21.17)	0.8178	(17.57)	0.5534	(14.06)	0.5504	(12.69)	
(5.20)	0.0700	(1.22)	0.1115	(1.57)	0.1244	(2.39)	0.3583	(5.55)	
(12.55)	0.6584	(8.56)	0.8295	(7.88)	0.2529	(3.97)	0.6782	(6.97)	
(−12.80)	−0.9620	(−9.93)	−1.5412	(−13.16)	−0.7358	(−6.74)	−1.0932	(−9.97)	
(−6.26)	−0.1458	(−6.15)	−0.2810	(−8.97)	−0.2408	(−7.36)	−0.1615	(−5.57)	
(3.55)	−0.0568	(−4.05)	0.0823	(4.91)	−0.0692	(−4.51)	−0.0798	(−4.64)	
	98.46		74.65		51.44		48.48		
	0.236		0.182		0.131		0.125		

(Continued on following page)

54 THE INTERDEPENDENCIES AMONG CORPORATE FINANCIAL POLICIES

TABLE 4.2 (Continued)

	1987		1988		1989		1990	
SDIV	0.2727	(7.79)	0.3803	(6.95)	0.1346	(4.55)	0.1250	(2.27)	0.1088
SCE	0.7677	(16.11)	0.5680	(8.99)	0.5158	(10.73)	0.3778	(5.78)	0.4808
SRD	0.2842	(4.41)	0.7987	(9.95)	0.4017	(7.74)	0.4455	(5.76)	0.1610
KD	0.1903	(2.53)	0.8224	(5.78)	0.6951	(5.57)	0.6358	(4.14)	0.3716
SDEP	−1.0787	(−9.71)	−1.0433	(−8.00)	−0.5083	(−4.47)	0.0345	(0.28)	0.0353
SNI	−0.3403	(−12.53)	−0.4163	(−10.75)	−0.1810	(−6.36)	−0.1408	(−3.15)	−0.1663
DE	−0.1494	(−8.68)	−0.1017	(−4.99)	−0.1348	(−7.38)	−0.1581	(−7.87)	−0.0764
F−Value	102.10		93.14		79.44		29.47		34.90
adjusted r^2	0.238		0.221		0.192		0.080		0.094

OLS   SRD

	1978		1979		1980		1981	
SLRD	1.0044	(135.16)	1.1629	(90.77)	1.1916	(140.44)	1.1044	(87.77)	1.1477
SCE	0.0192	(7.18)	0.0110	(2.76)	0.0063	(2.48)	0.0054	(1.36)	−0.0061
SDIV	−0.0018	(−1.04)	0.0052	(1.23)	0.0107	(3.86)	0.0140	(3.37)	−0.0153
SEF	−0.0180	(−14.35)	−0.0051	(−1.81)	0.0056	(3.69)	0.0161	(6.61)	0.0125
SNI	0.0021	(0.97)	−0.0052	(−3.49)	−0.0116	(−4.70)	−0.0137	(−3.57)	0.0141
F−Value	3669.48		1658.69		3958.86		1707.91		2902.25
adjusted r^2	0.889		0.782		0.894		0.779		0.852

	1987		1988		1989		1990	
SLRD	0.9010	(79.87)	0.8331	(83.90)	1.0391	(87.39)	1.0104	(136.05)	0.9947
SCE	0.0051	(0.65)	0.0055	(0.73)	−0.0010	(−0.11)	0.0018	(0.33)	−0.0021
SDIV	0.0587	(9.70)	−0.0164	(−2.14)	0.0249	(4.16)	−0.0190	(−3.64)	0.0299
SEF	0.0177	(4.98)	0.0014	(0.51)	0.0201	(4.95)	0.0006	(0.32)	0.0058
SNI	−0.0685	(−15.22)	0.0095	(1.73)	−0.0390	(−6.90)	0.0158	(3.81)	−0.0468
F−Value	1951.96		2350.28		3311.84		4018.48		2337.53
adjusted r^2	0.811		0.838		0.878		0.898		0.836

2SLS   SCE

	1978		1979		1980		1981	
SDIV	−0.0058	(−0.37)	−0.0355	(−1.37)	−0.3431	(−10.43)	−0.3293	(−11.82)	−0.3031
SEF	0.0432	(2.55)	0.6694	(15.83)	0.5284	(11.41)	0.4541	(15.14)	0.6886
SRD	0.0975	(1.63)	−0.2845	(−3.98)	0.0804	(1.13)	0.0903	(1.40)	0.1097
SNI	0.0199	(1.09)	0.0505	(5.36)	0.3587	(11.06)	0.3482	(12.96)	0.2992
SLIQ	−0.0753	(−11.31)	−0.0323	(−3.95)	−0.0511	(−5.72)	−0.1104	(−14.80)	−0.1193
SLSAL	0.0281	(5.19)	0.0046	(0.68)	−0.0219	(−3.25)	−0.0024	(−0.43)	−0.0279
F−Value	30.55		67.90		48.38		84.53		81.42
adjusted r^2	0.072		0.148		0.108		0.171		0.161

	1987		1988		1989		1990	
SDIV	0.0078	(0.28)	−0.0384	(−1.41)	−0.0012	(−0.04)	−0.1848	(−5.42)	−0.0916
SEF	0.0185	(0.60)	−0.0574	(−2.43)	0.1067	(3.50)	0.0738	(3.93)	0.0824
SRD	−0.0501	(−1.28)	0.0967	(2.51)	−0.0778	(−2.27)	0.1237	(3.79)	0.0541
SNI	−0.0269	(−1.11)	−0.0040	(−0.19)	−0.0119	(−0.52)	0.1366	(5.70)	0.0479
SLIQ	−0.0511	(−6.71)	−0.0469	(−6.46)	−0.0478	(−5.66)	−0.0662	(−8.82)	−0.0473
SLSAL	0.0329	(4.92)	0.0394	(5.93)	0.0326	(4.31)	0.0202	(3.32)	0.0155
F−Value	16.06		13.13		17.55		23.71		17.07
adjusted r^2	0.038		0.031		0.041		0.056		0.041

SIMULTANEOUS EQUATION ESTIMATION RESULTS

1991		1992		1993		1994		1995	
(3.03)	0.3331	(6.21)	0.2159	(4.71)	0.1895	(3.45)	0.2420	(4.40)	
(8.36)	0.5187	(6.25)	0.5849	(7.80)	0.5185	(7.17)	0.3568	(4.58)	
(2.90)	0.5542	(7.33)	−0.0029	(−0.04)	−0.1947	(−2.70)	−0.0109	(−0.14)	
(3.92)	0.7208	(4.96)	0.0214	(0.68)	0.0320	(0.94)	1.0390	(5.65)	
(0.38)	−0.6573	(−4.51)	−0.7329	(−5.55)	−0.6495	(−4.49)	−0.4895	(−2.94)	
(−5.99)	−0.4123	(−10.09)	−0.2804	(−7.47)	−0.2533	(−5.30)	−0.2797	(−5.75)	
(−4.51)	0.0340	(1.63)	−0.0182	(−0.92)	−0.0558	(−2.71)	−0.0345	(−1.54)	
	43.92		18.97		14.12		16.84		
	0.120		0.056		0.044		0.052		

1982		1983		1984		1985		1986	
(119.37)	1.2414	(116.17)	1.2043	(117.64)	1.2608	(81.65)	1.0558	(126.36)	
(−2.16)	0.0073	(1.57)	0.0075	(1.47)	0.0144	(1.87)	0.0024	(0.52)	
(−4.12)	0.0369	(12.68)	0.0186	(4.14)	0.0473	(6.18)	−0.0160	(−3.70)	
(6.88)	0.0160	(5.35)	0.0080	(3.40)	0.0199	(4.58)	0.0126	(5.22)	
(4.74)	−0.0482	(−15.11)	−0.0189	(−5.60)	−0.0462	(−7.11)	0.0149	(4.39)	
	3430.22		3228.62		1719.86		3765.63		
	0.886		0.874		0.786		0.890		

1991		1992		1993		1994		1995	
(91.11)	1.0269	(123.08)	1.0826	(83.68)	1.1708	(63.17)	0.6415	(52.46)	
(−0.21)	0.0126	(1.62)	0.0214	(1.88)	0.0229	(1.88)	0.0133	(1.04)	
(4.49)	−0.0255	(−4.47)	0.0735	(10.08)	0.1557	(16.07)	0.0912	(8.79)	
(1.49)	0.0068	(3.03)	0.0078	(2.16)	−0.0044	(−1.01)	−0.0058	(−1.33)	
(−9.63)	0.0238	(5.55)	−0.0795	(−13.93)	−0.1609	(−20.28)	−0.1014	(−11.49)	
	3575.30		1774.17		1289.08		769.74		
	0.890		0.807		0.765		0.656		

1982		1983		1984		1985		1986	
(−8.31)	−0.1946	(−8.27)	−0.1451	(−5.82)	−0.2415	(−7.28)	−0.0393	(−1.31)	
(16.48)	0.5635	(12.83)	0.2927	(6.77)	0.3612	(5.99)	−0.3399	(−5.23)	
(1.46))	0.3221	(6.39)	0.1975	(5.10)	0.2156	(5.52)	0.2212	(4.60)	
(9.69)	0.2153	(8.91)	0.1216	(5.51)	0.2173	(7.05)	−0.0199	(−0.77)	
(−12.36)	−0.0790	(−9.64)	−0.0945	(−13.29)	−0.0689	(−8.77)	−0.1013	(−9.83)	
(−3.88)	−0.0284	(−4.31)	0.0072	(1.08)	0.0079	(1.60)	0.0671	(6.09)	
	52.86		38.09		25.71		20.80		
	0.124		0.087		0.060		0.048		

1991		1992		1993		1994		1995	
(−4.50)	−0.0629	(−3.20)	−0.0544	(−2.90)	−0.1044	(−4.41)	−0.1191	(−5.24)	
(4.33)	0.0152	(0.66)	−0.0214	(−0.69)	0.0238	(1.32)	0.0278	(1.90)	
(2.03)	0.0130	(0.51)	0.0961	(3.56)	0.1399	(4.21)	0.1064	(2.97)	
(3.45)	0.0298	(1.81)	0.0283	(1.64)	0.0834	(4.04)	0.0985	(4.95)	
(−6.93)	−0.0412	(−7.01)	−0.0446	(−6.60)	−0.0460	(−5.93)	−0.0457	(−5.50)	
(2.64)	0.0257	(4.45)	0.0354	(4.95)	0.0261	(3.63)	0.0212	(2.82)	
	13.22		13.61		10.81		10.43		
	0.032		0.034		0.029		0.027		

(Continued on following page)

TABLE 4.2 (Continued)

2SLS	SDIV								
		1978		1979		1980		1981	
SLDIV	0.7925	(79.28)	1.2702	(229.34)	1.0785	(86.44)	0.6896	(66.38)	1.0142
SCE	−0.0015	(−0.14)	−0.0093	(−1.47)	−0.0366	(−3.41)	−0.0336	(−2.61)	0.0088
SEF	−0.0189	(−1.81)	0.0222	(2.71)	0.0493	(3.59)	0.0343	(2.83)	−0.0218
SRD	−0.0067	(−0.48)	0.0040	(0.35)	0.0167	(0.90)	0.0465	(2.08)	0.0362
SNI	0.0765	(15.98)	0.0016	(1.08)	0.0925	(11.64)	0.1788	(20.63)	0.1315
SLNI	−0.0208	(−2.04)	0.0191	(5.88)	0.0362	(6.62)	−0.0720	(−6.69)	−0.0723
SLIQ	−0.0009	(−0.51)	−0.0018	(−1.41)	−0.0078	(−3.37)	−0.0151	(−5.16)	−0.0087
F−Value	57713.99		10752.87		10494.09		8157.74		6590.90
adjusted r^2	0.994		0.970		0.969		0.959		0.948

		1987		1988		1989		1990	
SLDIV	0.7194	(50.12)	0.7300	(83.74)	0.5191	(20.23)	0.2973	(27.91)	0.9171
SCE	0.0715	(2.44)	−0.0113	(−0.85)	−0.0724	(−1.76)	−0.0760	(−3.13)	−0.0473
SEF	0.0579	(3.01)	−0.0254	(−2.67)	−0.0906	(−2.46)	0.0297	(2.09)	−0.0145
SRD	0.0999	(3.67)	0.1149	(7.61)	0.3612	(9.44)	0.3186	(13.12)	0.1456
SNI	0.1935	(14.91)	0.0550	(7.46)	0.0761	(3.05)	0.2680	(19.88)	0.1397
SLNI	−0.0852	(−6.26)	0.0024	(0.35)	0.1853	(7.55)	0.1891	(14.35)	−0.0170
SLIQ	−0.0120	(−2.05)	−0.0173	(−6.03)	−0.0640	(−6.33)	−0.0671	(−11.86)	−0.0207
F−Value	1369.59		3364.55		514.03		949.47		808.86
adjusted r^2	0.809		0.912		0.609		0.744		0.712

2SLS	SEF								
		1978		1979		1980		1981	
SDIV	−0.1974	(−6.86)	−0.1133	(−3.66)	0.2735	(6.77)	0.4246	(11.60)	0.1616
SCE	−0.0511	(−0.47)	0.5467	(10.30)	0.3379	(4.13)	0.8313	(12.23)	0.8346
SRD	0.6180	(5.33)	0.5338	(6.83)	0.2322	(2.44)	0.7098	(7.77)	0.3371
KD	2.1594	(16.35)	0.1398	(3.15)	0.3132	(6.20)	0.7078	(9.10)	0.6125
SDEP	0.4757	(2.44)	−0.3711	(−3.38)	−0.0644	(−0.38)	−1.1052	(−7.53)	−1.2531
SNI	0.2862	(7.90)	0.0356	(2.88)	−0.3298	(−9.16)	−0.4866	(−14.48)	−0.2047
DE	−0.1907	(−11.62)	−0.1547	(−13.98)	−0.1605	(−10.79)	0.0146	(0.96)	0.0414
F−Value	91.27		71.40		70.77		93.39		57.22
adjusted r^2	0.217		0.176		0.172		0.211		0.135

		1987		1988		1989		1990	
SDIV	0.2962	(6.97)	0.3345	(5.50)	0.1779	(3.54)	−0.0411	(−0.49)	0.0645
SCE	0.9025	(7.51)	0.6119	(4.62)	0.4206	(4.64)	0.2310	(2.02)	0.1565
SRD	0.2022	(2.65)	0.8882	(9.46)	0.3380	(4.79)	0.5093	(5.99)	0.2306
KD	0.1309	(1.73)	0.7434	(5.21)	0.6480	(5.16)	0.6340	(4.13)	0.3594
SDEP	−1.2076	(−7.86)	−1.0860	(−6.22)	−0.4026	(−2.88)	0.1796	(1.17)	0.2664
SNI	−0.3657	(−11.93)	−0.3831	(−8.97)	−0.2175	(−5.38)	−0.0368	(−0.63)	−0.1348
DE	−0.1427	(−8.14)	−0.1121	(−5.41)	−0.1318	(−7.11)	−0.1672	(−8.15)	−0.0779
F−Value	68.25		82.11		62.40		24.94		25.17
adjusted r^2	0.172		0.200		0.157		0.068		0.069

2SLS	SRD								
		1978		1979		1980		1981	
SLRD	1.0250	(126.38)	1.1471	(84.38)	1.1907	(139.23)	1.1053	(84.64)	1.1399
SCE	0.0323	(7.43)	−0.0169	(−2.58)	−0.0055	(−1.38)	0.0082	(1.32)	−0.0298
SDIV	−0.0036	(−1.88)	0.0049	(1.09)	0.0064	(1.84)	0.0071	(1.39)	−0.0237
SEF	−0.0400	(−17.08)	0.0301	(3.74)	0.0131	(2.82)	0.0193	(3.09)	0.0553
SNI	0.0043	(1.84)	−0.0029	(−1.84)	−0.0072	(−2.14)	−0.0079	(−1.59)	0.0241
F−Value	3256.48		1553.35		3894.55		1691.75		2374.88
adjusted r^2	0.877		0.771		0.892		0.777		0.825

1982		1983		1984		1985		1986	
(86.69)	0.8030	(49.19)	0.7768	(118.86)	0.6763	(47.55)	0.9090	(101.15)	
(0.66)	−0.0121	(−0.56)	−0.0355	(−3.32)	−0.0472	(−3.00)	−0.0102	(−0.73)	
(−1.54)	0.0364	(1.62)	0.0330	(3.25)	0.0520	(2.08)	0.0471	(3.09)	
(1.82)	0.1973	(7.95)	0.0400	(3.65)	0.1172	(6.58)	0.0513	(3.14)	
(17.97)	0.3770	(46.56)	−0.0119	(−2.00)	0.1898	(18.81)	0.1163	(15.13)	
(−6.78)	−0.3093	(−20.58)	0.0935	(13.39)	−0.0266	(−2.09)	−0.0481	(−6.75)	
(−2.88)	−0.0236	(−5.13)	−0.0130	(−5.89)	−0.0199	(−5.64)	−0.0099	(−2.72)	
	7859.56		12295.64		5709.08		4759.62		
	0.961		0.974		0.945		0.934		

1991		1992		1993		1994		1995	
(45.24)	0.8957	(83.13)	0.8011	(83.46)	0.8209	(92.20)	0.8180	(101.85)	
(−1.63)	−0.0352	(−1.86)	−0.0009	(−0.05)	−0.0162	(−1.28)	−0.0154	(−1.34)	
(−0.80)	0.0004	(0.04)	0.0068	(0.45)	−0.0078	(−1.08)	−0.0030	(−0.54)	
(5.59)	0.0896	(6.26)	0.0845	(5.69)	0.0597	(4.49)	0.0238	(1.63)	
(11.10)	0.0813	(9.29)	0.1001	(12.87)	0.0813	(9.65)	0.0454	(5.86)	
(−1.00)	0.0041	(0.49)	0.0235	(2.85)	−0.0154	(−1.96)	−0.0074	(−1.07)	
(−3.04)	−0.0223	(−6.64)	−0.0170	(−4.43)	−0.0118	(−3.69)	−0.0135	(−4.20)	
	3661.14		5041.94		6908.68		8684.43		
	0.921		0.943		0.961		0.968		

1982		1983		1984		1985		1986	
(3.62)	0.0987	(4.32)	0.2503	(5.99)	0.2470	(5.58)	0.1144	(2.84)	
(14.38)	0.7384	(11.04)	0.7027	(7.68)	0.3326	(4.25)	0.2337	(2.83)	
(3.76)	0.0008	(0.01)	0.0590	(0.76)	0.0120	(0.19)	0.2986	(4.24)	
(9.06)	0.4778	(6.16)	0.7297	(6.87)	0.2158	(3.33)	0.6540	(6.63)	
(−9.44)	−0.9879	(−7.87)	−1.4171	(−9.81)	−0.4492	(−3.21)	−0.7015	(−5.01)	
(−5.71)	−0.1493	(−5.87)	−0.2995	(−9.12)	−0.2726	(−7.17)	−0.1522	(−4.82)	
(2.68)	−0.0651	(−4.57)	0.0827	(4.86)	−0.0573	(−3.59)	−0.0774	(−4.41)	
	50.46		38.87		23.69		24.17		
	0.136		0.102		0.064		0.065		

1991		1992		1993		1994		1995	
(1.32)	0.3224	(5.43)	0.1742	(3.46)	0.1916	(3.12)	0.1611	(2.65)	
(1.64)	0.3271	(1.97)	0.2941	(2.28)	0.3973	(3.44)	0.3992	(3.10)	
(3.54)	0.5847	(7.09)	−0.0183	(−0.23)	−0.2044	(−2.29)	0.2393	(2.39)	
(3.76)	0.7061	(4.88)	0.0164	(0.52)	0.0334	(0.99)	1.0733	(5.81)	
(2.47)	−0.5047	(−2.67)	−0.5043	(−3.29)	−0.5331	(−3.16)	−0.5545	(−2.73)	
(−4.13)	−0.3995	(−9.17)	−0.2574	(−6.33)	−0.2558	(−4.75)	−0.1962	(−3.63)	
(−4.45)	0.0355	(1.64)	−0.0103	(−0.50)	−0.0501	(−2.31)	−0.0579	(−2.45)	
	38.23		10.10		8.22		15.21		
	0.106		0.029		0.025		0.047		

1982		1983		1984		1985		1986	
(106.36)	1.2634	(107.62)	1.1977	(106.59)	1.2641	(79.58)	1.0574	(121.86)	
(−5.47)	−0.0231	(−2.89)	0.0067	(0.77)	−0.0090	(−0.75)	−0.0174	(−2.45)	
(−5.23)	0.0240	(6.83)	0.0121	(2.30)	0.0430	(4.24)	−0.0245	(−4.98)	
(9.10)	0.0651	(7.18)	0.0560	(6.59)	0.0280	(1.49)	0.0403	(4.89)	
(6.59)	−0.0328	(−8.27)	−0.0083	(−1.94)	−0.0424	(−4.58)	0.0227	(5.77)	
	3040.66		2712.31		1704.85		3558.33		
	0.873		0.854		0.784		0.884		

(Continued on following page)

TABLE 4.2 (Continued)

	1987		1988		1989		1990	
SLRD	0.9034	(77.47)	0.8643	(73.09)	1.0219	(76.27)	1.0099	(131.72)	0.9917
SCE	−0.0521	(−3.67)	0.0213	(1.67)	−0.0170	(−1.26)	0.0106	(1.35)	−0.0306
SDIV	0.0464	(5.84)	0.0043	(0.46)	0.0435	(4.27)	−0.0090	(−1.16)	0.0301
SEF	0.0157	(1.57)	−0.0545	(−6.34)	0.0230	(1.83)	−0.0076	(−1.61)	0.0543
SNI	−0.0645	(−10.20)	−0.0153	(−2.18)	−0.0497	(−5.89)	0.0087	(1.61)	−0.0383
F−Value	1884.92		2007.18		3288.61		3980.60		2192.90
adjusted r²	0.806		0.815		0.877		0.897		0.827

3SLS   SCE

	1978		1979		1980		1981	
SDIV	−0.0110	(−0.70)	−0.0223	(−0.86)	−0.3624	(−11.49)	−0.3589	(−13.62)	−0.2145
SEF	−0.0132	(−0.79)	0.8016	(22.56)	0.6612	(16.77)	0.6473	(25.85)	0.8586
SRD	0.1378	(2.31)	−0.4046	(−5.90)	−0.0066	(−0.09)	−0.1706	(−2.76)	−0.1641
SNI	0.0271	(1.49)	0.0538	(5.83)	0.3878	(12.72)	0.3936	(15.65)	0.2314
SLIQ	−0.0801	(−12.23)	−0.0043	(−0.75)	−0.0267	(−3.86)	−0.0639	(−10.21)	−0.0498
SLSAL	0.0291	(5.48)	0.0002	(0.05)	−0.0101	(−1.96)	0.0034	(0.80)	−0.0036

	1987		1988		1989		1990	
SDIV	−0.0423	(−1.54)	−0.0293	(−1.08)	−0.0157	(−0.57)	−0.1683	(−4.95)	−0.0941
SEF	0.0891	(2.89)	−0.0914	(−3.90)	0.1688	(5.61)	0.1108	(5.96)	0.1167
SRD	−0.0261	(−0.67)	0.1203	(3.13)	−0.0889	(−2.60)	0.0984	(3.02)	0.0471
SNI	0.0234	(0.98)	−0.0177	(−0.84)	0.0076	(0.33)	0.1299	(5.44)	0.0539
SLIQ	−0.0525	(−7.00)	−0.0498	(−6.91)	−0.0422	(−5.07)	−0.0574	(−7.74)	−0.0475
SLSAL	0.0133	(2.04)	0.0380	(5.77)	0.0295	(3.98)	0.0241	(4.03)	0.0175

3SLS   SDIV

	1978		1979		1980		1981	
SLDIV	0.7885	(79.09)	1.2700	(229.52)	1.0585	(86.06)	0.6808	(66.15)	1.0175
SCE	−0.0063	(−0.61)	−0.0244	(−3.85)	−0.0891	(−8.50)	−0.0756	(−5.93)	0.0274
SEF	−0.0232	(−2.23)	0.0353	(4.34)	0.0870	(6.47)	0.0684	(5.70)	−0.0413
SRD	−0.0016	(−0.11)	−0.0032	(−0.28)	0.0131	(0.71)	0.0284	(1.28)	0.0437
SNI	0.0768	(16.06)	0.0024	(1.64)	0.1117	(14.19)	0.1972	(22.90)	0.1282
SLNI	−0.0174	(−1.71)	0.0191	(5.91)	0.0347	(6.51)	−0.0757	(−7.13)	−0.0727
SLIQ	−0.0021	(−1.21)	−0.0016	(−1.27)	−0.0082	(−3.60)	−0.0163	(−5.64)	−0.0090

	1987		1988		1989		1990	
SLDIV	0.7206	(50.72)	0.7272	(83.82)	0.5320	(21.10)	0.2859	(27.14)	0.9051
SCE	0.1887	(6.52)	−0.0272	(−2.06)	−0.0267	(−0.65)	−0.1726	(−7.18)	−0.1260
SEF	0.0757	(3.96)	−0.0400	(−4.25)	−0.1421	(−3.88)	0.0464	(3.26)	−0.0228
SRD	0.0928	(3.41)	0.1350	(8.99)	0.3679	(9.69)	0.3213	(13.26)	0.1584
SNI	0.1974	(15.33)	0.0438	(5.98)	0.0614	(2.47)	0.2648	(19.74)	0.1326
SLNI	−0.0843	(−6.36)	0.0129	(1.92)	0.1828	(7.66)	0.2054	(15.87)	0.0006
SLIQ	0.0005	(0.09)	−0.0191	(−6.77)	−0.0608	(−6.18)	−0.0653	(−11.69)	−0.0216

3SLS   SEF

	1978		1979		1980		1981	
SDIV	−0.1841	(−6.41)	−0.0063	(−0.21)	0.4074	(10.36)	0.4746	(13.33)	0.1970
SCE	−0.3753	(−3.61)	1.0177	(25.90)	0.9670	(15.10)	1.0927	(18.79)	0.9925
SRD	0.5083	(4.39)	0.5253	(6.75)	0.1563	(1.65)	0.5449	(6.04)	0.2941
KD	2.4950	(20.17)	0.2041	(6.93)	0.1176	(3.07)	0.3871	(6.47)	0.2499
SDEP	0.4831	(2.64)	−0.1047	(−1.42)	0.0825	(0.64)	−0.1624	(−1.40)	0.0151
SNI	0.2645	(7.33)	−0.0379	(−3.25)	−0.4530	(−13.03)	−0.5274	(−16.31)	−0.2233
DE	−0.1484	(−9.48)	−0.0389	(−4.32)	−0.0586	(−4.66)	0.0485	(4.06)	0.0504

1991		1992		1993		1994		1995	
(86.07)	1.0000	(96.59)	1.0942	(76.19)	1.1738	(61.86)	0.6437	(51.67)	
(−2.16)	−0.0230	(−1.70)	−0.0039	(−0.22)	0.0185	(1.12)	0.0136	(0.79)	
(3.27)	−0.0480	(−6.40)	0.0535	(6.03)	0.1579	(15.20)	0.0851	(7.59)	
(6.14)	0.0629	(7.29)	0.0839	(5.64)	0.0095	(1.01)	0.0264	(3.18)	
(−6.47)	0.0486	(8.04)	−0.0560	(−7.55)	−0.1605	(−18.86)	−0.0928	(−9.76)	
	2784.79		1466.97		1279.65		750.30		
	0.863		0.775		0.764		0.650		

1982		1983		1984		1985		1986	
(−6.17)	−0.1602	(−7.73)	−0.1264	(−5.24)	−0.2747	(−8.61)	−0.0416	(−1.43)	
(26.86)	0.7620	(21.43)	0.3623	(8.99)	0.5464	(10.13)	−0.4625	(−8.23)	
(−2.30)	0.2271	(4.72)	0.1183	(3.10)	0.1966	(5.10)	0.2316	(4.93)	
(8.10)	0.1982	(9.00)	0.1122	(5.33)	0.2586	(8.82)	−0.0258	(−1.05)	
(−7.53)	−0.0342	(−5.56)	−0.0696	(−10.56)	−0.0492	(−6.91)	−0.0856	(−9.69)	
(−0.83)	−0.0109	(−2.39)	0.0172	(3.03)	0.0071	(1.67)	0.0266	(2.88)	

1991		1992		1993		1994		1995	
(−4.62)	−0.0694	(−3.54)	−0.0473	(−2.53)	−0.1028	(−4.35)	−0.1200	(−5.29)	
(6.16)	0.0248	(1.08)	−0.0630	(−2.05)	0.0344	(1.92)	0.0518	(3.54)	
(1.77)	0.0112	(0.44)	0.0927	(3.44)	0.1426	(4.30)	0.1009	(2.82)	
(3.88)	0.0364	(2.20)	0.0177	(1.03)	0.0837	(4.06)	0.1014	(5.10)	
(−6.99)	−0.0437	(−7.45)	−0.0420	(−6.26)	−0.0438	(−5.67)	−0.0406	(−4.91)	
(3.00)	0.0228	(3.95)	0.0349	(4.92)	0.0303	(4.22)	0.0223	(2.98)	

1982		1983		1984		1985		1986	
(87.13)	0.7986	(49.07)	0.7711	(120.01)	0.6632	(47.44)	0.9084	(101.52)	
(2.07)	−0.0414	(−1.90)	−0.0798	(−7.59)	−0.1146	(−7.40)	0.0248	(1.80)	
(−2.95)	0.0572	(2.55)	0.0623	(6.24)	0.1090	(4.45)	0.0833	(5.56)	
(2.21)	0.2026	(8.17)	0.0388	(3.56)	0.1245	(7.04)	0.0412	(2.53)	
(17.55)	0.3805	(47.03)	−0.0048	(−0.81)	0.2094	(20.94)	0.1218	(15.97)	
(−6.85)	−0.3068	(−20.49)	0.0933	(13.88)	−0.0294	(−2.37)	−0.0498	(−7.13)	
(−2.97)	−0.0252	(−5.51)	−0.0140	(−6.49)	−0.0201	(−5.84)	−0.0086	(−2.40)	

1991		1992		1993		1994		1995	
(44.78)	0.8932	(83.10)	0.8011	(83.48)	0.8169	(91.94)	0.8152	(101.61)	
(−4.35)	−0.0582	(−3.07)	0.0060	(0.33)	−0.0231	(−1.83)	−0.0290	(−2.54)	
(−1.26)	0.0027	(0.21)	0.0049	(0.32)	−0.0119	(−1.67)	−0.0078	(−1.42)	
(6.09)	0.0890	(6.22)	0.0832	(5.60)	0.0614	(4.62)	0.0304	(2.08)	
(10.56)	0.0823	(9.41)	0.0983	(12.65)	0.0781	(9.28)	0.0434	(5.60)	
(0.03)	0.0050	(0.59)	0.0253	(3.07)	−0.0090	(−1.15)	−0.0032	(−0.47)	
(−3.19)	−0.0230	(−6.86)	−0.0159	(−4.15)	−0.0115	(−3.60)	−0.0132	(−4.13)	

1982		1983		1984		1985		1986	
(4.61)	0.1547	(6.95)	0.2765	(6.74)	0.3474	(7.99)	0.0780	(1.97)	
(21.11)	1.1209	(21.07)	1.2378	(14.72)	0.9210	(13.26)	−0.5909	(−8.75)	
(3.35)	−0.1718	(−2.83)	−0.0265	(−0.35)	−0.1274	(−2.04)	0.3103	(4.48)	
(6.67)	0.1078	(2.05)	0.5125	(6.30)	0.1195	(2.16)	0.7146	(9.26)	
(0.17)	−0.2304	(−2.53)	−0.6593	(−5.15)	−0.1794	(−1.49)	−0.5517	(−4.95)	
(−6.64)	−0.2106	(−8.59)	−0.3030	(−9.57)	−0.3532	(−9.52)	−0.1446	(−4.71)	
(5.51)	0.0109	(1.05)	0.1216	(8.95)	−0.0016	(−0.11)	−0.0434	(−2.89)	

(Continued on following page)

TABLE 4.2 (Continued)

	1987		1988		1989		1990	
SDIV	0.3033	(7.14)	0.3167	(5.22)	0.1656	(3.30)	0.0089	(0.11)	0.0920
SCE	1.0715	(9.01)	0.1773	(1.39)	0.7755	(8.73)	0.6098	(5.39)	0.4115
SRD	0.2082	(2.74)	0.8584	(9.22)	0.3648	(5.19)	0.4712	(5.55)	0.1976
KD	0.1168	(1.58)	0.8079	(6.08)	0.6120	(5.00)	0.5849	(3.86)	0.2568
SDEP	−1.1952	(−7.90)	−0.8730	(−5.30)	−0.3631	(−2.66)	0.1896	(1.25)	0.2017
SNI	−0.3666	(−11.98)	−0.3769	(−8.90)	−0.2004	(−4.97)	−0.0753	(−1.28)	−0.1633
TE	−0.1372	(−7.90)	−0.1243	(−6.37	−0.1168	(−6.42)	−0.1546	(−7.62)	−0.0317

3SLS	SRD

	1978		1979		1980		1981	
SLRD	1.0281	(126.81)	1.1229	(82.86)	1.1895	(139.09)	1.1001	(84.28)	1.1216
SCE	0.0418	(9.62)	−0.0645	(−10.24)	−0.0191	(−4.86)	0.0032	(0.52)	−0.0772
SDIV	−0.0043	(−2.29)	0.0039	(0.87)	0.0017	(0.50)	0.0039	(0.77)	−0.0316
SEF	−0.0519	(−22.75)	0.0703	(9.00)	0.0224	(4.82)	0.0266	(4.26)	0.1008
SNI	0.0054	(2.27)	−0.0002	(−0.13)	−0.0022	(−0.66)	−0.0042	(−0.85)	0.0334

	1987		1988		1989		1990	
SLRD	0.8983	(77.07)	0.8769	(74.48)	1.0159	(75.84)	1.0110	(131.87)	0.9844
SCE	−0.1106	(−7.90)	0.0086	(0.68)	−0.0486	(−3.61)	0.0247	(3.13)	−0.0711
SDIV	0.0440	(5.54)	0.0159	(1.71)	0.0422	(4.14)	−0.0066	(−0.85)	0.0235
SEF	0.0223	(2.23)	−0.0860	(−10.34)	0.0361	(2.88)	−0.0137	(−2.92)	0.1036
SNI	−0.0629	(−9.95)	−0.0306	(−4.40)	−0.0476	(−5.65)	0.0063	(1.16)	−0.0281

TABLE 4.3 Estimated System Equations, U.S. WorldScope Universe

OLS	SCE

	1982 (t)		1983		1984	
SDIV	−0.2249	(−6.23)	−0.1956	(−11.27)	−0.2180	(−6.40)	−0.7639
SEF	0.1166	(10.63)	0.1939	(14.13)	0.1539	(14.87)	0.2004
SRD	0.1983	(3.73)	0.1265	(3.21)	0.0862	(2.20)	0.1728
SNI	0.1978	(6.55)	0.2705	(12.47)	0.1711	(8.38)	0.7252
SLIQ	−0.0919	(−10.20)	−0.0826	(−12.88)	−0.0615	(−10.39)	−0.0972
SLSAL	−0.0032	(−0.57)	−0.0130	(−3.78)	0.0053	(1.24)	−0.0590
SDEP	1.6677	(24.56)	1.3219	(29.13)	1.2435	(27.16)	2.0147
F−Value	130.07		186.02		148.56		4391.91
adjusted r^2	0.428		0.505		0.432		0.954

	1989		1990		1991	
SDIV	−0.1064	(−8.84)	−0.3522	(−18.03)	−0.4748	(−22.36)	−0.7344
SEF	0.1138	(11.62)	0.1635	(13.03)	0.2172	(15.19)	0.3050
SRD	0.0061	(0.21)	0.2756	(8.52)	0.3659	(10.59)	0.3495
SNI	0.1228	(10.28)	0.4416	(33.82)	0.5693	(46.33)	0.8311
SLIQ	−0.0353	(−6.58)	−0.1364	(−22.32)	−0.1774	(−28.70)	−0.1838
SLSAL	0.0036	(1.70)	−0.0033	(−0.74)	−0.0837	(−28.91)	−0.0856
SDEP	1.0996	(29.70)	0.6210	(16.08)	0.5504	(11.23)	0.5235
F−Value	157.14		305.42		557.98		2444.96
adjusted r^2	0.333		0.487		0.628		0.878

SIMULTANEOUS EQUATION ESTIMATION RESULTS

1991		1992		1993		1994		1995	
(1.88)	0.3472	(5.88)	0.1929	(3.86)	0.2011	(3.28)	0.1824	(3.00)	
(4.39)	0.3282	(2.13)	−0.1131	(−0.91)	0.5037	(4.37)	0.5832	(4.61)	
(3.05)	0.5460	(6.75)	−0.0762	(−0.97)	−0.2109	(−2.36)	0.2761	(2.76)	
(2.80)	0.2091	(1.68)	0.0015	(0.05)	0.0270	(0.80)	0.8954	(4.95)	
(1.95)	−0.4832	(−2.97)	−0.6099	(−4.35)	−0.5460	(−3.24)	−0.5045	(−2.53)	
(−5.02)	−0.4347	(−10.13)	−0.2890	(−7.20)	−0.2633	(−4.89)	−0.2152	(−3.99)	
(−1.88)	0.0707	(3.81)	0.0622	(3.33)	−0.0460	(−2.13)	−0.0268	(−1.15)	

1982		1983		1984		1985		1986	
(105.06)	1.2735	(108.60)	1.1927	(106.24)	1.2644	(79.60)	1.0529	(121.42)	
(−15.23)	−0.0731	(−9.43)	−0.0230	(−2.67)	−0.0078	(−0.66)	−0.0003	(−0.05)	
(−6.99)	0.0173	(4.95)	0.0055	(1.04)	0.0421	(4.16)	−0.0260	(−5.29)	
(18.16)	0.1107	(12.59)	0.0897	(11.11)	0.0315	(1.67)	0.0622	(7.70)	
(9.21)	−0.0237	(−6.02)	−0.0004	(−0.09)	−0.0414	(−4.47)	0.0259	(6.62)	

1991		1992		1993		1994		1995	
(85.51)	0.9656	(96.02)	1.0943	(76.28)	1.1708	(61.72)	0.6392	(51.40)	
(−5.04)	−0.0515	(−3.81)	0.0287	(1.58)	0.0570	(3.48)	0.0250	(1.46)	
(2.56)	−0.0645	(−8.72)	0.0451	(5.11)	0.1602	(15.42)	0.0798	(7.12)	
(12.12)	0.1181	(15.81)	0.1543	(11.36)	0.0135	(1.43)	0.0586	(7.16)	
(−4.76)	0.0693	(11.91)	−0.0406	(−5.57)	−0.1615	(−18.98)	−0.0854	(−9.00)	

1985		1986		1987		1988	
(−29.45)	−0.7039	(−36.01)	−0.7570	(−38.76)	−0.1175	(−5.80)	
(10.97)	0.3550	(22.70)	0.2140	(20.80)	0.1067	(10.70)	
(3.15)	0.1240	(2.07)	0.1800	(3.22)	0.0178	(0.47)	
(82.08)	0.7163	(81.60)	0.7623	(93.13)	0.1160	(8.29)	
(−10.80)	−0.1391	(−13.90)	−0.1143	(−12.39)	−0.0279	(−4.18)	
(−10.96)	−0.0650	(−9.37)	−0.0241	(−4.78)	0.0032	(1.16)	
(31.24)	2.0002	(29.38)	1.5361	(29.32)	0.5984	(16.94)	
	4585.01		3333.90		62.32		
	0.954		0.932		0.178		

1992		1993		1994		1995	
(−33.80)	−0.5881	(−23.32)	−0.6049	(−25.11)	−0.5172	(−25.04)	
(17.18)	0.0889	(4.57)	0.2091	(11.74)	0.2757	(14.21)	
(7.94)	0.4078	(8.57)	0.3485	(7.47)	0.5619	(14.77)	
(110.00)	0.6648	(65.31)	0.7304	(86.96)	0.5741	(54.37)	
(−23.74)	−0.1838	(−18.11)	−0.1846	(−19.35)	−0.1448	(−16.38)	
(−12.97)	−0.0457	(−5.83)	−0.0662	(−8.85)	−0.1199	(−18.32)	
(17.10)	0.5727	(13.06)	1.2036	(24.33)	0.9781	(18.03)	
	804.58		1810.23		907.60		
	0.701		0.839		0.721		

(Continued on following page)

TABLE 4.3 (Continued)

OLS	SDIV							
		1982		1983		1984		
SLDIV	1.4361	(146.68)	0.5633	(287.01)	0.2995	(19.13)	1.3383	
SCE	−0.0208	(−4.86)	0.0099	(1.88)	−0.0665	(−4.27)	−0.0613	
SEF	0.0117	(5.70)	−0.0064	(−1.97)	−0.0025	(−0.34)	0.0096	
SRD	0.0341	(3.51)	−0.0257	(−2.70)	0.0054	(0.19)	0.0379	
SNI	0.0644	(12.44)	0.0499	(10.54)	0.3014	(26.56)	0.0551	
SLIQ	−0.0128	(−7.56)	−0.0008	(−0.51)	−0.0309	(−7.19)	−0.0054	
F−Value	21012.14		95938.48		448.21		3150.46	
adjusted r^2	0.990		0.998		0.664		0.927	
		1989		1990		1991		
SLDIV	0.6966	(55.02)	0.4436	(37.92)	0.6997	(42.06)	0.6190	
SCE	−0.1109	(−5.39)	−0.1119	(−6.88)	−0.1025	(−7.34)	−0.1736	
SEF	0.1467	(13.22)	0.1137	(11.35)	0.0150	(1.30)	0.0481	
SRD	0.0696	(2.15)	0.0757	(2.88)	0.0481	(1.76)	0.0793	
SNI	0.2223	(16.75)	0.2109	(17.74)	0.1285	(16.40)	0.1871	
SLIQ	−0.0297	(−4.80)	−0.0249	(−4.67)	−0.0186	(−3.36)	−0.0305	
F−Value	1713.07		689.65		545.22		1261.526	
adjusted r^2	0.825		0.648		0.586		0.762	

OLS	SEF							
		1982		1983		1984		
SDIV	0.2023	(2.17)	0.1251	(4.03)	−0.0463	(−0.55)	0.1077	
SCE	0.6721	(9.91)	0.6357	(12.30)	0.8484	(13.66)	0.2642	
SRD	0.4318	(3.29)	0.2645	(3.34)	0.4522	(4.67)	0.1161	
SNI	−0.2045	(−2.64)	−0.2009	(−4.52)	−0.1405	(−2.75)	−0.1571	
TE	0.0419	(1.68)	0.0143	(0.99)	0.0905	(5.65)	−0.0062	
SDEP	−1.2231	(−5.77)	−1.0104	(−8.48)	−1.9410	(−14.15)	−0.8110	
F−Value	20.62		28.58		53.86		12.69	
adjusted r^2	0.089		0.115		0.189		0.045	
		1989		1990		1991		
SDIV	0.2059	(7.97)	0.3153	(9.43)	0.0109	(0.39)	0.2037	
SCE	0.5126	(11.65)	0.4065	(13.10)	0.2511	(10.91)	0.2372	
SRD	0.2222	(3.74)	0.1238	(2.30)	0.3237	(6.58)	0.0730	
SNI	−0.1987	(−7.48)	−0.2182	(−8.12)	−0.0578	(−3.85)	−0.2390	
DE	−0.0295	(−2.73)	−0.0361	(−3.44)	−0.0009	(−0.10)	0.0545	
SDEP	−0.8706	(−9.48)	−0.4048	(−6.15)	−0.5338	(−8.09)	−0.3738	
F−Value	39.16		38.85		43.76		48.18	
adjusted r^2	0.096		0.092		0.100		0.107	

SIMULTANEOUS EQUATION ESTIMATION RESULTS

1985		1986		1987		1988	
(99.47)	0.9516	(123.43)	0.4990	(33.43)	0.7245	(60.76)	
(−10.87)	−0.0249	(−4.16)	−0.2901	(−23.18)	−0.0292	(−2.11)	
(1.80)	0.0170	(3.26)	0.0768	(10.68)	0.0338	(5.17)	
(2.33)	−0.0057	(−0.31)	0.1343	(3.31)	−0.0252	(−1.02)	
(10.16)	0.0273	(4.89)	0.3011	(26.19)	0.0537	(5.71)	
(−1.98)	−0.0037	(−1.14)	−0.0353	(−5.12)	−0.0036	(−0.82)	
	5915.45		754.90		1154.23		
	0.958		0.727		0.777		

1992		1993		1994		1995	
(56.46)	0.9431	(135.17)	0.7876	(82.35)	0.8342	(141.16)	
(−19.17)	−0.0386	(−7.95)	−0.0332	(−4.66)	−0.0275	(−4.95)	
(5.32)	−0.0055	(−1.15)	0.0538	(8.07)	0.0091	(1.51)	
(3.46)	0.0067	(0.56)	0.0070	(0.38)	0.0182	(1.51)	
(23.02)	0.0353	(8.67)	0.0408	(6.42)	0.0356	(8.83)	
(−7.07)	0.0019	(0.71)	−0.0045	(−1.12)	−0.0070	(−2.45)	
	4818.14		1795.63		5124.25		
	0.923		0.816		0.926		

1985		1986		1987		1988	
(2.38)	0.4600	(12.93)	0.7435	(13.46)	0.2389	(5.32)	
(7.85)	0.6100	(20.30)	1.0414	(24.05)	0.5086	(10.27)	
(1.51)	0.0816	(0.95)	0.5049	(4.22)	0.0971	(1.16)	
(−5.36)	−0.4886	(−18.66)	−0.7503	(−18.64)	−0.2314	(−7.29)	
(−0.45)	0.0166	(1.11)	−0.2046	(−11.26)	0.0309	(2.15)	
(−6.89)	−1.2610	(−10.45)	−1.6680	(−12.29)	−0.5983	(−7.12)	
	73.03		151.00		27.07		
	0.218		0.346		0.073		

1992		1993		1994		1995	
(7.10)	−0.0086	(−0.30)	0.1660	(5.38)	0.1129	(5.16)	
(12.09)	0.0858	(4.29)	0.1912	(8.97)	0.2190	(12.45)	
(1.46)	0.0332	(0.67)	0.1379	(2.56)	0.0426	(1.08)	
(−12.17)	−0.0074	(−0.42)	−0.1119	(−5.78)	−0.1363	(−10.03)	
(6.56)	−0.0224	(−2.53)	−0.0068	(−0.73)	−0.0240	(−3.06)	
(−10.63)	−0.3618	(−7.74)	−0.6515	(−10.36)	−0.5193	(−9.06)	
	13.38		24.80		34.56		
	0.030		0.055		0.076		

(Continued on following page)

TABLE 4.3 (Continued)

OLS	SRD							
		1982		1983		1984		
SLRD	1.2283	(51.84)	1.1325	(83.03)	1.1156	(99.81)	1.2058	
SCE	0.0118	(1.59)	0.0111	(1.80)	0.0081	(1.49)	−0.0056	
SDIV	0.1147	(10.45)	−0.0233	(−5.81)	−0.0524	(−6.25)	−0.0128	
SEF	0.0203	(5.56)	0.0254	(6.42)	0.0163	(6.37)	0.0112	
SNI	−0.1063	(−11.99)	0.0351	(6.26)	0.0366	(7.76)	0.0071	
F−Value	564.92		1414.10		2084.43		2075.02	
adjusted r^2	0.700		0.848		0.885		0.875	
		1989		1990		1991		
SLRD	1.0783	(124.67)	1.0722	(106.13)	1.0658	(138.66)	1.0141	
SCE	0.0119	(2.33)	0.0114	(2.21)	−0.0049	(−1.51)	−0.0037	
SDIV	−0.0157	(−4.70)	0.0033	(0.59)	−0.0112	(−2.75)	−0.0059	
SEF	0.0154	(5.49)	0.0043	(1.19)	0.0240	(7.95)	0.0094	
SNI	0.0187	(5.69)	−0.0094	(−2.21)	0.0027	(1.27)	0.0048	
F−Value	3154.31		2326.48		3964.20		3831.26	
adjusted r^2	0.878		0.838		0.896		0.890	

2SLS	SCE							
		1982		1983		1984		
SLRD	−0.3658	(−3.89)	−0.4962	(−9.23)	−0.6535	(−4.10)	−0.6001	
SEF	0.9382	(4.58)	1.0754	(8.88)	0.8423	(7.47)	2.6728	
SRD	0.0468	(0.30)	0.1257	(1.40)	−0.0856	(−0.93)	0.0568	
SNI	0.3730	(4.41)	0.5902	(9.54)	0.4750	(5.74)	0.6950	
SLIQ	−0.1288	(−5.33)	−0.1250	(−8.57)	−0.1456	(−7.94)	−0.0486	
SLSAL	−0.0517	(−2.87)	−0.0915	(−7.23)	−0.0671	(−4.66)	−0.2159	
SDEP	1.6523	(10.16)	1.4032	(14.80)	1.7313	(13.98)	2.4592	
F−Value	22.85		47.74		33.84		332.76	
adjusted r^2	0.112		0.205		0.145		0.611	
		1989		1990		1991		
SDIV	−0.0829	(−5.02)	−0.6543	(−7.26)	−0.4574	(−4.25)	−0.9176	
SEF	0.2818	(5.20)	1.9175	(8.21)	2.1821	(11.74)	1.7752	
SRD	−0.0457	(−1.37)	0.0442	(0.38)	−0.2353	(−1.86)	0.1061	
SNI	0.1135	(7.38)	0.5909	(12.15)	0.5622	(11.74)	0.8551	
SLIQ	−0.0312	(−5.28)	−0.0877	(−4.32)	−0.1405	(−7.35)	−0.1810	
SLSAL	0.0016	(0.64)	−0.1097	(−5.80)	−0.0967	(−9.90)	−0.1977	
SDEP	1.1668	(26.67)	0.9950	(7.62)	1.6178	(9.33)	0.9047	
F−Value	121.52		42.02		83.95		662.62	
adjusted r^2	0.278		0.113		0.201		0.662	

SIMULTANEOUS EQUATION ESTIMATION RESULTS

1985		1986		1987		1988	
(101.47)	1.1082	(100.05)	1.0010	(104.36)	1.0142	(76.44)	
(−1.70)	−0.0077	(−2.52)	0.0098	(3.11)	−0.0052	(−0.77)	
(−2.26)	−0.0101	(−2.41)	0.0091	(2.09)	−0.0256	(−3.98)	
(3.42)	0.0130	(4.51)	0.0025	(1.42)	0.0206	(6.45)	
(2.14)	0.0085	(2.81)	−0.0099	(−3.19)	0.0221	(5.01)	
	2011.95		2201.66		1177.82		
	0.867		0.866		0.748		

1992		1993		1994		1995	
(137.65)	1.0232	(149.50)	0.9207	(135.79)	0.9429	(94.56)	
(−1.33)	−0.0024	(−0.93)	−0.0014	(−0.52)	0.1182	(29.83)	
(−1.41)	−0.0050	(−1.29)	−0.0039	(−0.94)	−0.0325	(−6.00)	
(3.27)	0.0048	(1.79)	0.0018	(0.69)	−0.0008	(−0.16)	
(1.76)	0.0040	(1.76)	0.0022	(0.88)	0.0400	(11.96)	
	4573.45		3833.05		2416.55		
	0.905		0.887		0.831		

1985		1986		1987		1988	
(−5.69)	−0.7675	(−26.53)	−1.0316	(−39.88)	−0.2903	(−4.90)	
(6.52)	0.9871	(21.97)	0.3054	(18.45)	0.8382	(5.13)	
(0.26)	0.2635	(2.81)	0.1488	(2.28)	0.2171	(2.22)	
(20.94)	0.7674	(59.12)	0.8155	(87.83)	0.2841	(6.15)	
(−1.42)	−0.1327	(−9.12)	−0.1083	(−10.76)	−0.0840	(−4.64)	
(−6.69)	−0.0903	(−8.95)	−0.0084	(−1.36)	−0.0373	(−3.58)	
(9.83)	1.7658	(17.76)	1.3900	(24.05)	0.7840	(9.82)	
	2258.00		2910.79		16.12		
	0.911		0.923		0.051		

1992		1993		1994		1995	
(−17.77)	−0.6259	(−23.10)	−0.7466	(−12.34)	−0.6206	(−10.62)	
(16.78)	0.3003	(3.63)	1.9475	(11.29)	2.6493	(12.10)	
(1.12)	0.4221	(8.17)	−0.0561	(−0.47)	0.2972	(2.50)	
(54.89)	0.6681	(62.48)	0.7206	(37.38)	0.7038	(22.61)	
(−11.73)	−0.1801	(−17.00)	−0.1294	(−5.90)	−0.0484	(−1.91)	
(−13.18)	−0.0569	(−6.36)	−0.2777	(−10.68)	−0.2016	(−10.61)	
(−13.31)	0.6269	(12.18)	1.8497	(14.35)	1.6907	(10.45)	
	769.60		386.77		144.53		
	0.691		0.526		0.290		

(Continued on following page)

TABLE 4.3 (Continued)

2SLS	SDIV							
		1982		1983		1984		
SLDIV	1.4127	(85.86)	0.5643	(272.86)	0.2991	(18.69)	1.3210	
SCE	−0.0662	(−6.34)	0.0175	(2.37)	−0.0424	(−1.89)	−0.0747	
SEF	0.1003	(5.12)	−0.0177	(−2.26)	−0.0343	(−1.55)	0.0204	
SRD	0.0075	(0.39)	−0.0145	(−1.36)	0.0457	(1.49)	0.0449	
SNI	0.0809	(9.01)	0.0472	(9.40)	0.2988	(25.83)	0.0672	
SLIQ	−0.0194	(−6.26)	−0.0004	(−0.25)	−0.0283	(−5.63)	−0.0080	
F−Value	8241.65		94877.94		439.14		3134.30	
adjusted r^2	0.976		0.998		0.659		0.927	
		1989		1990		1991		
SLDIV	0.7035	(54.15)	0.4457	(36.17)	0.6860	(28.35)	0.6220	
SCE	0.0130	(0.40)	−0.0924	(−4.23)	0.0568	(2.01)	−0.1684	
SEF	0.0711	(1.38)	0.0262	(0.56)	−0.5641	(−7.99)	0.0526	
SRD	0.0847	(2.38)	0.1020	(3.42)	0.1275	(2.96)	0.0905	
SNI	0.2091	(15.07)	0.2051	(15.40)	0.1070	(9.00)	0.1829	
SLIQ	−0.0275	(−4.30)	−0.0254	(−4.45)	−0.0041	(−0.49)	−0.0302	
F−Value	1627.49		645.60		269.32		1246.35	
adjusted r^2	0.817		0.632		0.411		0.759	

2SLS	SEF							
		1982		1983		1984		
SDIV	0.1253	(1.28)	0.1298	(3.96)	−0.4407	(−2.64)	−0.0350	
SCE	0.2842	(1.97)	0.6100	(6.38)	0.5065	(4.18)	0.0872	
SRD	0.2408	(1.50)	0.1373	(1.58)	0.3361	(3.18)	0.0827	
SDEP	−0.5386	(−1.74)	−0.9623	(−5.91)	−1.5041	(−8.13)	−0.4109	
SNI	−0.1515	(−1.86)	−0.2076	(−4.41)	0.0498	(0.60)	−0.0232	
DE	0.0447	(1.72)	0.0207	(1.40)	0.0848	(4.94)	−0.0160	
F−Value	2.38		8.31		23.25		2.83	
adjusted r^2	0.007		0.033		0.089		0.007	
		1989		1990		1991		
SDIV	−0.0508	(−1.47)	−0.0001	(−0.00)	−0.1405	(−3.66)	0.1648	
SCE	0.3486	(3.47)	0.2592	(6.43)	0.1928	(7.61)	0.1814	
SRD	0.2110	(3.14)	0.2043	(3.34)	0.2572	(4.88)	0.0749	
SDEP	−0.6449	(−4.82)	−0.3157	(−4.58)	−0.5061	(−7.58)	−0.3509	
SNI	0.0033	(0.10)	−0.0344	(−0.93)	−0.0133	(−0.80)	−0.1919	
DE	−0.0578	(−4.91)	−0.0503	(−4.47)	0.0079	(0.83)	0.0563	
F−Value	9.63		11.24		33.81		34.16	
adjusted r^2	0.023		0.027		0.079		0.078	

SIMULTANEOUS EQUATION ESTIMATION RESULTS

1985		1986		1987		1988	
(91.50)	0.9416	(103.30)	0.4863	(31.07)	0.7285	(58.74)	
(−11.42)	−0.0359	(−4.47)	−0.3080	(−21.50)	−0.0008	(−0.03)	
(1.05)	0.0363	(2.92)	0.1001	(10.25)	0.0089	(0.32)	
(2.55)	0.0042	(0.20)	0.1318	(3.00)	−0.0168	(−0.54)	
(10.81)	0.0376	(5.01)	0.3185	(24.46)	0.0484	(4.56)	
(−2.80)	−0.0060	(−1.74)	−0.0366	(−5.16)	−0.0020	(−0.38)	
	5862.47		738.07		1139.89		
	0.958		0.722		0.775		
1992		1993		1994		1995	
(55.01)	0.9400	(129.00)	0.7879	(80.65)	0.8506	(109.69)	
(−16.56)	−0.0419	(−7.76)	−0.0312	(−4.07)	0.0001	(0.01)	
(2.35)	−0.0605	(−3.67)	0.0170	(0.99)	−0.2261	(−7.78)	
(3.68)	0.0078	(0.60)	0.0092	(0.46)	0.0087	(0.49)	
(20.52)	0.0389	(8.72)	0.0400	(5.93)	0.0128	(2.22)	
(−6.83)	0.0004	(0.14)	−0.0041	(−1.00)	−0.0088	(−2.33)	
	4567.33		1762.84		3178.27		
	0.919		0.813		0.886		
1985		1986		1987		1988	
(−0.64)	0.4560	(11.53)	1.4064	(15.73)	0.1361	(2.39)	
(1.90)	0.5807	(16.32)	1.4242	(22.63)	0.2232	(1.81)	
(1.00)	−0.0481	(−0.52)	0.4496	(3.34)	−0.1275	(−1.27)	
(−3.00)	−1.1567	(−8.97)	−2.1278	(−13.76)	−0.4110	(−3.88)	
(−0.61)	−0.4766	(−15.93)	−1.1379	(−19.17)	−0.1875	(−5.26)	
(−1.13)	0.0234	(1.53)	−0.1436	(−7.15)	0.0504	(3.19)	
	49.08		135.71		7.63		
	0.157		0.322		0.020		
1992		1993		1994		1995	
(4.43)	−0.0441	(−1.43)	0.0110	(0.30)	0.0593	(2.54)	
(8.03)	0.0207	(0.96)	0.0956	(4.04)	0.1448	(7.38)	
(1.40)	0.0496	(0.94)	0.1980	(3.42)	0.0107	(0.24)	
(−9.89)	−0.3305	(−7.01)	−0.5631	(−8.75)	−0.4232	(−7.20)	
(−8.36)	0.0347	(1.88)	−0.0233	(−1.08)	−0.0918	(−6.35)	
(6.71)	−0.0186	(−2.09)	−0.0100	(−1.08)	−0.0174	(−2.18)	
	10.40		14.33		15.70		
	0.023		0.032		0.035		

(Continued on following page)

TABLE 4.3 (Continued)

2SLS	SRD	1982		1983		1984	
SLRD	1.2000	(27.98)	1.1331	(82.61)	1.1138	(95.47)	1.2104
SCE	−0.0246	(−1.29)	0.0029	(0.35)	−0.0002	(−0.02)	−0.0159
SDIV	0.0911	(4.47)	−0.0231	(−5.61)	−0.0814	(−4.78)	−0.0223
SEF	0.2068	(5.91)	0.0359	(3.86)	0.0214	(2.96)	0.0539
SNI	−0.0792	(−4.76)	0.0352	(6.10)	0.0491	(6.33)	0.0168
F−Value	178.92		1397.17		2048.12		1862.38
adjusted r²	0.424		0.846		0.883		0.863

		1989		1990		1991	
SLRD	1.0795	(122.63)	1.0694	(93.06)	1.0522	(110.56)	1.0130
SCE	0.0043	(0.54)	−0.0048	(−0.62)	−0.0272	(−5.19)	−0.0110
SDIV	−0.0182	(−4.34)	0.0125	(1.25)	0.0002	(0.03)	−0.0131
SEF	0.0367	(2.88)	0.0900	(5.02)	0.1201	(7.61)	0.0457
SNI	0.0219	(5.74)	−0.0083	(−1.32)	0.0034	(1.20)	0.0112
F−Value	3066.65		1825.57		2752.13		3592.92
adjusted r²	0.875		0.802		0.856		0.883

3SLS	SCE	1982		1983		1984	
SDIV	−0.3947	(−4.75)	−0.3322	(−7.00)	0.3070	(2.26)	−0.5431
SEF	1.2849	(15.13)	1.4101	(13.63)	0.6962	(10.11)	3.3587
SRD	−0.2445	(−1.75)	−0.0841	(−0.97)	−0.3074	(−3.61)	−0.0907
SNI	0.3833	(5.58)	0.4374	(7.63)	−0.0539	(−0.82)	0.6751
SLIQ	−0.0726	(−6.24)	−0.0632	(−4.91)	−0.0187	(−1.58)	−0.0471
SLSAL	−0.0260	(−4.08)	−0.0425	(−4.55)	0.0133	(1.85)	−0.0749
SDEP	0.6945	(5.72)	1.4600	(15.52)	1.6766	(15.84)	2.4467

		1989		1990		1991	
SDIV	−0.0698	(−4.28)	−0.4895	(−5.56)	0.5473	(7.09)	−0.9211
SEF	0.4426	(8.90)	2.4429	(16.92)	2.6589	(29.25)	1.9194
SRD	−0.0791	(−2.40)	−0.1644	(−1.55)	−0.4263	(−3.93)	−0.1312
SNI	0.1103	(7.26)	0.5021	(10.69)	0.1250	(4.11)	0.8664
SLIQ	−0.0224	(−4.12)	−0.0447	(−4.68)	−0.0463	(−5.87)	−0.0650
SLSAL	0.0006	(0.29)	−0.0330	(−4.03)	−0.0172	(−4.05)	−0.0357
SDEP	1.2201	(28.47)	0.8573	(7.45)	0.3969	(3.04)	0.8180

3SLS	SDIV	1982		1983		1984	
SLDIV	1.3756	(90.66)	0.5647	(273.30)	0.2635	(16.81)	1.3167
SCE	−0.1279	(−15.18)	0.0191	(2.59)	−0.0666	(−2.98)	−0.0778
SEF	0.1879	(14.67)	−0.0259	(−3.32)	−0.0568	(−2.61)	0.0275
SRD	−0.0170	(−0.95)	−0.0140	(−1.32)	0.0779	(2.55)	0.0468
SNI	0.1031	(12.72)	0.0461	(9.19)	0.3172	(27.76)	0.0699
SLIQ	−0.0216	(−10.44)	−0.0002	(−0.13)	−0.0363	(−7.45)	−0.0090

SIMULTANEOUS EQUATION ESTIMATION RESULTS

1985		1986		1987		1988	
(96.22)	1.1092	(99.26)	1.0014	(104.09)	1.0388	(62.96)	
(−3.86)	−0.0124	(−3.12)	0.0055	(1.26)	−0.0242	(−1.59)	
(−3.28)	−0.0151	(−3.01)	0.0038	(0.56)	−0.0372	(−3.83)	
(4.31)	0.0135	(2.04)	0.0013	(0.51)	0.1179	(8.11)	
(4.08)	0.0129	(3.27)	−0.0060	(−1.36)	0.0366	(5.97)	
	2006.29		2193.37		805.82		
	0.866		0.866		0.670		

1992		1993		1994		1995	
(132.80)	1.0291	(137.21)	0.9205	(133.44)	0.9359	(85.55)	
(−3.19)	−0.0048	(−1.59)	−0.0036	(−1.26)	0.1295	(26.47)	
(−2.34)	−0.0052	(−1.15)	−0.0058	(−1.18)	−0.0272	(−4.30)	
(6.29)	0.0619	(6.30)	0.0232	(3.35)	−0.1064	(−5.39)	
(3.31)	0.0042	(1.57)	0.0037	(1.36)	0.0305	(7.46)	
	3865.17		3736.88		2020.96		
	0.889		0.885		0.804		

1985		1986		1987		1988	
(−5.39)	−0.7667	(−26.70)	−1.1902	(−48.19)	−0.2190	(−4.34)	
(14.30)	1.1274	(40.95)	0.4326	(36.52)	0.9555	(11.25)	
(−0.45)	0.0215	(0.24)	0.1027	(1.63)	0.0809	(0.96)	
(21.18)	0.7769	(63.36)	0.9053	(108.12)	0.2369	(6.99)	
(−3.32)	−0.0411	(−5.51)	−0.0375	(−5.08)	−0.0350	(−3.59)	
(−4.09)	−0.0221	(−5.09)	−0.0093	(−2.67)	−0.0078	(−1.51)	
(10.77)	1.9055	(20.00)	0.8745	(18.47)	0.5829	(8.66)	

1992		1993		1994		1995	
(−17.96)	−0.6223	(−22.98)	−0.6668	(−11.09)	−0.4794	(−8.51)	
(30.80)	0.4537	(5.62)	2.3188	(24.99)	2.9650	(32.97)	
(−1.45)	0.4212	(8.17)	−0.2060	(−1.90)	0.5908	(5.38)	
(57.96)	0.6637	(62.20)	0.6901	(37.07)	0.6142	(23.12)	
(−7.68)	−0.1761	(−16.92)	−0.0651	(−5.42)	−0.0123	(−1.40)	
(−4.71)	−0.0535	(−6.16)	−0.0992	(−7.26)	−0.0333	(−4.58)	
(13.60)	0.6303	(12.36)	2.0417	(18.08)	1.1831	(9.29)	

1985		1986		187		1988	
(91.32)	0.9225	(102.03)	0.2763	(20.52)	0.7285	(58.75)	
(−11.92)	−0.0555	(−6.99)	−0.4846	(−39.19)	−0.0015	(−0.06)	
(1.42)	0.0640	(5.21)	0.2078	(23.01)	0.0007	(0.03)	
(2.66)	0.0106	(0.52)	0.1613	(3.72)	−0.0157	(−0.51)	
(11.27)	0.0568	(7.66)	0.4922	(43.56)	0.0478	(4.51)	
(−3.15)	−0.0062	(−1.81)	−0.0317	(−4.95)	−0.0025	(−0.49)	

(Continued on following page)

TABLE 4.3 (Continued)

	1989		1990		1991		
SLDIV	0.7043	(54.31)	0.4332	(35.61)	0.5380	(24.79)	0.5925
SCE	0.0540	(1.66)	−0.0912	(−4.21)	0.2522	(10.70)	−0.2032
SEF	0.0296	(0.58)	−0.0373	(−0.81)	−1.0635	(−19.36)	0.0936
SRD	0.0867	(2.44)	0.1176	(3.96)	0.2156	(5.27)	0.0976
SNI	0.2057	(14.85)	0.2109	(15.94)	0.0921	(8.13)	0.2126
SLIQ	−0.0302	(−4.76)	−0.0273	(−4.85)	−0.0041	(−0.73)	−0.0334

3SLS   SEF

	1982		1983		1984		
SDIV	0.2807	(3.15)	0.1160	(3.91)	−0.5892	(−3.73)	0.1458
SCE	0.6575	(10.03)	0.5775	(6.26)	1.1232	(14.39)	0.2823
SRD	0.2942	(1.96)	0.1531	(1.81)	0.4662	(4.58)	0.0438
SDEP	−0.1186	(−0.87)	−0.8873	(−5.74)	−2.1082	(−14.62)	−0.7483
SNI	−0.2654	(−3.68)	−0.1864	(−4.46)	0.1318	(1.76)	−0.1962
DE	0.0341	(3.77)	0.0098	(1.32)	0.0168	(2.07)	0.0186

	1989		1990		1991		
SDIV	0.0340	(1.00)	0.1790	(3.65)	−0.3197	(−8.90)	0.4274
SCE	1.0739	(11.90)	0.3764	(11.97)	0.3501	(19.38)	0.4609
SRD	0.1978	(2.97)	0.0693	(1.20)	0.1664	(3.34)	0.1083
SDEP	−1.4693	(−11.93)	−0.3218	(−5.35)	0.0368	(0.77)	−0.3984
SNI	−0.1050	(−3.31)	−0.2066	(−7.44)	0.0012	(0.08)	−0.4162
DE	−0.0092	(−0.84)	0.0241	(4.12)	0.0039	(1.50)	0.0165

3SLS   SRD

	1982		1983		1984		
SLRD	1.1222	(27.43)	1.1316	(82.50)	1.1123	(95.38)	1.2095
SCE	−0.1438	(−9.09)	0.0011	(0.14)	−0.0029	(−0.37)	−0.0208
SDIV	0.0412	(2.05)	−0.0238	(−5.76)	−0.0820	(−4.82)	−0.0253
SEF	0.3737	(18.96)	0.0475	(5.12)	0.0249	(3.45)	0.0906
SNI	−0.0330	(−2.03)	0.0363	(6.29)	0.0495	(6.39)	0.0213

	1989		1990		1991		
SLRD	1.0798	(122.68)	1.0635	(93.00)	1.0323	(110.17)	1.0119
SCE	−0.0035	(−0.44)	−0.0332	(−4.60)	−0.0639	(−13.50)	−0.0263
SDIV	−0.0186	(−4.41)	0.0043	(0.44)	0.0337	(4.88)	−0.0291
SEF	0.0479	(3.80)	0.1707	(10.48)	0.2241	(16.44)	0.0724
SNI	0.0230	(6.02)	0.0031	(0.51)	0.0032	(1.15)	0.0252

SIMULTANEOUS EQUATION ESTIMATION RESULTS

1992		1993		1994		1995	
(53.04)	0.9326	(128.16)	0.7856	(80.42)	0.8740	(114.93)	
(−20.39)	−0.0513	(−9.56)	−0.0347	(−4.55)	0.0618	(8.61)	
(4.19)	−0.1150	(−7.20)	−0.0094	(−0.55)	−0.4101	(−16.80)	
(3.98)	0.0140	(1.07)	0.0181	(0.91)	−0.0511	(−3.00)	
(24.32)	0.0465	(10.47)	0.0428	(6.34)	−0.0234	(−4.35)	
(−7.57)	−0.0035	(−1.30)	−0.0076	(−1.87)	−0.0096	(−3.15)	

1985		1986		1987		1988	
(3.22)	0.6611	(19.77)	2.2811	(30.72)	0.1901	(3.41)	
(10.05)	0.8611	(36.79)	1.9673	(40.42)	0.8712	(13.34)	
(0.55)	0.0450	(0.51)	0.0538	(0.41)	−0.0581	(−0.61)	
(−7.09)	−1.7503	(−15.88)	−2.0956	(−15.86)	−0.4758	(−6.06)	
(−8.37)	−0.6763	(−32.95)	−1.7635	(−39.46)	−0.2243	(−6.66)	
(3.12)	0.0157	(2.23)	−0.0100	(−0.69)	0.0291	(3.44)	

1992		1993		1994		1995	
(13.03)	0.0671	(2.18)	0.2661	(7.94)	0.1512	(6.92)	
(32.04)	0.1895	(8.97)	0.4071	(28.24)	0.3302	(33.21)	
(2.10)	−0.0255	(−0.49)	0.1067	(1.89)	−0.2222	(−5.28)	
(−11.74)	−0.4523	(−10.56)	−0.9089	(−15.36)	−0.3487	(−7.53)	
(−26.11)	−0.1054	(−5.80)	−0.2888	(−19.98)	−0.1996	(−18.18)	
(3.88)	0.0230	(2.92)	0.0175	(3.54)	0.0044	(1.57)	

1985		1986		1987		1988	
(96.18)	1.1102	(99.35)	1.0025	(104.21)	1.0433	(63.32)	
(−5.16)	−0.0141	(−3.57)	0.0015	(0.34)	−0.0820	(−5.82)	
(−3.76)	−0.0168	(−3.36)	−0.0016	(−0.23)	−0.0488	(−5.03)	
(7.61)	0.0154	(2.32)	0.0030	(1.20)	0.1960	(15.81)	
(5.26)	0.0147	(3.71)	−0.0020	(−0.44)	0.0503	(8.32)	

1992		1993		1994		1995	
(132.74)	1.0335	(137.83)	0.9227	(133.88)	0.9074	(83.90)	
(−7.86)	−0.0152	(−5.00)	−0.0123	(−4.34)	0.1572	(34.15)	
(−5.26)	−0.0101	(−2.24)	−0.0126	(−2.58)	−0.0121	(−1.93)	
(10.21)	0.1029	(11.20)	0.0392	(5.75)	−0.1896	(−10.30)	
(7.60)	0.0102	(3.84)	0.0107	(3.96)	0.0109	(2.77)	

(Continued on following page)

Dividends are positively associated with rising net income and last year's dividends, supporting the positions of Lintner (1956), Fama and Babiak (1968), and Switzer (1984).

The rejection of the perfect markets hypothesis is found in (1) the influence of dividends, R&D expenditures, and new debt financing on the investment decision, (2) evidence suggesting that increasing capital expenditures lower dividends whereas new debt financing and R&D are positively associated with increasing dividends, and (3) the interdependence between investment and new debt variables. The empirical evidence of the 3,000-firm universe confirms the necessity of using simultaneous equations to model econometrically the interdependencies of financial decisions. The evidence is more supportive of the existence of imperfect markets in the larger universe than in the original Guerard and McCabe (1990) 303-firm sample.

In the international analysis using the WorldScope database for the 1982–1995 period, one finds almost complete agreement with the 3,000-

TABLE 4.4 Estimated System Equations, Japan WorldScope Universe

OLS	SCE	1982 (t)		1983		1984		
SDIV	0.7779	(1.90)	0.8991	(2.14)	0.8734	(2.08)	0.4346
SEF	0.1444	(4.11)	0.0520	(2.13)	−0.0099	(−0.54)	0.0335
SRD	0.6289	(5.37)	0.4668	(4.89)	0.5104	(5.38)	0.2067
SNI	0.1788	(1.83)	0.2448	(2.27)	0.3927	(4.00)	0.2325
SLIQ	−0.0465	(−3.10)	−0.0444	(−3.24)	−0.0473	(−3.92)	−0.0154
SLSAL	0.0099	(0.54)	−0.0121	(−1.30)	−0.0041	(−0.66)	−0.0146
SDEP	0.8476	(10.29)	0.8825	(12.01)	0.6116	(9.34)	1.2806
F−Value	36.00		44.67		36.13		56.52
adjusted r^2	0.405		0.436		0.351		0.401
		1989		1990		1991		
SDIV	1.5949	(5.81)	0.1365	(0.70)	0.0235	(0.12)	−0.1181
SEF	−0.0067	(−0.80)	−0.0068	(−1.59)	−0.0031	(−0.60)	0.0107
SRD	0.1313	(1.92)	0.2360	(4.80)	0.3390	(6.61)	0.3410
SNI	0.0184	(0.47)	0.0178	(0.58)	0.0266	(1.53)	0.0126
SLIQ	−0.0165	(−2.60)	0.0032	(0.77)	−0.0034	(−0.77)	−0.0017
SLSAL	−0.0007	(−0.22)	−0.0046	(−3.19)	−0.0057	(−2.49)	−0.0026
SDEP	0.7439	(13.40)	0.6530	(17.84)	0.5374	(14.65)	0.5120
F−Value	42.32		69.11		48.46		50.82
adjusted r^2	0.188		0.235		0.159		0.155

stock Compustat universe results (see Table 4.3). The Japanese estimated system equations, shown in Table 4.4, do not provide as strong a case for interdependent financial decisions as the U.S. equations; although R&D and new debt financing positively affect capital expenditures, there is a positive relationship between dividends and capital expenditures in Japan. Japanese new debt financing is associated only with higher capital expenditures, not with R&D or dividends. Japanese dividends are associated with higher new debt financing and are independent of capital expenditures and R&D. New debt is associated with higher capital expenditures in Japan, not with R&D, and often negatively with dividends. Dividends and capital expenditures do not affect Japanese R&D; only net debt financing and net income are significantly associated with R&D. The G5 (European) estimated system equations are shown in Table 4.5 and are more consistent with the estimated Japanese relationships than the U.S. estimates. New debt financing and capital expenditures constitute the only violation of the imperfect markets hypothesis in the G5 estimated system equations.

(Text continued on page 92)

1985		1986		1987		1988	
(0.98)	0.4026	(1.33)	0.2927	(0.98)	1.5094	(5.25)	
(2.01)	0.0552	(3.23)	0.0003	(0.08)	−0.0019	(−0.25)	
(1.98)	0.1232	(1.39)	0.2674	(3.17)	0.0069	(0.09)	
(2.47)	0.0659	(1.67)	0.1091	(2.35)	0.0154	(0.57)	
(−1.27)	−0.0154	(−1.82)	−0.0171	(−2.49)	−0.0221	(−3.20)	
(−1.98)	0.0111	(1.73)	0.0085	(1.85)	−0.0028	(−0.66)	
(15.69)	1.0731	(16.43)	0.7342	(11.01)	0.6485	(10.08)	
	53.90		28.92		22.84		
	0.311		0.161		0.119		
1992		1993		1994		1995	
(−0.60)	0.0393	(0.29)	−0.0411	(−0.35)	0.0026	(0.03)	
(1.36)	0.0228	(2.78)	0.0084	(1.30)	0.0155	(2.95)	
(7.29)	0.2230	(6.01)	0.1824	(5.78)	0.1858	(6.40)	
(0.93)	0.0164	(1.11)	0.0228	(1.75)	0.0178	(1.80)	
(−0.40)	−0.0068	(−2.07)	−0.0048	(−1.84)	−0.0063	(−2.73)	
(−0.61)	0.0071	(1.85)	0.0062	(1.42)	0.0091	(2.16)	
(15.45)	0.4195	(16.36)	0.3418	(15.91)	0.3170	(14.16)	
	54.53		49.17		43.96		
	0.163		0.149		0.136		

(Continued on following page)

74 THE INTERDEPENDENCIES AMONG CORPORATE FINANCIAL POLICIES

TABLE 4.4 (Continued)

OLS	SDIV							
		1982 (t)		1983		1984		
SLDIV	0.9308	(29.74)	0.6677	(23.85)	0.7545	(22.85)	0.8369
SCE	0.0015	(0.45)	0.0060	(1.81)	−0.0012	(−0.34)	−0.0006
SEF	0.0022	(0.89)	0.0054	(3.24)	0.0057	(4.35)	0.0096
SRD	−0.0055	(−0.64)	−0.0099	(−1.31)	−0.0080	(−1.07)	−0.0092
SNI	0.0423	(6.61)	0.0425	(5.41)	0.0626	(8.92)	0.0351
SLIQ	−0.0017	(−1.54)	0.0023	(2.18)	0.0017	(1.77)	0.0012
F−Value	259.28		206.24		230.50		551.78
adjusted r^2	0.812		0.757		0.752		0.851
		1989		1990		1991		
SLDIV	0.7151	(38.98)	0.7866	(40.73)	0.8433	(52.43)	0.8151
SCE	0.0076	(4.19)	−0.0028	(−1.32)	−0.0003	(−0.18)	0.0010
SEF	0.0050	(9.41)	0.0085	(22.88)	0.0055	(15.15)	0.0032
SRD	−0.0005	(−0.11)	0.0117	(2.60)	−0.0001	(−0.02)	0.0014
SNI	0.0168	(6.26)	0.0269	(9.84)	0.0067	(5.08)	0.0071
SLIQ	0.0020	(4.60)	−0.0000	(−0.05)	0.0014	(4.03)	0.0021
F−Value	429.76		446.17		598.47		855.18
adjusted r^2	0.674		0.632		0.671		0.729

OLS	SEF							
		1982		1983		1984		
SDIV	−1.5796	(−2.33)	−1.5313	(−1.27)	−0.7604	(−0.56)	−0.7052
SCE	0.2979	(3.88)	0.2062	(1.73)	−0.1803	(−1.37)	0.0691
SRD	−0.1782	(−0.99)	0.1491	(0.65)	0.2124	(0.78)	−0.1300
SNI	−0.0983	(−0.59)	−0.3361	(−1.10)	0.6735	(2.21)	0.6811
DE	0.0668	(2.56)	0.0899	(2.13)	0.0260	(0.61)	0.1209
SDEP	0.0940	(0.68)	−0.2302	(−1.13)	−0.0533	(−0.26)	−0.6064
F−Value	5.98		1.65		2.48		6.25
adjusted r^2	0.077		0.010		0.019		0.052
		1989		1990		1991		
SDIV	−5.1617	(−4.13)	−6.6236	(−4.24)	−3.2427	(−2.62)	0.7199
SCE	−0.2202	(−2.03)	−0.6980	(−3.99)	−0.2722	(−2.12)	0.0873
SRD	−0.4283	(−1.65)	−1.6663	(−4.95)	−0.4718	(−1.71)	−0.1228
SNI	−0.3198	(−2.03)	−1.0200	(−4.57)	−0.1186	(−1.24)	0.0378
DE	0.2498	(8.34)	0.5037	(14.69)	0.1068	(4.15)	−0.0045
SDEP	0.0482	(0.21)	−0.3024	(−1.09)	0.1503	(0.72)	0.1071
F−Value	12.52		49.17		4.23		1.09
adjusted r^2	0.052		0.157		0.011		0.000

SIMULTANEOUS EQUATION ESTIMATION RESULTS

1985		1986		1987		1988	
(39.17)	0.8905	(61.64)	0.8388	(54.15)	0.8206	(54.96)	
(−0.37)	0.0028	(1.92)	0.0023	(1.47)	0.0048	(3.13)	
(13.94)	0.0099	(12.69)	0.0007	(2.75)	0.0037	(9.49)	
(−1.85)	−0.0026	(−0.62)	0.0028	(0.65)	−0.0030	(−0.75)	
(7.96)	0.0199	(10.65)	0.0365	(16.05)	0.0235	(16.70)	
(2.07)	0.0009	(2.06)	−0.0009	(−2.30)	0.0012	(3.32)	
	969.57		883.03		718.67		
	0.876		0.839		0.793		
1992		1993		1994		1995	
(61.11)	0.9479	(75.24)	0.8175	(73.43)	0.9837	(83.26)	
(0.67)	0.0015	(0.81)	0.0018	(0.81)	−0.0005	(−0.18)	
(6.44)	0.0000	(0.06)	0.0012	(1.83)	0.0017	(2.81)	
(0.43)	−0.0071	(−2.23)	−0.0009	(−0.28)	−0.0007	(−0.22)	
(7.66)	0.0151	(13.03)	0.0212	(17.69)	0.0104	(9.58)	
(7.31)	0.0008	(2.80)	0.0008	(3.22)	−0.0003	(−0.98)	
	1302.89		1323.71		1504.12		
	0.802		0.805		0.826		

1985		1986		1987		1988	
(−0.43)	−0.1187	(−0.16)	5.3525	(2.40)	0.2970	(0.20)	
(0.51)	0.2933	(3.87)	0.0449	(0.21)	−0.1267	(−1.02)	
(−0.39)	−0.1010	(−0.53)	−0.4858	(−0.87)	−0.6891	(−2.22)	
(2.09)	−0.5554	(−6.47)	0.2749	(0.88)	−0.1962	(−1.70)	
(2.57)	0.1153	(5.30)	−0.1320	(−2.32)	0.1163	(3.05)	
(−1.92)	−0.5831	(−3.56)	0.5320	(1.12)	−0.4258	(−1.52)	
	13.41		1.99		3.92		
	0.083		0.006		0.015		
1992		1993		1994		1995	
(0.97)	−0.2671	(−0.64)	−0.1187	(−0.25)	1.2381	(2.73)	
(1.18)	0.1621	(2.55)	0.1089	(1.31)	0.3067	(3.03)	
(−0.81)	−0.0846	(−0.82)	−0.1486	(−1.30)	−0.0452	(−0.35)	
(0.86)	−0.0665	(−0.75)	−0.0285	(−0.63)	−0.2209	(−5.20)	
(−0.31)	−0.0079	(−0.88)	0.0103	(1.08)	0.0132	(1.30)	
(0.95)	−0.0418	(−0.55)	−0.1429	(−1.72)	−0.1798	(−1.72)	
	2.61		1.00		6.89		
	0.005		−0.000		0.018		

(Continued on following page)

TABLE 4.4 (Continued)

OLS	SRD	1982		1983		1984		
SLRD	1.1126	(46.68)	1.1123	(37.61)	0.9999	(34.06)	1.0169
SCE	0.0397	(5.36)	0.0401	(3.95)	0.0179	(1.61)	0.0174
SDIV	−0.0346	(−0.52)	−0.0931	(−0.92)	−0.1772	(−1.66)	0.0119
SEF	−0.0097	(−1.64)	0.0063	(1.21)	0.0143	(3.26)	0.0036
SNI	0.0268	(1.86)	0.0418	(1.81)	0.0377	(1.53)	0.0247
F−Value	526.44		357.70		288.72		301.17
adjusted r^2	0.879		0.818		0.760		0.721
		1989		1990		1991		
SLRD	0.8630	(49.59)	0.9865	(68.92)	1.0350	(102.61)	1.0701
SCE	0.0141	(2.16)	0.0147	(2.43)	0.0106	(2.64)	−0.0030
SDIV	0.0879	(1.36)	0.0259	(0.53)	0.0133	(0.41)	0.0290
SEF	0.0073	(4.06)	0.0033	(3.70)	0.0055	(7.00)	0.0050
SNI	0.0131	(1.37)	0.0177	(2.35)	0.0046	(1.56)	−0.0008
F−Value	527.08		1055.79		2264.92		6651.87
adjusted r^2	0.678		0.772		0.865		0.946

2SLS	SCE	1982		1983		1984		
SDIV	1.0116	(1.89)	1.1412	(1.58)	2.0601	(3.15)	3.3777
SEF	0.2395	(1.60)	0.5118	(3.24)	−0.0132	(−0.23)	0.5296
SRD	0.3967	(3.07)	0.3807	(2.52)	0.6984	(5.85)	0.5021
SNI	0.1787	(1.74)	0.4369	(2.59)	0.2462	(2.12)	−0.1750
SLIQ	−0.0482	(−2.70)	−0.0741	(−3.39)	−0.0586	(−3.89)	−0.1125
SLSAL	0.0059	(0.30)	−0.1000	(−3.11)	−0.0034	(−0.34)	−0.1501
SDEP	0.8391	(8.04)	0.8269	(7.80)	0.5603	(8.20)	1.2273
F−Value	30.24		23.79		37.18		22.49
adjusted r^2	0.363		0.287		0.358		0.206
		1989		1990		1991		
SDIV	2.7034	(6.37)	0.2018	(0.74)	−0.4503	(−1.49)	0.1479
SEF	0.0955	(2.65)	−0.0252	(−3.37)	−0.0580	(−2.87)	0.1124
SRD	0.2311	(2.33)	0.2223	(3.84)	0.2852	(4.86)	0.3820
SNI	0.0014	(0.03)	0.0058	(0.18)	0.0143	(0.77)	0.0158
SLIQ	−0.0413	(−4.09)	0.0078	(1.60)	0.0059	(1.05)	−0.0100
SLSAL	−0.0182	(−2.65)	−0.0010	(−0.54)	0.0069	(1.37)	−0.0261
SDEP	0.6758	(10.66)	0.6501	(17.27)	0.5655	(14.32)	0.4826
F−Value	38.51		69.27		45.28		47.63
adjusted r^2	0.174		0.235		0.150		0.146

SIMULTANEOUS EQUATION ESTIMATION RESULTS

1985		1986		1987		1988	
(36.08)	1.0697	(84.73)	0.9477	(57.58)	0.9200	(37.85)	
(2.15)	0.0112	(2.74)	0.0119	(2.14)	−0.0049	(−0.62)	
(0.12)	−0.0197	(−0.52)	0.0791	(1.49)	0.2160	(2.92)	
(1.22)	0.0055	(2.60)	−0.0002	(−0.22)	−0.0034	(−1.75)	
(1.13)	0.0077	(1.45)	−0.0002	(−0.02)	0.0010	(0.13)	
	1535.19		705.21		300.10		
	0.903		0.776		0.571		

1992		1993		1994		1995	
(176.35)	0.9855	(203.48)	0.9455	(181.11)	0.9811	(168.23)	
(−1.17)	0.0009	(0.31)	0.0013	(0.35)	0.0086	(1.97)	
(1.35)	0.0296	(1.80)	0.0109	(0.59)	0.0357	(1.99)	
(5.87)	−0.0009	(−0.85)	0.0026	(2.40)	0.0009	(0.82)	
(−0.53)	0.0020	(1.11)	0.0087	(4.10)	0.0017	(0.87)	
	8668.75		6840.27		5917.05		
	0.957		0.947		0.939		

1985		1986		1987		1988	
(2.70)	0.9435	(2.37)	0.3453	(0.47)	3.7357	(2.79)	
(2.94)	0.3218	(4.07)	0.2820	(2.60)	0.8349	(2.87)	
(2.28)	0.1917	(1.74)	0.4521	(2.01)	0.3715	(0.97)	
(−0.84)	0.1596	(2.98)	0.0228	(0.21)	0.1499	(1.42)	
(−2.90)	−0.0450	(−3.49)	−0.0240	(−1.58)	−0.1337	(−2.92)	
(−2.99)	−0.0304	(−2.18)	−0.0095	(−0.79)	−0.1472	(−2.83)	
(8.94)	1.0754	(14.25)	0.5786	(3.66)	0.9563	(3.81)	
	42.42		7.16		3.06		
	0.261		0.041		0.013		

1992		1993		1994		1995	
(0.54)	0.3388	(1.52)	0.3084	(1.54)	−0.0042	(−0.04)	
(2.63)	0.3194	(5.05)	0.2297	(3.98)	0.0363	(3.45)	
(7.43)	0.2677	(5.37)	0.2636	(5.61)	0.1977	(6.55)	
(1.10)	0.0436	(2.17)	0.0451	(2.54)	0.0238	(2.31)	
(−1.74)	−0.0161	(−3.33)	−0.0210	(−3.86)	−0.0077	(−3.19)	
(−2.45)	0.0054	(1.06)	−0.0131	(−1.75)	0.0068	(1.57)	
(13.08)	0.4028	(11.95)	0.3379	(12.24)	0.3161	(14.01)	
	35.58		32.08		44.14		
	0.112		0.102		0.137		

(Continued on following page)

TABLE 4.4 (Continued)

2SLS	SDIV							
		1982		1983		1984		
SLDIV	0.9459	(26.76)	0.6998	(22.73)	0.7589	(20.47)	0.9031
SCE	−0.0012	(−0.24)	0.0031	(0.70)	0.0033	(0.76)	0.0004
SEF	0.0118	(1.25)	0.0161	(4.78)	0.0083	(3.11)	0.0154
SRD	−0.0068	(−0.72)	−0.0071	(−0.78)	−0.0162	(−1.75)	−0.0063
SNI	0.0425	(6.40)	0.0414	(4.93)	0.0591	(7.93)	0.0255
SLIQ	−0.0023	(−1.79)	0.0014	(1.26)	0.0018	(1.73)	−0.0002
F−Value	249.15		187.96		226.35		485.40
adjusted r^2	0.805		0.739		0.749		0.834
		1989		1990		1991		
SLDIV	0.8071	(32.57)	0.9150	(38.17)	0.9473	(46.95)	0.8237
SCE	0.0100	(2.72)	−0.0009	(−0.25)	0.0003	(0.12)	0.0016
SEF	0.0146	(11.25)	0.0136	(22.17)	0.0126	(16.54)	0.0061
SRD	0.0018	(0.26)	0.0199	(3.48)	0.0036	(0.75)	0.0006
SNI	0.0155	(5.11)	0.0244	(8.36)	0.0068	(4.66)	0.0070
SLIQ	−0.0001	(−0.21)	−0.0020	(−4.37)	0.0001	(0.15)	0.0019
F−Value	349.57		400.70		503.33		838.22
adjusted r^2	0.627		0.607		0.632		0.725

2SLS	SEF							
		1982		1983		1984		
SDIV	−2.6177	(−3.03)	−4.3096	(−2.49)	−8.3060	(−3.78)	−9.9790
SCE	0.0439	(0.34)	0.1794	(1.08)	−0.2251	(−1.24)	−0.3516
SRD	−0.0331	(−0.17)	0.0051	(0.02)	−0.3198	(−0.92)	−0.3728
SDEP	0.3489	(1.99)	−0.1262	(−0.54)	0.1047	(0.47)	0.0102
SNI	−0.0503	(−0.30)	−0.3615	(−1.17)	1.0518	(3.24)	1.1686
DE	0.0854	(3.00)	0.1475	(2.97)	0.1542	(2.93)	0.2796
F−Value	4.19		1.93		4.98		9.70
adjusted r^2	0.051		0.014		0.050		0.083
		1989		1990		1991		
SDIV	−18.5434	(−9.32)	−41.8216	(−13.51)	−15.2204	(−8.66)	−2.0958
SCE	−0.4453	(−1.83)	−2.2826	(−5.68)	−0.4932	(−2.72)	0.0657
SRD	−1.4665	(−4.29)	−1.6055	(−3.44)	−0.8255	(−2.67)	−0.2106
SDEP	0.6201	(2.11)	1.4532	(3.66)	0.5435	(2.41)	0.1716
SNI	−0.2271	(−1.36)	−0.8103	(−3.08)	−0.1496	(−1.52)	0.0523
DE	0.4647	(12.05)	0.9483	(18.58)	0.2559	(8.43)	0.0295
F−Value	25.58		72.21		17.45		1.72
adjusted r^2	0.106		0.215		0.053		0.002

SIMULTANEOUS EQUATION ESTIMATION RESULTS

1985		1986		1987		1988
(35.04)	0.9245	(51.18)	0.8379	(46.78)	0.9196	(31.73)
(0.17)	0.0032	(1.17)	0.0029	(0.96)	0.0025	(0.53)
(12.29)	0.0244	(11.59)	0.0053	(2.00)	0.0219	(9.72)
(−0.94)	0.0013	(0.24)	−0.0011	(−0.17)	0.0088	(0.95)
(5.09)	0.0249	(10.62)	0.0347	(12.46)	0.0260	(10.63)
(−0.29)	−0.0007	(−1.36)	−0.0009	(−1.92)	−0.0019	(−2.47)
	682.08		663.26		254.25	
	0.832		0.796		0.574	
1992		1993		1994		1995
(59.75)	0.9372	(65.59)	0.7948	(57.56)	0.9834	(82.71)
(0.83)	0.0001	(0.04)	0.0038	(1.24)	−0.0007	(−0.22)
(5.63)	−0.0101	(−2.27)	−0.0137	(−3.62)	0.0044	(3.75)
(0.18)	−0.0098	(−2.78)	−0.0084	(−2.10)	−0.0027	(−0.77)
(7.48)	0.0141	(10.95)	0.0203	(14.72)	0.0110	(9.86)
(6.42)	0.0011	(3.40)	0.0020	(4.87)	−0.0004	(−1.43)
	1177.45		1037.89		1488.85	
	0.785		0.764		0.824	

1985		1986		1987		1988
(−4.72)	−3.3303	(−3.76)	3.2481	(1.22)	−5.4999	(−3.09)
(−1.39)	0.5250	(3.50)	0.3348	(1.00)	−0.0715	(−0.30)
(−0.89)	−0.3177	(−1.53)	−0.5551	(−0.85)	−0.6338	(−1.50)
(0.02)	−0.7726	(−3.55)	0.3277	(0.63)	−0.4366	(−1.37)
(3.40)	−0.4884	(−5.56)	0.3523	(1.09)	−0.1767	(−1.52)
(5.24)	0.1690	(7.24)	−0.1042	(−1.72)	0.2112	(4.99)
	15.07		1.40		4.86	
	0.093		0.002		0.020	
1992		1993		1994		1995
(−2.14)	−0.5392	(−1.06)	−1.2250	(−2.12)	1.3920	(2.66)
(0.67)	0.0890	(1.14)	0.1115	(1.04)	0.3141	(2.37)
(−1.33)	−0.0508	(−0.48)	−0.2292	(−1.93)	−0.0654	(−0.49)
(1.45)	−0.0080	(−0.10)	−0.1208	(−1.40)	−0.1820	(−1.68)
(1.18)	−0.0593	(−1.54)	−0.0008	(−0.02)	−0.2234	(−5.23)
(1.82)	−0.0057	(−0.59)	0.0232	(2.27)	0.0117	(1.11)
	1.84		1.89		6.25	
	0.003		0.003		0.016	

(Continued on following page)

TABLE 4.4 (Continued)

2SLS	SRD							
		1982		1983		1984		
SLRD	1.1428	(43.72)	1.1249	(37.18)	0.9991	(32.91)	1.0103
SCE	−0.0077	(−0.68)	0.0245	(1.97)	0.0203	(1.46)	0.0309
SDIV	0.0095	(0.11)	−0.0373	(−0.30)	−0.3207	(−2.25)	0.1080
SEF	0.0160	(0.79)	0.0063	(0.65)	0.0215	(2.60)	0.0104
SNI	0.0310	(1.85)	0.0396	(1.58)	0.0523	(1.76)	−0.0031
F−Value	457.42		352.78		285.90		297.74
adjusted r^2	0.864		0.816		0.758		0.719
		1989		1990		1991		
SLRD	0.8689	(48.61)	0.9921	(66.82)	1.0453	(100.64)	1.0748
SCE	0.0152	(1.30)	0.0147	(1.48)	−0.0010	(−0.20)	−0.0075
SDIV	0.0246	(0.30)	0.0328	(0.51)	0.0289	(0.72)	0.0241
SEF	0.0123	(3.78)	0.0057	(4.70)	0.0093	(6.88)	0.0130
SNI	0.0153	(1.53)	0.0155	(1.93)	0.0047	(1.55)	−0.0011
F−Value	522.01		1051.77		2222.55		6348.02
adjusted r^2	0.676		0.771		0.863		0.943

3SLS	SCE							
		1982		1983		1984		
SDIV	1.2309	(2.33)	1.0064	(1.41)	1.9725	(3.04)	3.5795
SEF	0.3829	(2.66)	0.4535	(3.15)	−0.0667	(−1.19)	0.6212
SRD	0.4034	(3.13)	0.3574	(2.39)	0.6913	(5.82)	0.5109
SNI	0.1667	(1.64)	0.1949	(1.22)	0.2507	(2.16)	−0.4522
SLIQ	−0.0550	(−3.30)	−0.0454	(−2.62)	−0.0504	(−3.44)	−0.1017
SLSAL	0.0009	(0.05)	−0.0172	(−0.59)	0.0056	(0.59)	−0.1011
SDEP	0.7460	(7.50)	0.8728	(8.48)	0.5449	(8.24)	1.2964
		1989		1990		1991		
SDIV	2.6751	(6.36)	0.1115	(0.41)	−0.5985	(−2.01)	0.2483
SEF	0.0872	(2.46)	−0.0320	(−4.47)	−0.0852	(−4.40)	0.1983
SRD	0.2305	(2.33)	0.1694	(2.94)	0.2466	(4.22)	0.3993
SNI	−0.0052	(−0.12)	0.0129	(0.41)	0.0183	(1.00)	0.0114
SLIQ	−0.0392	(−4.00)	0.0121	(2.65)	0.0088	(1.63)	−0.0119
SLSAL	−0.0074	(−1.10)	−0.0059	(−3.34)	0.0012	(0.26)	−0.0257
SDEP	0.6929	(10.97)	0.6503	(17.40)	0.5806	(14.85)	0.4641

3SLS	SDIV							
		1982		1983		1984		
SLDIV	0.9478	(26.84)	0.7113	(23.14)	0.7186	(19.52)	0.9218
SCE	−0.0015	(−0.29)	−0.0027	(−0.62)	0.0153	(3.61)	−0.0022
SEF	0.0139	(1.48)	0.0235	(7.17)	0.0058	(2.17)	0.0189
SRD	−0.0071	(−0.76)	−0.0036	(−0.40)	−0.0288	(−3.14)	−0.0020
SNI	0.0421	(6.35)	0.0426	(5.11)	0.0565	(7.61)	0.0222
SLIQ	−0.0023	(−1.80)	0.0010	(0.88)	0.0026	(2.64)	−0.0007

SIMULTANEOUS EQUATION ESTIMATION RESULTS

1985		1986		1987		1988	
(34.68)	1.0670	(82.02)	0.9679	(44.04)	0.9270	(37.39)	
(2.75)	0.0167	(2.73)	−0.0069	(−0.62)	−0.0271	(−1.97)	
(0.93)	−0.0340	(−0.83)	0.0147	(0.20)	0.2967	(3.54)	
(2.31)	0.0086	(1.98)	0.0205	(2.04)	0.0012	(0.22)	
(−0.13)	0.0090	(1.62)	−0.0004	(−0.04)	−0.0000	(−0.00)	
	1526.45		456.28		296.78		
	0.903		0.691		0.568		

1992		1993		1994		1995	
(169.72)	0.9865	(200.38)	0.9483	(177.89)	0.9813	(166.59)	
(−2.25)	−0.0002	(−0.05)	−0.0085	(−1.84)	0.0095	(1.71)	
(0.94)	0.0154	(0.80)	−0.0102	(−0.48)	0.0165	(0.82)	
(7.17)	0.0040	(0.68)	0.0023	(0.55)	0.0013	(0.63)	
(−0.66)	0.0029	(1.56)	0.0099	(4.56)	0.0023	(1.18)	
	8573.24		6809.77		5912.19		
	0.957		0.947		0.939		

1985		1986		1987		1988	
(3.19)	0.3272	(0.87)	−0.0971	(−0.14)	0.9002	(0.82)	
(4.18)	0.3069	(4.46)	0.4620	(6.25)	0.6552	(3.84)	
(2.40)	0.1589	(1.48)	0.6350	(2.96)	0.6909	(2.02)	
(−2.34)	0.1623	(3.17)	−0.0088	(−0.08)	0.0638	(0.65)	
(−3.16)	−0.0238	(−2.38)	−0.0269	(−2.37)	−0.0316	(−1.42)	
(−2.44)	0.0157	(1.34)	−0.0061	(−1.01)	0.0138	(0.46)	
(9.63)	1.0600	(14.26)	0.3659	(2.84)	−0.1151	(−0.64)	

1992		1993		1994		1995	
(0.91)	0.3506	(1.65)	0.3419	(1.80)	−0.0686	(−0.64)	
(4.82)	0.4923	(9.91)	0.3854	(8.22)	0.0704	(6.78)	
(7.79)	0.2626	(5.33)	0.2833	(6.28)	0.1962	(6.50)	
(0.80)	0.0539	(2.80)	0.0397	(2.32)	0.0308	(2.99)	
(−2.12)	−0.0099	(−2.43)	−0.0170	(−3.62)	−0.0074	(−3.09)	
(−2.49)	0.0032	(0.84)	−0.0101	(−1.71)	0.0069	(1.62)	
(12.65)	0.3927	(11.93)	0.3320	(12.67)	0.3204	(14.22)	

1985		1986		1987		1988	
(35.93)	0.9255	(52.04)	0.8372	(47.58)	0.9412	(34.22)	
(−0.87)	−0.0033	(−1.21)	−0.0086	(−3.20)	−0.0195	(−5.75)	
(15.42)	0.0336	(17.86)	0.0118	(5.01)	0.0391	(25.90)	
(−0.30)	0.0073	(1.34)	0.0089	(1.47)	0.0245	(2.75)	
(4.45)	0.0290	(12.60)	0.0341	(12.40)	0.0283	(11.73)	
(−1.02)	−0.0012	(−2.44)	−0.0012	(−2.98)	−0.0028	(−5.19)	

(Continued on following page)

82 THE INTERDEPENDENCIES AMONG CORPORATE FINANCIAL POLICIES

TABLE 4.4 (Continued)

	1989		1990		1991		
SLDIV	0.8299	(33.71)	0.9067	(37.83)	0.9708	(48.40)	0.8266
SCE	0.0099	(2.72)	0.0036	(0.98)	0.0097	(3.86)	0.0003
SEF	0.0184	(14.53)	0.0134	(21.76)	0.0168	(22.77)	0.0083
SRD	0.0057	(0.85)	0.0170	(2.97)	0.0025	(0.52)	0.0016
SNI	0.0154	(5.07)	0.0242	(8.30)	0.0067	(4.57)	0.0069
SLIQ	−0.0009	(−1.59)	−0.0018	(−4.05)	−0.0004	(−0.96)	0.0019

3SLS   SEF

	1982		1983		1984		
SDIV	−2.7183	(−3.16)	−4.0830	(−2.44)	−7.7829	(−3.55)	−9.2881
SCE	0.2758	(2.18)	1.1126	(8.28)	−0.3513	(−1.94)	0.7397
SRD	−0.1183	(−0.60)	−0.3470	(−1.32)	−0.2344	(−0.67)	−0.5897
SDEP	0.1424	(0.83)	−0.9530	(−4.55)	0.2145	(0.96)	−1.3054
SNI	−0.0833	(−0.49)	−0.4493	(−1.49)	1.1397	(3.52)	1.0215
DE	0.0861	(3.07)	0.1336	(2.88)	0.1320	(2.52)	0.2371

	1989		1990		1991		
SDIV	−20.5434	(−10.38)	−37.5032	(−12.23)	−15.1524	(−8.80)	−2.5196
SCE	0.3931	(1.67)	−4.6916	(−12.40)	−2.0647	(−12.03)	0.9101
SRD	−1.5404	(−4.51)	−0.8507	(−1.83)	−0.3536	(−1.15)	−0.5380
SDEP	−0.0705	(−0.25)	2.9908	(7.73)	1.4560	(6.78)	−0.2650
SNI	−0.2493	(−1.50)	−0.6829	(−2.60)	−0.1383	(−1.41)	0.0316
DE	0.4887	(12.83)	0.8718	(17.34)	0.2641	(9.10)	0.0491

3SLS   SRD

	1982		1983		1984		
SLRD	1.1622	(44.57)	1.1225	(37.10)	0.9655	(31.90)	1.0099
SCE	−0.0429	(−3.87)	0.0276	(2.22)	0.0609	(4.45)	0.0338
SDIV	0.0733	(0.86)	−0.0409	(−0.33)	−0.4090	(−2.87)	0.1270
SEF	0.0488	(2.43)	0.0062	(0.64)	0.0260	(3.51)	0.0176
SNI	0.0282	(1.68)	0.0390	(1.56)	0.0437	(1.48)	−0.0155

	1989		1990		1991		
SLRD	0.8668	(48.49)	0.9884	(66.58)	1.0457	(100.68)	1.0801
SCE	0.0223	(1.92)	0.0272	(2.74)	0.0027	(0.05)	−0.0157
SDIV	0.0003	(0.00)	0.0176	(0.27)	0.0323	(0.81)	0.0265
SEF	0.0118	(3.62)	0.0067	(5.50)	0.0118	(8.80)	0.0207
SNI	0.0158	(1.59)	0.0149	(1.86)	0.0045	(1.48)	−0.0014

1992		1993		1994		1995	
(60.03)	0.9263	(65.55)	0.7796	(58.25)	0.9792	(82.46)	
(0.16)	0.0125	(5.40)	0.0284	(9.76)	−0.0035	(−1.11)	
(7.77)	−0.0209	(−4.97)	−0.0295	(−8.85)	0.0079	(6.92)	
(0.46)	−0.0142	(−4.05)	−0.0180	(−4.57)	−0.0018	(−0.51)	
(7.41)	0.0130	(10.10)	0.0193	(14.11)	0.0119	(10.59)	
(6.24)	0.0012	(3.81)	0.0023	(6.18)	−0.0005	(−1.72)	

1985		1986		1987		1988	
(−4.63)	−2.3263	(−2.74)	−1.2566	(−0.51)	−2.9956	(−1.94)	
(3.54)	1.3217	(11.77)	2.6294	(15.93)	0.9595	(8.02)	
(−1.41)	−0.3979	(−1.94)	−1.5473	(−2.45)	−0.9276	(−2.28)	
(−3.44)	−1.4149	(−8.14)	−1.0963	(−2.99)	0.1193	(0.60)	
(3.01)	−0.4942	(−5.64)	−0.1239	(−0.39)	−0.1458	(−1.27)	
(5.11)	0.1178	(5.62)	0.1186	(2.43)	0.0992	(3.46)	

1992		1993		1994		1995	
(−2.63)	−0.5858	(−1.23)	−0.9209	(−1.70)	1.4475	(2.78)	
(9.65)	1.4221	(24.20)	1.8021	(21.88)	0.9949	(7.61)	
(−3.41)	−0.3650	(−3.49)	−0.5228	(−4.46)	−0.1885	(−1.41)	
(−2.32)	−0.5369	(−7.52)	−0.6118	(−8.31)	−0.4128	(−3.86)	
(0.71)	−0.0952	(−2.48)	−0.0439	(−0.96)	−0.2340	(−5.48)	
(3.17)	0.0037	(0.47)	0.0188	(2.22)	0.0128	(1.22)	

1985		1986		1987		1988	
(34.67)	1.0654	(81.91)	1.0072	(46.78)	0.9286	(37.46)	
(3.01)	0.0195	(3.19)	−0.0578	(−5.98)	−0.0338	(−2.47)	
(1.10)	−0.0382	(−0.93)	−0.0264	(−0.36)	0.3040	(3.63)	
(3.92)	0.0088	(2.02)	0.0493	(5.79)	0.0036	(0.65)	
(−0.64)	0.0090	(1.62)	−0.0012	(−0.11)	0.0002	(0.03)	

1992		1993		1994		1995	
(170.58)	0.9876	(200.60)	0.9485	(177.92)	0.9792	(166.25)	
(−4.71)	−0.0032	(−0.94)	−0.0095	(−2.08)	0.0180	(3.26)	
(1.03)	0.0187	(0.97)	−0.0110	(−0.51)	0.0152	(0.76)	
(11.69)	0.0079	(1.36)	0.0016	(0.38)	0.0012	(0.61)	
(−0.82)	0.0032	(1.73)	0.0100	(4.59)	0.0022	(1.12)	

(Continued on following page)

TABLE 4.5 Estimated System Equations, G5 (European) WorldScope Universe

OLS	SCE						
		1982 (t)		1983		1984	
SDIV	0.3077	(1.08)	0.8657	(4.01)	0.6781	(3.89)	0.9045
SEF	0.0833	(2.94)	0.0562	(2.49)	0.0201	(1.42)	0.0632
SRD	0.1782	(1.55)	0.1220	(1.33)	0.0699	(0.87)	0.1531
SNI	0.3608	(4.79)	0.2097	(4.61)	0.2143	(5.12)	0.1937
SLIQ	−0.0924	(−5.89)	−0.0820	(−6.60)	−0.0638	(−6.36)	−0.0874
SLSAL	0.0132	(1.61)	0.0029	(0.42)	0.0050	(1.33)	0.0064
SDEP	0.8268	(11.01)	0.8923	(13.92)	0.9599	(17.85)	0.7720
F−Value	31.46		47.12		66.71		27.98
adjusted r^2	0.309		0.389		0.446		0.231
		1989		1990		1991	
SDIV	0.0545	(0.92)	−0.0238	(−0.37)	0.0038	(0.06)	−0.0191
SEF	0.0368	(4.43)	0.1021	(10.95)	0.0728	(8.18)	0.0629
SRD	0.0887	(1.04)	−0.0229	(−0.39)	−0.0514	(−0.77)	−0.0161
SNI	0.1097	(5.64)	0.0261	(2.35)	0.0402	(4.48)	0.0535
SLIQ	−0.0880	(−10.27)	−0.0503	(−6.33)	−0.0549	(−7.88)	−0.0339
SLSAL	0.0066	(2.24)	0.0145	(3.92)	0.0048	(1.19)	0.0051
SDEP	0.9415	(23.02)	0.7368	(19.71)	0.9441	(25.73)	0.9677
F−Value	96.52		84.59		110.75		132.48
adjusted r^2	0.219		0.185		0.231		0.251

OLS	SDIV						
		1982		1983		1984	
SLDIV	0.0719	(5.73)	0.8737	(31.77)	0.8469	(29.70)	0.9449
SCE	0.0195	(3.03)	0.0050	(1.12)	0.0129	(2.58)	0.0199
SEF	0.0142	(3.25)	0.0067	(2.57)	0.0121	(5.92)	0.0073
SRD	0.0303	(1.66)	−0.0373	(−3.47)	0.0319	(2.67)	−0.0197
SNI	0.1510	(15.69)	0.0444	(8.99)	0.0481	(8.28)	0.0264
SLIQ	0.0128	(5.12)	−0.0008	(−0.51)	0.0000	(0.01)	0.0014
F−Value	87.56		355.69		289.51		350.46
adjusted r^2	0.522		0.808		0.752		0.770
		1989		1990		1991	
SLDIV	0.7118	(18.41)	0.5775	(41.58)	0.7484	(56.73)	1.1134
SCE	0.0090	(1.49)	−0.0043	(−0.98)	0.0019	(0.50)	0.0006
SEF	0.0068	(2.52)	0.0089	(4.02)	0.0088	(4.64)	0.0125
SRD	−0.0109	(−0.40)	0.0204	(1.49)	0.0010	(0.07)	−0.0069
SNI	0.0767	(12.62)	0.0337	(13.15)	0.0205	(10.94)	0.0237
SLIQ	−0.0018	(−0.62)	0.0049	(2.56)	0.0027	(1.77)	−0.0039
F−Value	107.06		376.35		589.85		783.73
adjusted r^2	0.211		0.467		0.580		0.631

SIMULTANEOUS EQUATION ESTIMATION RESULTS

1985		1986		1987		1988	
(3.86)	0.7946	(3.57)	0.4060	(2.29)	0.0529	(0.39)	
(4.69)	0.0468	(2.48)	0.1077	(5.06)	−0.0006	(−2.80)	
(1.26)	0.0718	(0.51)	0.1046	(0.70)	0.0678	(0.63)	
(3.87)	0.1966	(2.98)	0.2199	(3.65)	0.2117	(6.93)	
(−5.55)	−0.1207	(−7.00)	−0.1195	(−6.96)	−0.0755	(−7.36)	
(1.35)	0.0042	(0.55)	−0.0010	(−0.20)	0.0048	(1.72)	
(9.46)	0.7466	(8.33)	0.8002	(8.52)	0.7225	(12.86)	
	24.05		23.00		35.35		
	0.164		0.134		0.136		

1992		1993		1994		1995	
(−0.52)	0.0571	(1.90)	0.0193	(0.44)	0.1300	(2.62)	
(8.60)	0.0180	(2.89)	0.0634	(10.44)	0.0557	(8.67)	
(−0.25)	−0.0614	(−1.23)	−0.0605	(−1.17)	−0.0349	(−0.65)	
(7.46)	0.0051	(1.17)	0.0543	(8.07)	0.0554	(8.59)	
(−5.24)	−0.0135	(−3.08)	−0.0045	(−3.51)	−0.0116	(−7.81)	
(2.08)	0.0220	(8.46)	−0.0066	(−7.21)	0.0045	(1.67)	
(28.82)	0.8944	(33.15)	0.8116	(28.32)	0.7128	(25.39)	
	169.67		130.16		102.48		
	0.301		0.250		0.219		

1985		1986		1987		1988	
(38.85)	0.6239	(19.55)	0.9089	(28.93)	0.4958	(23.49)	
(5.89)	0.0103	(2.37)	−0.0032	(−0.77)	0.0035	(0.90)	
(5.96)	0.0060	(2.45)	0.0138	(4.86)	−0.0002	(−5.72)	
(−1.77)	−0.0131	(−0.73)	0.0038	(0.19)	−0.0205	(−1.21)	
(5.95)	0.1045	(13.73)	0.0751	(10.02)	0.0654	(14.38)	
(0.92)	−0.0023	(−1.01)	−0.0019	(−0.80)	0.0002	(0.15)	
	145.06		229.99		173.08		
	0.513		0.581		0.403		

1992		1993		1994		1995	
(66.94)	0.6360	(47.33)	0.5354	(52.29)	0.8252	(59.12)	
(0.12)	0.0235	(3.12)	0.0082	(1.59)	0.0117	(2.53)	
(5.59)	0.0077	(2.67)	0.0035	(3.11)	0.0094	(5.75)	
(−0.34)	0.0163	(0.70)	0.0455	(2.88)	0.0039	(0.28)	
(11.12)	0.0135	(6.71)	0.0194	(10.16)	0.0122	(7.53)	
(−1.90)	0.0058	(2.80)	−0.0011	(−2.91)	−0.0022	(−5.89)	
	400.12		509.19		625.40		
	0.466		0.529		0.597		

(Continued on following page)

TABLE 4.5 (Continued)

SLIQ	−0.1655	(−3.59)	−0.1014	(−4.36)	−0.0292	(−1.61)	−0.0815
SLSAL	0.0744	(2.57)	−0.0460	(−2.19)	0.0360	(3.28)	−0.0038
SDEP	0.1340	(0.46)	1.0145	(8.04)	0.9265	(11.65)	0.8793
F−Value	6.04		15.77		32.15		21.05
adjusted r^2	0.069		0.169		0.276		0.183
		1989		1990		1991	
SDIV	−0.0323	(−0.17)	−0.0729	(−0.71)	−0.0545	(−0.67)	−0.0594
SEF	−0.3110	(−6.14)	−0.0411	(−0.92)	0.0050	(0.25)	0.0428
SRD	−0.0597	(−0.47)	−0.0531	(−0.75)	−0.0922	(−1.22)	−0.0261
SNI	−0.0232	(−0.71)	0.0047	(0.34)	0.0052	(0.39)	0.0422
SLIQ	−0.0890	(−7.83)	−0.0554	(−6.54)	−0.0505	(−7.08)	−0.0316
SLSAL	0.0339	(6.28)	0.0290	(4.93)	0.0110	(2.51)	0.0071
SDEP	0.8109	(14.28)	0.7080	(17.59)	0.9364	(25.08)	0.9621
F−Value	60.09		61.94		99.05		122.86
adjusted r^2	0.148		0.142		0.212		0.237

2SLS   SDIV

		1982		1983		1984	
SLDIV	0.0765	(3.79)	0.8861	(25.14)	0.9182	(25.21)	1.0145
SCE	0.0659	(4.39)	0.0075	(1.02)	0.0103	(1.50)	0.0172
SEF	0.1262	(5.65)	0.0496	(4.03)	0.0357	(5.46)	0.0404
SRD	0.0470	(1.55)	−0.0330	(−2.13)	0.0477	(3.22)	−0.0162
SNI	0.1407	(8.71)	0.0419	(6.68)	0.0469	(7.16)	0.0203
SLIQ	0.0176	(4.22)	−0.0008	(−0.39)	−0.0029	(−1.55)	0.0023
F−Value	39.58		231.09		233.23		163.80
adjusted r^2	0.327		0.731		0.709		0.609
		1989		1990		1991	
SLDIV	0.8097	(15.91)	0.5875	(40.65)	0.7529	(55.94)	1.1097
SCE	0.0487	(4.46)	−0.0069	(−0.84)	−0.0016	(−0.24)	−0.0096
SEF	0.0452	(4.04)	0.0273	(4.06)	0.0268	(6.95)	0.0446
SRD	−0.0033	(−0.10)	0.0114	(0.71)	0.0138	(0.85)	0.0047
SNI	0.0819	(11.94)	0.0356	(13.03)	0.0300	(11.55)	0.0415
SLIQ	0.0001	(0.04)	0.0057	(2.88)	0.0015	(0.93)	−0.0075
F−Value	99.85		365.98		573.61		745.76
adjusted r^2	0.199		0.460		0.573		0.619

2SLS   SEF

		1982		1983		1984	
SDIV	2.4902	(1.42)	−0.6953	(−1.01)	−3.5730	(−3.96)	−3.5201
SCE	−0.0206	(−0.18)	0.1658	(1.32)	−0.3006	(−1.59)	0.1410
SRD	−0.3491	(−0.72)	−0.4231	(−2.16)	−0.1426	(−0.50)	−0.4021
SDEP	−0.3758	(−2.34)	−0.2888	(−1.69)	0.2919	(1.15)	−0.4279
SNI	−0.5974	(−2.67)	−0.0345	(−0.35)	0.1680	(1.12)	−0.1876
DE	0.0257	(0.50)	0.0541	(1.84)	0.1319	(3.35)	0.3042
F−Value	4.59		1.98		3.74		7.03
adjusted r^2	0.043		0.011		0.028		0.055

(−4.72)	−0.1213	(−6.98)	−0.1104	(−4.83)	−0.0756	(−7.31)
(−0.58)	0.0046	(0.59)	−0.0142	(−1.90)	0.0052	(1.87)
(9.29)	0.7070	(7.67)	1.0879	(7.53)	0.6886	(11.89)
	23.14		15.19		35.84	
	0.159		0.091		0.137	

1992		1993		1994		1995
(−1.27)	0.0413	(0.93)	0.0242	(0.39)	0.2261	(3.51)
(3.01)	−0.0138	(−1.53)	−0.0091	(−0.63)	0.0526	(4.48)
(−0.35)	−0.0708	(−1.25)	−0.1133	(−1.93)	−0.0750	(−1.24)
(4.02)	−0.0044	(−0.89)	0.0223	(2.44)	0.0519	(5.18)
(−4.79)	−0.0125	(−2.83)	−0.0025	(−1.86)	−0.0109	(−5.33)
(2.57)	0.0240	(9.07)	0.0021	(1.16)	0.0044	(1.57)
(28.40)	0.8855	(32.53)	0.8038	(27.22)	0.7129	(25.02)
	166.73		109.15		95.62	
	0.297		0.218		0.207	

1985		1986		1987		1988
(26.64)	0.6355	(19.47)	0.9866	(23.72)	0.4872	(22.65)
(1.87)	0.0067	(1.17)	−0.0058	(−1.03)	0.0280	(3.57)
(6.99)	0.0114	(2.77)	0.0564	(4.42)	−0.0002	(−5.41)
(−0.94)	−0.0058	(−0.31)	0.0153	(0.64)	−0.0270	(−1.38)
(2.97)	0.1048	(13.60)	0.0648	(7.33)	0.0619	(13.14)
(1.00)	−0.0026	(−1.12)	−0.0015	(−0.56)	0.0020	(1.12)
	143.46		187.19		170.66	
	0.510		0.529		0.399	

1992		1993		1994		1995
(64.32)	0.6364	(47.28)	0.5358	(52.26)	0.8182	(57.29)
(−0.98)	0.0330	(2.72)	0.0198	(2.05)	0.0405	(4.98)
(11.37)	0.0116	(2.83)	0.0034	(2.49)	0.0207	(7.36)
(0.19)	−0.0036	(−0.14)	0.0417	(2.38)	−0.0023	(−0.14)
(14.73)	0.0147	(6.70)	0.0191	(9.89)	0.0205	(8.60)
(−3.42)	0.0060	(2.87)	−0.0011	(−2.86)	−0.0039	(−7.54)
	398.90		507.34		608.51	
	0.465		0.528		0.590	

1985		1986		1987		1988
(−3.39)	−1.2702	(−1.52)	−1.8541	(−4.53)	−6.0635	(−6.93)
(0.53)	0.0398	(0.44)	0.0368	(0.62)	−0.3367	(−1.70)
(−1.08)	0.0104	(0.04)	−0.2574	(−1.05)	−0.9520	(−2.29)
(−1.30)	−0.3267	(−1.72)	−0.4431	(−2.92)	0.4704	(1.93)
(−1.19)	0.0596	(0.41)	0.1733	(1.73)	0.0575	(0.47)
(5.07)	0.0840	(2.22)	0.1287	(5.37)	0.2377	(1317.52)
	2.01		10.85		395467.41	
	0.007		0.056		0.999	

(Continued on following page)

TABLE 4.5 (Continued)

	1989		1990		1991		
SDIV	−2.8346	(−7.22)	−0.8371	(−4.07)	−0.9050	(−5.27)	0.1560
SCE	−0.2620	(−2.37)	0.4010	(4.42)	0.1920	(2.04)	0.3303
SRD	−0.5901	(−2.34)	−0.0115	(−0.08)	−0.1731	(−1.12)	−0.2529
SDEP	0.1517	(1.01)	−0.4409	(−4.31)	−0.2237	(−1.93)	−0.3297
SNI	0.1677	(2.29)	−0.0160	(−0.61)	−0.0811	(−3.26)	−0.1069
DE	−0.0008	(−7.29)	−0.0007	(−11.28)	−0.0008	(−23.14)	−0.0010
F−Value	24.07		34.12		293.21		561.53
adjusted r²	0.055		0.072		0.407		0.550

2SLS   SRD

	1982		1983		1984		
SLRD	1.1407	(109.77)	1.0864	(46.55)	1.0463	(52.59)	1.1350
SCE	−0.0069	(−1.59)	−0.0029	(−0.29)	0.0052	(0.58)	0.0005
SDIV	0.0532	(1.33)	0.0170	(0.32)	0.0413	(0.83)	0.0068
SEF	0.0017	(0.21)	0.0417	(2.36)	0.0086	(1.16)	0.0011
SNI	−0.0000	(−0.01)	−0.0020	(−0.20)	0.0033	(0.34)	−0.0011
F−Value	2862.95		498.66		561.84		1748.42
adjusted r²	0.968		0.831		0.831		0.933

	1989		1990		1991		
SLRD	1.0953	(95.78)	1.2293	(88.21)	0.9657	(96.97)	0.8948
SCE	−0.0036	(−1.06)	0.0189	(3.31)	0.0036	(0.87)	0.0123
SDIV	−0.0060	(−0.37)	0.0391	(2.34)	0.0373	(3.37)	0.0012
SEF	0.0031	(1.00)	−0.0003	(−0.05)	0.0047	(1.91)	0.0040
SNI	0.0023	(0.93)	−0.0117	(−5.67)	0.0085	(5.12)	0.0028
F−Value	1892.41		1640.96		1895.67		1340.18
adjusted r²	0.799		0.761		0.787		0.709

3SLS   SCE

	1982		1983		1984		
SDIV	8.5461	(5.32)	0.7822	(1.70)	−0.0757	(−0.20)	0.5144
SEF	−1.5719	(−5.89)	0.9708	(5.71)	−0.3627	(−3.76)	0.2166
SRD	−0.6816	(−2.20)	0.4684	(2.44)	−0.0545	(−0.42)	0.2036
SNI	−1.3104	(−3.89)	0.1886	(2.14)	0.2340	(3.57)	0.2199
SLIQ	−0.0486	(−1.80)	−0.0384	(−2.91)	−0.0218	(−1.69)	−0.0850
SLSAL	−0.0032	(−0.20)	−0.0107	(−0.84)	0.0056	(0.59)	0.0055
SDEP	0.1030	(0.82)	0.8950	(7.84)	0.9375	(12.19)	0.7994

	1989		1990		1991		
SDIV	0.1319	(0.71)	−0.0744	(−0.72)	−0.0574	(−0.71)	−0.0651
SEF	−0.3757	(−9.06)	−0.0449	(−1.00)	−0.0011	(−0.05)	0.0407
SRD	−0.2229	(−1.78)	−0.0531	(−0.75)	−0.0904	(−1.20)	−0.0316
SNI	−0.0490	(−1.57)	0.0048	(0.35)	0.0023	(0.17)	0.0397
SLIQ	−0.0500	(−5.66)	−0.0535	(−6.36)	−0.0504	(−7.08)	−0.0280
SLSAL	0.0105	(2.63)	0.0253	(4.33)	0.0086	(1.96)	0.0091
SDEP	0.7823	(14.39)	0.7148	(17.82)	0.9375	(25.11)	0.9627

1992		1993		1994		1995	
(1.51)	−0.1275	(−1.25)	−0.3340	(−1.09)	−0.7446	(−4.27)	
(2.76)	0.1089	(1.00)	0.5562	(1.71)	0.0562	(0.44)	
(−1.51)	0.0060	(0.05)	0.2152	(0.75)	−0.1148	(−0.70)	
(−2.40)	−0.2372	(−2.06)	−0.4136	(−1.39)	−0.1260	(−1.09)	
(−5.84)	−0.0667	(−6.46)	−0.0457	(−1.29)	0.0017	(0.17)	
(−34.62)	−0.0010	(−47.09)	−0.0008	(−15.51)	−0.0010	(−57.94)	
	564.67		65.60		968.11		
	0.552		0.125		0.696		

1985		1986		1987		1988	
(92.09)	1.0584	(86.42)	0.9629	(73.04)	1.0425	(75.54)	
(0.08)	−0.0010	(−0.29)	−0.0003	(−0.10)	0.0075	(1.43)	
(0.29)	−0.0101	(−0.35)	−0.0335	(−1.42)	−0.0088	(−0.31)	
(0.30)	−0.0004	(−0.14)	−0.0092	(−1.39)	0.0000	(0.15)	
(−0.25)	0.0115	(1.88)	0.0102	(1.76)	−0.0007	(−0.18)	
	1499.15		1111.22		1151.44		
	0.901		0.848		0.790		

1992		1993		1994		1995	
(81.48)	0.9669	(99.65)	0.9420	(110.40)	0.9399	(101.51)	
(2.54)	0.0084	(1.82)	0.0037	(0.75)	−0.0062	(−1.23)	
(0.16)	0.0063	(0.79)	0.0088	(0.91)	0.0153	(1.48)	
(2.04)	0.0018	(1.12)	0.0008	(1.11)	0.0007	(0.69)	
(1.96)	0.0012	(1.40)	0.0024	(2.61)	0.0007	(1.14)	
	2001.42		2460.35		2074.40		
	0.785		0.819		0.804		

1985		1986		1987		1988	
(1.74)	1.4819	(4.09)	1.6037	(4.49)	0.7844	(3.18)	
(3.93)	0.0337	(1.02)	0.7812	(6.79)	−0.0004	(−1.69)	
(1.49)	0.0976	(0.66)	0.1912	(0.89)	0.1425	(1.17)	
(3.94)	0.0813	(1.03)	−0.0441	(−0.49)	0.1369	(3.78)	
(−5.31)	−0.1218	(−7.01)	−0.0679	(−3.97)	−0.0729	(−7.09)	
(0.92)	0.0112	(1.44)	−0.0013	(−0.23)	0.0089	(3.34)	
(8.63)	0.7098	(7.80)	1.0741	(7.74)	0.6688	(11.90)	

1992		1993		1994		1995	
(−1.40)	0.0431	(0.97)	0.0296	(0.47)	0.2449	(3.81)	
(2.87)	−0.0164	(−1.82)	−0.0077	(−0.54)	0.0483	(4.12)	
(−0.42)	−0.0705	(−1.25)	−0.1131	(−1.93)	−0.0724	(−1.20)	
(3.79)	−0.0052	(−1.06)	0.0228	(2.49)	0.0479	(4.79)	
(−4.27)	−0.0121	(−2.73)	−0.0025	(−1.86)	−0.0103	(−5.05)	
(3.28)	0.0233	(8.85)	0.0017	(0.92)	0.0063	(2.28)	
(28.45)	0.8825	(32.43)	0.8023	(27.17)	0.6894	(24.35)	

(Continued on following page)

TABLE 4.5 (Continued)

3SLS	SDIV							
		1982		1983		1984		
SLDIV	0.0144	(1.13)	0.8952	(25.85)	0.9346	(25.83)	1.0402	
SCE	0.1028	(10.23)	−0.0098	(−1.42)	0.0283	(4.26)	−0.0047	
SEF	0.1788	(11.45)	0.0805	(7.77)	0.0517	(8.28)	0.0561	
SRD	0.0757	(2.57)	−0.0202	(−1.34)	0.0494	(3.34)	−0.0046	
SNI	0.1545	(10.41)	0.0446	(7.18)	0.0427	(6.55)	0.0240	
SLIQ	0.0069	(2.88)	0.0001	(0.05)	−0.0036	(−2.02)	0.0002	
		1989		1990		1991		
SLDIV	0.7314	(14.92)	0.5890	(40.76)	0.7545	(56.09)	1.0989	
SCE	0.0845	(7.91)	−0.0062	(−0.76)	−0.0019	(−0.29)	−0.0152	
SEF	0.0401	(3.63)	0.0325	(4.84)	0.0316	(8.22)	0.0494	
SRD	−0.0144	(−0.43)	0.0126	(0.79)	0.0149	(0.92)	0.0043	
SNI	0.0789	(11.50)	0.0364	(13.30)	0.0325	(12.54)	0.0440	
SLIQ	0.0033	(1.07)	0.0057	(2.89)	0.0014	(0.91)	−0.0062	

3SLS	SEF							
		1982		1983		1984		
SDIV	4.6817	(3.95)	−0.8264	(−1.42)	−1.6058	(−1.87)	−3.8275	
SCE	−0.5321	(−8.12)	0.8408	(11.84)	−1.6632	(−11.53)	0.6029	
SRD	−0.4479	(−2.28)	−0.4286	(−2.20)	−0.1978	(−0.69)	−0.5117	
SDEP	0.0125	(0.15)	−0.6949	(−5.59)	1.5594	(7.02)	−0.2546	
SNI	−0.7967	(−4.32)	−0.1387	(−1.44)	0.4354	(2.96)	−0.2446	
DE	0.0102	(0.36)	0.0228	(1.34)	0.0453	(1.29)	0.2983	
		1989		1990		1991		
SDIV	−1.8788	(−4.91)	−0.8006	(−3.90)	−0.8951	(−5.21)	0.2732	
SCE	−1.3526	(−15.41)	0.2487	(2.75)	0.0518	(0.56)	0.5664	
SRD	−0.7763	(−3.08)	−0.0301	(−0.22)	−0.1890	(−1.23)	−0.2235	
SDEP	1.0936	(7.96)	−0.3418	(−3.36)	−0.0878	(−0.77)	−0.5841	
SNI	0.0468	(0.68)	−0.0173	(−0.65)	−0.0673	(−2.71)	−0.1283	
DE	−0.0003	(−3.55)	−0.0007	(−11.32)	−0.0009	(−24.09)	−0.0009	

3SLS	SRD							
		1982		1983		1984		
SLRD	1.1385	(109.62)	1.0994	(47.39)	1.0467	(52.61)	1.1351	
SCE	−0.0092	(−2.12)	−0.0199	(−2.06)	0.0103	(1.15)	0.0008	
SDIV	0.0752	(1.89)	0.0314	(0.59)	0.0451	(0.91)	0.0059	
SEF	−0.0021	(−0.27)	0.0670	(3.99)	0.0136	(1.85)	0.0011	
SNI	−0.0037	(−0.44)	0.0005	(0.05)	0.0020	(0.20)	−0.0011	
		1989		1990		1991		
SLRD	1.0953	(95.77)	1.2284	(88.16)	0.9660	(97.00)	0.8945	
SCE	−0.0040	(−1.15)	0.0272	(4.77)	0.0047	(1.12)	0.0152	
SDIV	−0.0083	(−0.52)	0.0390	(2.33)	0.0377	(3.40)	0.0017	
SEF	0.0029	(0.96)	−0.0024	(−0.52)	0.0057	(2.35)	0.0036	
SNI	0.0026	(1.02)	−0.0121	(−5.85)	0.0091	(5.45)	0.0025	

1985		1986		1987		1988	
(27.43)	0.6140	(18.84)	1.0049	(24.23)	0.4814	(22.55)	
(−0.53)	0.0249	(4.38)	−0.0199	(−3.54)	0.0557	(7.32)	
(11.37)	0.0112	(2.70)	0.0674	(5.37)	−0.0002	(−5.08)	
(−0.27)	−0.0105	(−0.56)	0.0209	(0.87)	−0.0383	(−1.96)	
(3.57)	0.1003	(13.03)	0.0649	(7.36)	0.0569	(12.16)	
(0.11)	−0.0004	(−0.17)	−0.0010	(−0.41)	0.0048	(2.81)	
1992		1993		1994		1995	
(63.85)	0.6361	(47.26)	0.5354	(52.22)	0.8119	(56.90)	
(−1.56)	0.0363	(2.99)	0.0223	(2.31)	0.0577	(7.12)	
(12.68)	0.0122	(2.96)	0.0039	(2.90)	0.0217	(7.77)	
(0.17)	−0.0041	(−0.16)	0.0414	(2.37)	−0.0030	(−0.19)	
(15.71)	0.0149	(6.79)	0.0193	(9.97)	0.0210	(8.88)	
(−2.99)	0.0059	(2.85)	−0.0011	(−2.89)	−0.0040	(−7.76)	

1985		1986		1987		1988	
(−3.84)	−1.3511	(−1.62)	−1.9261	(−4.77)	−6.6862	(−7.68)	
(2.68)	0.0765	(0.85)	0.7535	(17.04)	−0.4751	(−2.43)	
(−1.37)	0.0110	(0.04)	−0.2269	(−0.93)	−0.9633	(−2.32)	
(−1.03)	−0.3532	(−1.86)	−0.8819	(−6.25)	0.6259	(2.66)	
(−1.58)	0.0579	(0.40)	0.1242	(1.25)	0.1365	(1.13)	
(5.83)	0.0869	(2.30)	0.0580	(2.80)	0.2376	(1318.56)	
1992		1993		1994		1995	
(2.65)	−0.1214	(−1.19)	−0.3283	(−1.07)	−0.7545	(−4.33)	
(4.93)	−0.0020	(−0.02)	0.5153	(1.59)	0.2087	(1.63)	
(−1.34)	−0.0030	(−0.02)	0.2105	(0.73)	−0.0938	(−0.57)	
(−4.50)	−0.1429	(−1.25)	−0.3830	(−1.28)	−0.2768	(−2.41)	
(−7.06)	−0.0670	(−6.49)	−0.0442	(−1.25)	0.0005	(0.05)	
(−33.80)	−0.0010	(−47.01)	−0.0008	(−15.55)	−0.0010	(−57.98)	

1985		1986		1987		1988	
(92.10)	1.0581	(86.39)	0.9607	(72.89)	1.0414	(75.47)	
(0.14)	0.0005	(0.16)	0.0067	(2.14)	0.0133	(2.55)	
(0.25)	−0.0120	(−0.42)	−0.0474	(−2.02)	−0.0192	(−0.69)	
(0.30)	−0.0012	(−0.49)	−0.0138	(−2.12)	0.0000	(0.06)	
(−0.24)	0.0114	(1.87)	0.0113	(1.95)	−0.0002	(−0.06)	
1992		1993		1994		1995	
(81.45)	0.9668	(99.64)	0.9421	(110.42)	0.9399	(101.51)	
(3.13)	0.0104	(2.25)	0.0024	(0.49)	−0.0092	(−1.83)	
(0.22)	0.0061	(0.77)	0.0088	(0.91)	0.0164	(1.53)	
(1.85)	0.0018	(1.17)	0.0010	(1.42)	0.0006	(0.61)	
(1.79)	0.0012	(1.44)	0.0025	(2.70)	0.0007	(1.11)	

CONCLUSIONS

An econometric model has been developed to determine a firm's research, dividend, investment, and new capital issue interdependence. The interdependence of financial decisions is of considerable importance to the manager who is interested in integrating research and development decisions with dividend, investment, and financing decisions. We find significant relationships between the research, capital expenditure, dividend, and new debt decisions in the United States. We find only new debt and investment decision interdependencies in Japan and the G5 countries. Much more work is necessary in examining the interdependence of financial decisions and how these decisions impact corporate expenditures and stockholder wealth.

REFERENCES

Ben-Zion, U. "The R&D and Investment Decision and Its Relationship to the Firm's Market Value: Some Preliminary Results." In *R&D, Patents, and Productivity,* edited by Z. Griliches. Chicago: University of Chicago Press, 1984.

Damon, W. W., and R. Schramm. "A Simultaneous Decision Model for Production, Marketing and Finance." *Management Sci.* 18 (1972):161–172.

Dhrymes, P. J. *Econometrics: Statistical Foundations and Applications,* New York: Springer-Verlag, 1974.

Dhrymes, P. J. "Investment, Dividends, and External Finance Behavior of Firms." In *Determinants of Investment Behavior,* edited by Robert Ferber. New York: Columbia University Press, 1967.

Dhrymes, P. J., and M. Kurz. "On the Dividend Policy of Electric Utilities." *Rev. Economics and Statist.* 46 (1964):76–81.

Fama, E. F. "The Empirical Relationship Between the Dividend and Investment Decisions of Firms." *Amer. Economic Rev.* 63 (1974):304–318.

Grabowski, H. G. "The Determinants of Industrial Research and Development: A Study of the Chemical, Drug and Petroleum Industries." *Journal of Political Economy* 76 (1968):292–306.

Grabowski, H. G., and D. C. Mueller. "Managerial and Stockholder Welfare Models of Firm Expenditures." *Rev. Economics and Static.* 54 (1972):9–24.

Guerard, J. B., Jr., and G. M. McCabe. "The Integration of Research and Development Management into the Firm Decision Process." In *Management of R&D and Engineering,* edited by D. Kocaoglu. Amsterdam: North Holland, 1990.

Guerard, J. B., Jr., and B. K. Stone. "Strategic Planning and the Investment-Financing Behaviour of Major Industrial Companies." *Journal of the Operational Research Society* 38 (1987):1039–1050.

Guerard, J. B., Jr., A. S. Bean, and S. Andrews. "R&D Management and Corporate Financial Policy." *Management Science* 33 (1987):1419–1427.

Hambrick, D. C., I. C. MacMillan, and R. R. Barbosa. "Changes in Product R&D Budgets." *Management Sci.* 29 (1983):757–769.

Higgins, R. C. "The Corporate Dividend-Saving Decision." *J. Financial and Quantitative Anal.* 7 (1972):1527–1541.

Jalilvand, A., and R. S. Harris. "Corporate Behavior in Adjusting to Capital Structure and Dividend Targets: An Econometric Study." *J. Finance* 39 (1984):127–145.

Lintner, J. "Distributions of Incomes of Corporations Among Dividends, Retained Earnings and Taxes." *Amer. Economic. Rev.* 46 (1979):119–135.

Mansfield, E. "R&D and Innovation." In *R&D, Patents, and Productivity,* edited by Z. Griliches. Chicago: University of Chicago Press, 1984.

Mansfield, E. "How Economists See R&D." *Harvard Business Review* 59 (1981a):98–106.

Mansfield, E. "Composition of R&D Expenditures: Relationship to Size of Firm, Concentration, and Innovative Output." *Review of Economics and Statistics* 63 (1981b):610–615.

Mansfield, E. "Size of Firm, Market Structure and Innovation." *J. Political Economy* 71 (1963):556–576.

McCabe, G. M. "The Empirical Relationship Between Investment and Financing: A New Look." *J. Financial and Quantitative Anal.* 14 (1979):119–135.

McDonald, J. G., B. Jacquillat, and M. Nussenbaum. "Dividend, Investment, and Financial Decisions: Empirical Evidence on French Firms." *J. Financial and Quantitative Anal.* 10 (1975):741–755.

Meyer, J. R., and E. Kuh. *The Investment Decision.* Cambridge: Harvard University Press, 1957.

Miller, M., and F. Modigliani. "Dividend Policy, Growth, and the Valuation of Shares." *J. Business* 34 (1961):411–433.

Mueller, D. C. "The Firm Decision Process: An Econometric Investigation." *Quart. J. Economics* 81 (1967):58–87.

Peterson, P., and G. Benesh. "A Reexamination of the Empirical Relationship Between Investment and Financial Decisions." *J. Financial and Quantitative Anal.* 18 (1983):439–454.

Scherer, F. M. "Firm Size, Market Structure, Opportunity, and the Output of Patented Inventions." *Amer. Economic Rev.* 55 (1965):1104–1113.

Switzer, L. "The Determinants of Industrial R&D: A Funds Flow Simultaneous Equation Approach." *Rev. Economics and Statist.* 66 (1984):163–168.

Tinbergen, J. *A Method and Its Application to Investment Activity.* Geneva: League of Nations, 1939.

5
COMPARING CENSUS/NSF R&D DATA WITH COMPUSTAT R&D DATA

This chapter reviews the results of a project in which research and development (R&D) data drawn from National Science Foundation (NSF) and Census Bureau studies were substituted in several financial models for R&D data drawn from the 1975–1982 Compustat tapes. The result using the Compustat data did not differ significantly from that based on the NSF/Census data in the aggregate, but significant differences were observed for certain industries and certain years. This chapter discusses the financial models and determinants of corporate R&D expenditures using the different databases, and suggests further questions for research.

The general objective of this chapter is to review and summarize findings as to whether R&D expenditures should be included in a set of financial decisions that influence the market value of a firm, as reflected in the price of its stock (Weston and Copeland 1986). A rigorous theoretical position, known as the *perfect markets hypothesis,* has guided research on this issue for many years, as discussed in the preceding chapter. The perfect markets hypothesis asserts that the value of a firm's stock is determined by the firm's ability to invest in opportunities that will produce enhanced earnings, dividends, or cash flow. It further asserts that dividend

Reprinted from *Research Policy,* Vol 18, No. 4, Bean and Guerard, "A Comparison of Census/NSF R&D Data vs. Compustat R&D Data in a Financial Decision-Making Model," pp. 193–208, 1989 with kind permission from Elsevier Science-NL, Sara Burgerhartstraat 25, 1055 kV Amsterdam, The Netherlands.

policy is independent of these investment decisions; however, the investment decision and the decision to issue new capital stock are interdependent. Thus, the perfect markets hypothesis asserts that the stock market value of a firm is not affected by dividend policy, only by opportunities for further returns, as described in Chapter 4. This chapter considers whether federal financing of R&D influences a firm's financial decision-making process.

INNOVATION, R&D, AND STOCKHOLDER WEALTH

It is well known that there is substantial underinvestment in R&D in the United States economy, primarily because social benefits exceed private rates of return from innovative activities (Mansfield, Rappaport, Romeo, Wagner, and Beardsley 1977). Furthermore, underinvestment in R&D diminishes potential stockholder wealth because R&D activities have been associated with the market value of major industrial firms (Ben-Zion 1984; Guerard, Bean, and Stone 1990). It is helpful to model the R&D budgetary process of major industrial companies, since empirical evidence supports the hypothesis that decisions on R&D expenditures are made simultaneously with a firm's other financial decisions (Switzer 1984; Guerard and McCabe 1990), as discussed in the preceding chapter. A firm's decision to increase its R&D expenditures impacts its decisions on dividends, investments, and new debt issuance. Thus, there is a need to understand how corporate decisions on expenditure reallocation may affect a firms's stock price.

As discussed earlier, in Chapter 4, the empirical research on the perfect markets hypothesis has yielded mixed results since its formulation by Miller and Modigliani (1961), although the majority of studies published in the 1970s support the existence of imperfect markets. A review of this research led Guerard and Bean (1987) to formulate a revised model of the stock price valuation process that allows interdependencies that were precluded by the perfect markets hypothesis. One of the goals of this study was to revive the Mueller (1967) hypothesis that research and development expenditures should be included in tests of the perfect markets hypothesis. Guerard and Bean (1987) used data obtained from Compustat and U.S. Patent Office tapes to estimate such a model.

In July 1984, the National Science Foundation and the U.S. Bureau of the Census announced the availability of a longitudinal database that would enable researchers to explore the relationship between R&D and other economic variables on an enterprise basis. This chapter compares the results obtained from the NSF data with those derived from the Compustat/Patent Office data.

STUDY MODEL

The model used in the Guerard and McCabe (1992) study assumes interdependence between several financial decisions. It employs investment, dividends, and new capital financing equations to describe the firm's budget constraint. The manager may use available funds for research and development (R&D), capital investment (CE), or dividends (DIV), or to increase net working capital (LIQ). The sources of funds are represented by net income (NI), depreciation (DEP), and new debt financing (EF). Thus, the budget constraint is:

$$R\&D + CE + DIV + LIQ = NI + DEP + EF$$

Research and development (R&D) expenditures are modeled in terms of investments, dividends, and new capital issues (to reflect the imperfect markets hypothesis) and previous research and development expenditures (to serve as a surrogate for previous patents and R&D activities). We use Compustat R&D data as well as NSF/Census R&D data in this chapter. We also use a three-year lag on the R&D variable, as was done in Guerard, Bean, and Andrews (1987). There is little difference between the three-year lag structure and the use of a one-year lag, as was done in Chapter 4. This result is consistent with the Guerard, Bean, and McCabe (1986) results in which the authors found no significant differences between using contemporary and distributed lag variables of investment, dividends, and R&D. The investment equation (CE) uses the *rate of profit theory* (Tinbergen (1939) and Dhrymes and Kurz (1967)) in which net income positively affects investment. The accelerator position on investment is also examined through the two-year growth in sales (DSAL) variable. Depreciation is normally included in the investment analysis because depreciation describes the deterioration of capital in the productive process. This study uses cash flows (CF) to incorporate both net income and depreciation effects; moreover, other non-cash expenditures are included in the firm's cash flows. The investment, dividend, and external financing equations used here were discussed in Chapter 4. The variables are again divided by assets to reduce heteroscedasticity.

It is expected that the price of common stock (PCS) would be positively correlated with research and development expenditures, patents, lagged patents, book value of equity, and investment (Ben-Zion 1984). The study model is summarized in Figure 5.1.

Research and Development	(R & D)	= f(C̄E, D̄IV, ĒF⁺, R & D$_{t-1}^+$, R & D$_{t-2}^+$, R & D$_{t-3}^+$, Lagged Patents⁺, NĪ⁺, DĒP⁺)
Capital Expenditures	(CE)	= f(DSĀL⁺, CF⁺, ĒF⁺, D̄IV, R̄ & D̄, L̄IQ, tax rate)
Dividends	(DIV)	= f(CĒ⁺, NĪ⁺, L̄IQ, LDĪV, RŌE, ĒF⁺)
External Financing	(EF)	= f(DĪV⁺, CĒ⁺, R & D⁺, C̄F, K̄D, D̄E)
Common Stock Price	(PCS)	= f(R & D⁺, Patents⁺, Lagged Patents⁺, Book Value of Equity⁺, CĒ⁺, DĪV⁺)

Figure 5.1. Summary of hypothesized relationships.

FINANCIAL DECISION ESTIMATION RESULTS

Guerard and McCabe (1992) used a sample of 303 very large U.S. firms drawn from 12 industries to model a firm's financial decisions. This database covered the 1971–1982 time period and was constructed from Compustat and Patent Office data. Cross-sectional regressions were used to estimate the model (see Figure 5.2). Industry dummy variables, based on Standard Industrial Code (SIC) classifications, were developed to examine industry financial differences. Our econometric estimations of the R&D, capital expenditures, dividends, new debt, and stock price decisions use only contemporaneous variables because the use of distributed lags of the investment, dividend, and R&D variables did not enhance the regression models. Higgins (1972) and McCabe (1979) previously used models in which composite investments and dividend variables used three-to-four year weights with lagged variables having weights of 0.65, 0.35, and 0.10 for one-, two-, and three- year lags. The results of Guerard and McCabe (1987) are summarized in Figure 5.2.

The three-stage least squares regression estimates indicated that research and development (R&D) expenditures were positively associated with the previous year's R&D expenditures and (current) net income. Dividends were an alternative use to R&D funds in most years, whereas investments were most often positively associated with increasing R&D activities. The lack of statistically significant relationships between R&D, depreciation, and external funds were quite surprising; only the dividend variable significantly violated the perfect markets hypothesis in the R&D activities equa-

Research and Development	(R & D)	= f($\overset{+}{R \& D}_{t-1}$, $\overset{+}{\text{Net Income}}$, Dividends)
Capital Expenditures	(CE)	= f($\overset{+}{\text{Cash Flow}}$, $\overset{+}{R \& D}$, $\overset{+}{\text{New Debt}}$, $\overset{-}{\text{Net Working Capital}}$)
Dividends	(DIV)	= f($\overset{+}{\text{Dividends}_{t-1}}$, $\overset{+}{\text{Net Income}}$, $\overset{-}{\text{New Debt}}$)
External Financing	(EF)	= f($\overset{+}{\text{Capital Expenditures}}$, $\overset{-}{\text{Cash Flow}}$, $\overset{-}{\text{Dividends}}$)
Common Stock Price	(PCS)	= f($\overset{+}{\text{Dividends}}$, $\overset{+}{\text{Capital Expenditures}}$, $\overset{+}{R \& D}$, $\overset{+}{\text{Book Value}}$)

Figure 5.2. Summary of actual relationships.
(N = 303 firms; Compustat and Patent data)

tion. The electronics industry tended to spend more on R&D activities, holding everything else constant, than other industries.

The three-stage least squares estimate of the investment (CE) equation indicated that the dividend variable was generally positive in the investment equation (contrary to the imperfect markets hypothesis). The statistically significant positive coefficient of the new debt variable in the investment equation complemented the work of McCabe (1979), Peterson and Benesh (1983), and Guerard and McCabe (1987). Similarly, Dhrymes and Kurz (1967) did not always find a significant positive association between new debt and investments. R&D expenditures were positive in the investment equation; Mueller (1967) found an inverse relationship between investments and R&D in his earlier work, and no relationship between the variables in his later work with Grabowski (Mueller and Grabowski 1972). Decreases in net liquidity were associated with rising investment. Cash flow positively affected investment, while the tax rate had statistically significant negative coefficients in 1976 and 1977 in the investment equation. The positive coefficients on the R&D and dividend variables were counter to the imperfect markets hypothesis.

Dividends (DIV) were positively associated with rising net income and previous dividends, supporting the positions of Lintner (1956) and Fama and Babiak (1968). There was a slight tendency for rising investment to accompany increasing dividends. The hypothesized (negative) relationship between investments and dividends was never realized in the dividend equation estimates. Little support for the imperfect markets hypothesis was found in the dividend equation. R&D was often negatively associated with dividends, but the relationship was not statistically significant. The petroleum industry had a very unstable dividend policy (the industry dummy variable was usually significant, whether positive or negative).

While external funds (EF) are normally issued in response to increasing investments, the hypothesized (positive) relationships between new debt, dividends, and research and development were not found. The relationship found between new debt and investment was consistent with the perfect markets hypothesis. New debt financing was negatively associated with cash flow. Furthermore, debt-to-equity ratio coefficients were (unexpectedly) positive. One would expect that as cash flow increased, there would be less need to issue capital. In addition, a higher debt-to-equity ratio would raise the risk to the firm's creditors and normally reduce future capital issues. The rubber, machinery, chemical, and drug industries tended to be "new-debt-issues-intensive."

Support for the imperfect markets hypothesis was indicated by the positive interdependencies of new debt issues and investments (contrary to the perfect markets hypothesis). Moreover, the *alternative uses of funds* concept was found in the significant negative interdependencies between R&D expenditures and dividends. Furthermore, R&D expenditures and investment decisions were interrelated.

In a follow-up study by Guerard, Bean, and McCabe (1987), the stock price (PCS) was positively affected by dividends, investment, R&D, and the book value of common stock, in agreement with Ben-Zion (1984), who studied the relationship between R&D and the firm's market value for 157 firms during the 1969–1977 period. However, patents issued to a firm did not appear to increase the value of the firm's equity (only in 1982), as was found by Ben-Zion. Lagged patents were positively associated with the stock price in most years. The significance levels for the dividend, capital expenditures, and R&D variables tended to decline in the latter years of the study (1980–1982). The firm's measure of systematic risk (its beta) is incorrectly positive in the estimated stock price equation. A positive coefficient was also found by Ben-Zion (1984). It is noteworthy that the book value of equity variable dominates the stock price equation, as Ben-Zion found, except for 1981.[1] Common stock in the machinery industry appeared to be overpriced relative to that of other industries (see Figure 5.2).

These findings cast additional doubt on the perfect markets hypothesis. They also suggest that R&D expenditures (but not necessarily patents) are among the variables that influence the stock market value of a firm.

Against this background, the announced availability of the NSF/Census data series provided an opportunity to replicate the preceding analysis. Such a replication was seen as a way of reviewing the perfect markets hypothesis one more time, and of testing the sensitivity of the empirical model to alternative data. If the empirical results using the NSF/Census data set were very similar to the Compustat/Patent results, it would encourage further work on the linkages between the data sets (NSF 1985).

The current work attempts to replicate the previous (Compustat) study by substituting the R&D expenditure data from the NSF/Census database for the R&D expenditure data in the Compustat database over the 1975–1982 time period. Census Bureau employees performed the substitutions and provided results of the computer runs to the authors. Whereas the prior work covered 303 firms from 1972 to 1982, the present study uses a sample of 158–188 firms over the 1975–1982 time period. The matching procedure created by substituting the NSF/Census data for the Compustat data reduced the original 303-firm sample to 158–188 firms by eliminating (nonmanufacturing) firms not engaged in R&D activities.[2] This procedure permitted comparisons to be made at several different levels:

1. Annual comparisons of the aggregate R&D data across the whole sample;
2. Annual comparisons of industry-level data across several subsamples;
3. Examination of the sensitivity of the models to the substitution of NSF/Census data for Compustat data; and
4. Examination of the impact of the compressed time frame on results.

While there are several results that are worthy of detailed discussion, the overall results suggest that the substitution of the NSF/Census data had little effect on the model structure, thus indicating a good linkage between the NSF/Census data set and the more "business-oriented" Compustat data.[3]

Comparison of R&D Expenditure Data

When the process of "matching" was finished, samples ranging in number from 158 to 188 were obtained for the years 1975 through 1982. Examination of variances (Table 5.1) and means (Table 5.2) of the complete sample shows that the R&D expenditure data are not statistically different in six of the eight years, but that the 1978 and 1979 samples are significantly different. The variance (Table 5.1) of the Compustat data is 1.29 times greater than that of the NSF/Census data in 1978, which is significant at the 0.05 level (F test), while it is significantly less than that of the Compustat data in 1979. Clearly, the reversal of the direction of the differences and the level of significance attained (0.05 in 1978 and 0.01 in 1979) suggest that R&D

TABLE 5.1 Comparison of R&D Expenditure Data for Matched Samples of Firms: Compustat vs. NSF/Census Data 1975–1982

Test of Significance for differences in sample variances: F ratio

		1975	1976	1977	1978	1979	1980	1981	1982
Complete sample	$F=$	1.08	1.09	1.15	1.29**	2.55*	1.16	1.08	1.04
	$N=$	(188)	(188)	(188)	(188)	(180)	(177)	(162)	(158)
Industries									
Construction	$F=$	1.50	1.50	1.50	1.67	2.00	1.33	1.33	2.00
	$N=$	(16)	(16)	(16)	(16)	(16)	(16)	(17)	(17)
Petroleum	$F=$	11.5*	14.5*	16*	15*	18*	1.50	1.25	1.20
	$N=$	(12)	(12)	(12)	(12)	(12)	(11)	(11)	(11)
Machinery	$F=$	1.17	1.00	1.08	1.15	1.06	1.06	1.05	1.09
	$N=$	(16)	(16)	(16)	(16)	(16)	(16)	(15)	(15)
Electronics	$F=$	1.29	1.19	1.18	1.83	1.23	1.06	1.12	1.09
	$N=$	(21)	(21)	(21)	(21)	(22)	(22)	(17)	(15)
Drugs	$F=$	1.08	1.31	1.40	1.57	1.37	1.47	1.50	1.44
	$N=$	(25)	(25)	(25)	(25)	(25)	(24)	(23)	(23)
Chemicals	$F=$	1.00	1.09	1.05	1.23	1.18	1.28	1.24	1.02
	$N=$	(24)	(24)	(24)	(24)	(24)	(24)	(23)	(23)

*Significant at 0.01 level; Compustat variance < Census variance.
**Significant at 0.05 level; Compustat variance > Census variance.

TABLE 5.2 Comparison of R&D Expenditure Data for Matched Samples of Firms: Compustat vs. NSF/Census Data 1975–1982

Z scores for differences in means*

		1975	1976	1977	1978	1979	1980	1981	1982
Complete sample	Z =	0.004	0.014	0.315	0.308	0.858	0.395	0.200	0.266
	N =	(188)	(188)	(188)	(188)	(180)	(177)	(162)	(158)
Industries									
Construction	Z =	0.222	0.816	0.097	0.029	0.328	0.531	0.360	0.643
	N =	(16)	(16)	(16)	(16)	(16)	(16)	(17)	(17)
Petroleum	Z =	1.549	−1.229	−1.348	−1.382	−1.148	−1.344	−1.161	−1.500
	N =	(12)	(12)	(12)	(12)	(12)	(11)	(11)	(11)
Machinery	Z =	0.063	−0.031	−0.018	−0.362	−0.225	−0.490	−0.085	−0.236
	N =	(16)	(16)	(16)	(16)	(16)	(16)	(15)	(15)
Electronics	Z =	0.003	−0.134	0.364	0.578	0.424	0.375	0.150	0.443
	N =	(21)	(21)	(21)	(21)	(22)	(22)	(17)	(15)
Drugs	Z =	0.092	−0.027	0.192	0.377	0.319	0.460	0.383	0.313
	N =	(25)	(25)	(25)	(25)	(25)	(24)	(23)	(23)
Chemicals	Z =	1.076	0.825	1.230	1.008	0.616	0.752	0.793	0.063
	N =	(24)	(24)	(24)	(24)	(24)	(24)	(23)	(23)

*A positive score indicates that the Compustat mean exceeds the NSF/Census mean; a negative score indicates that the NSF/Census mean exceeds the Compustat mean.

expenditures were reported very differently by some firms in these two years.[4] The sample means (Table 5.2) further suggest that 1979 was "unusual" for R&D reporting purposes, for the sample as a whole. Firms tended to report higher R&D expenditures in the Compustat format than they did in the NSF/Census survey, except for 1979 and 1982. The size of the difference was particularly great in 1982. Since it is well known (NSF 1985) that the Securities and Exchange Commission (SEC) Forms 10-K and 10-Q data used to develop the Compustat data permit certain activities (e.g., engineering and technical service) to be included in R&D that are excluded under the NSF definitions, one would expect the Compustat means to be greater than the NSF/Census means.[5]

The industry-level data in Tables 5.1 and 5.2 suggest strong differences between the two data sources in the petroleum industry from 1975 to 1979. The differences in R&D reporting in the petroleum industry could result from the accounting treatment of exploration activities and other non-cash expenditures. The variances are significantly different for these years, with the Compustat data showing the greater variance but a lower mean value. Thus, the petroleum industry reported higher mean R&D expenditures under the more restrictive NSF/Census definitions than under the broader SEC definitions.[6] Other than for the petroleum group, the industry-level data are well behaved. Industry groupings with fewer than ten cases were not analyzed separately because of sample size. (There were only seven cases in the next largest industry after petroleum.)

In summary, the R&D expenditure data for the complete sample are comparable for all years except 1978 and 1979. When specific industry data are examined, the petroleum industry data have dissimilar variances in 1975–1979, and the mean values are uniformly different throughout. Moreover, the differences in the petroleum industry mean R&D expenditures are in the opposite direction from the complete sample differences in all years except 1979 and 1982. This suggests that the industry consistently underreported its R&D expenditures on SEC Forms 10-K and 10-Q, while the firms in the complete sample tended to overreport, relative to the NSF/Census figures.

Comparison of Regression Results

The question addressed by the regression results is whether the support for the imperfect markets hypothesis is strengthened or weakened by substituting the NSF/Census data for the Compustat data. Because of the losses in sample size due to the pairwise matching of firms and the reduced time frame (1978–1982 vs. 1975–1982), it is necessary to examine changes in the results due to sample size and time frame reduction, as well as those due to differences in data sources. These results are presented in Tables 5.3 through 5.6.

1. When the R&D expenditures equation (Table 5.3) is reestimated using the reduced Compustat sample (a decrease from 303 firms to 158–188 firms), the model that most closely fits the data is:

$$\text{RCOMP(78–82): R\&D} = f(\overset{+}{\text{R\&D}}_{t-1}, \overset{+}{\text{R\&D}}_{t-2}, \overset{+}{\text{R\&D}}_{t-3})$$

The rationale for this judgment is that, in order to be retained in the model, a variable must be statistically significant and have the same sign in at least three of the five periods studied. Applying the same criteria to the NSF/Census data yields the following result:

$$\text{CENS (78–82): R\&D} = f(\overset{+}{\text{R\&D}})_{t-1}$$

Recall that the 303-firm Compustat sample supported the following model for the 1975–1982 period, as well as for the shorter period.[7]

$$\text{COMP303(75–82): R\&D} = f(\overset{+}{\text{R\&D}}_{t-1}, \overset{+}{\text{NI}}, \overset{-}{\text{DIV}})$$

One important factor may be that the 303-firm sample is highly diversified and includes firms that reported no R&D expenditures. The NSF/Census sample, which determines the reduced Compustat (RCOMP) sample, does not contain such (predominately nonmanufacturing) firms.[8]

On the positive side, all three models strongly support the importance of last year's R&D expenditures as a prediction of this year's R&D.

2. When the investment (CE) equation was reestimated, it was perfectly stable. As can be seen by inspecting Table 5.4, nothing changes. The tax rate variable is significant in three of the five years, but changes signs (as is the case with dividends).[9] The investment equation is:

$$\text{CE} = f(\overset{+}{\text{CF}}, \overset{-}{\text{R\&D}}, \overset{+}{\text{EF}}, \overset{-}{\text{LIQ}})$$

3. The dividend (DIV) equation (Table 5.5) is also stable for the 1978–1982 period. However, it is noteworthy that the new debt variable (EF), which was significantly and negatively related to dividends in 1975–1977,

(Text continues on page 110.)

TABLE 5.3 Comparative Two-Stage Regression Results. Dependent Variable: R&D expenditures (R&D)

	Constant	RD_{t-1}	RD_{t-2}	RD_{t-3}	NI	DEP	EF	CE	DIV
1978: 303	0.000	1.454*	−0.050	−0.317*	−0.036	−0.004	−0.012*	0.013*	−0.134*
188 Cens	0.000	0.511*	0.502*	−0.107	0.001	0.020	−0.002	−0.012	0.056
188 Comp	0.002**	0.956*	0.093	−0.076	0.008	0.028	−0.001	0.001	−0.012
1979: 303	0.000	1.490*	−0.223	−0.190	0.042*	0.012	−0.010	0.033*	−0.177*
180 Cens	0.018	0.968**	−0.213	0.267	−0.171**	−0.302*	−0.144**	0.198*	0.120
180 Comp	−0.001	0.759*	−0.551*	0.845*	0.002	−0.010	−0.013	0.023*	−0.020
1980: 303	−0.001	1.378*	−0.046	−0.272	0.074*	0.002	−0.008	−0.005	−0.169*
177 Cens	−0.001	0.890*	0.022	0.127*	−0.001	0.048*	−0.026*	−0.015	−0.014
177 Comp	−0.061	−0.537	0.546*	0.994	0.012	0.038*	−0.018**	−0.013	0.009
1981: 303	0.000	1.476*	−0.238	−0.242*	0.028*	0.020	−0.000	0.001	−0.041
161 Cens	0.002	0.653*	0.376*	0.069	−0.003	−0.010	−0.050	0.012	−0.066*
161 Comp	−0.001	1.119*	−0.397*	0.294*	−0.005	0.019	−0.019*	−0.011	0.014
1982: 303	−0.019	2.382*	−2.379*	0.030*	−0.061*	0.225*	0.030	−0.084*	0.853*
158 Cens	−0.000	0.630*	0.108	0.265	−0.008	0.020	−0.024	−0.003	0.102
158 Comp	−0.001	1.088*	−0.573	0.469*	−0.016	0.097	−0.009	0.012	0.091*

* = Significant at the 5% level
** = Significant at the 10% level
Cens = Census R&D database used in analysis
Comp = Compustat R&D database used in analysis

TABLE 5.4 Two-Stage Least Squares Results. Dependent Variable: Investment (CE)

	Constant	CF	LIQ	Tax Rate	DIV	RD	EF
1978: 303	0.069*	0.336*	−0.194*	−0.008	−0.036	0.187**	0.457*
188 Cens	0.079*	0.338*	−0.181*	−0.012	−0.257	0.228*	0.208*
188 Comp	0.076*	0.320*	−0.182*	−0.010	−0.224	0.231*	0.218*
1979: 303	0.071*	0.345*	−0.200*	−0.001	−0.186**	0.422*	0.353*
180 Cens	0.073*	0.425*	−0.188*	−0.059*	−0.201	0.153*	0.423*
180 Comp	0.084*	0.365*	−0.213*	−0.054*	−0.126	0.464*	0.405*
1980: 303	0.075*	0.348*	−0.185*	0.000	0.093	0.312*	0.609*
177 Cens	0.073*	0.293*	−0.181*	−0.001	−0.131	0.508*	0.291*
177 Comp	0.077*	0.282*	−0.184*	−0.001	−0.151	0.461*	0.295*
1981: 303	0.070*	0.473*	−0.213*	−0.015	0.346*	0.398*	0.157*
161 Cens	0.080*	0.509*	−0.188*	−0.065**	−0.401	0.166	0.428*
161 Comp	0.081*	0.508*	−0.190*	−0.066**	−0.421	0.165	0.426*
1982: 303	0.070*	0.250*	−0.164*	0.028*	0.346*	0.398*	0.157*
158 Cens	0.061*	0.258*	−0.163*	0.030*	0.243	0.241*	0.151*
158 Comp	0.062*	0.257*	−0.165*	0.030*	0.200	0.274*	0.149*

*   = Significant at the 5% level
** = Significant at the 10% level
Cens = Census R&D database used in analysis
Comp = Compustat R&D database used in analysis

107

TABLE 5.5 Two-Stage Least Squares Regression Results. Dependent Variable: Dividend (DIV)

	Constant	LDIV	NI	ROE	LIQ	CE	RD	EF
1978: 303	0.002	0.965*	0.053*	−0.006	−0.005	0.008	−0.026*	−0.010
188 Cens	0.002	0.959*	0.080*	−0.002	−0.045*	−0.009	−0.027*	−0.006
188 Comp	0.001	0.958*	0.080*	−0.002	−0.005*	−0.009	−0.017**	−0.007
1979: 303	0.003	0.955*	0.067*	−0.017*	−0.002	−0.015*	0.005	−0.012**
180 Cens	0.001	0.982*	0.064*	−0.004	−0.004	−0.010	0.001	−0.008
180 Comp	0.001	0.981*	0.066*	−0.004	−0.003	−0.009	−0.005	−0.008
1980: 303	0.000	1.008*	0.055*	0.001*	0.007*	−0.003	−0.001	0.001
177 Cens	0.001	1.029*	0.065*	0.012*	−0.005*	−0.014**	−0.008	0.006
177 Comp	0.001	1.028*	0.065*	0.011*	−0.005*	−0.015*	−0.004	−0.006
1981: 303	0.001	1.073*	0.024*	−0.002	−0.004**	0.007**	−0.014**	−0.003
161 Cens	0.001	0.085*	0.077*	−0.015*	−0.029	0.012*	−0.021*	−0.612**
161 Comp	0.000	0.986*	0.079*	−0.015*	−0.003	0.012*	−0.017**	−0.012**
1982: 303	−0.008	0.429*	0.174*	0.000	−0.055*	0.034	0.651*	0.126*
158 Cens	−0.001*	0.995*	0.022*	0.003*	0.034*	0.010*	−0.005	−0.002
158 Comp	−0.014*	0.996*	0.022*	0.003*	0.034*	0.010*	−0.004	−0.014

*   = Significant at the 5% level
** = Significant at the 10% level
Cens = Census R&D database used in analysis
Comp = Compustat R&D database used in analysis

TABLE 5.6 Two-Stage Least Squares Results. Dependent Variable: New Debt (EF)

	Constant	CF	KD	DE	CE	RD	DIV
1978: 303	0.002	−0.155*	−0.027*	0.019*	0.322*	0.514*	0.272*
188 Cens	0.023*	0.012	0.023	0.014**	0.082	0.185	−0.751*
188 Comp	0.026*	0.024	0.003	0.011	0.088	0.023	−0.801*
1979: 303	−0.008	−0.085**	−0.020**	0.013*	0.313*	0.329*	0.381*
180 Cens	0.013	0.024	0.005	0.014*	0.203*	−0.045	−0.725*
180 Comp	0.011	0.017	0.004	0.016*	0.188*	0.097	−0.709*
1980: 303	0.011	−0.029	−0.004	−0.007*	0.472*	0.042	−0.820*
177 Cens	0.034*	−0.033	0.017	−0.012*	0.226*	−0.053	−0.505*
177 Comp	0.034*	−0.031	0.002	−0.012*	0.225*	−0.061	−0.504*
1981: 303	0.023*	0.055	0.005	0.003	0.347*	−0.072	−1.291*
161 Cens	−0.005	−0.061	0.008	0.036*	0.278*	0.157	−0.251
161 Comp	−0.005	−0.064	0.009	0.036*	0.281*	0.183**	−0.276
1982: 303	0.009	−0.192	0.032*	−0.000	0.302*	0.117	0.533*
158 Cens	0.038*	−0.113	0.002*	−0.004	0.348*	−0.190	−0.514
158 Comp	0.037*	−0.114	−0.001	−0.004	0.348*	−0.171	0.488

*   = Significant at the 5% level
** = Significant at the 10% level
Cens = Census R&D database used in analysis
Comp = Compustat R&D database used in analysis

109

did not hold up in the 1978–1982 period. Thus, the data support the following dividend equation in all cases:

$$DIV = f(\overset{+}{DIV}, \overset{+}{NI}_{t-1})$$

4. The new debt (EF) equation (Table 5.6) must also be modified because of the time frame covered in this study. While both cash flow and dividends were significantly and negatively related to new debt over the 1975–1982 time frame, cash flow does not hold up between 1978 and 1982, and dividends waver somewhat. For the 1978–1982 period, the Compustat 303 data support the following model:

$$COMP303(78-82): EF = f(\overset{+}{CE}, \overset{+}{DIV})$$

The longer Compustat period (1975–1982) model is:

$$COMP303(75-82): EF = f(\overset{+}{CE}, \overset{+}{DIV}, \overset{-}{CF})$$

Using the reduced Compustat sample, the equation for the 1978–1982 would become:

$$RCOMP(78-82): EF = f(\overset{+}{CE}, \overset{-}{DIV})$$

Using the NSF/Census data in the reduced sample, the 1978–1982 equation is:

$$CENS(78-82): EF = f(\overset{+}{CE}, \overset{+}{DE}, \overset{-}{DIV})$$

These results provide a mixed picture of the new debt equation. Clearly, new debt is issued to finance capital expenditure in all cases. However, the direction of the dividend relationship seems sensitive to the time frame issue. It is significantly and negatively related to new debt except for the 1978–1982 time period using the Compustat 303-firm sample, when it is marginally positive. For the reduced Compustat sample, the dividend variable becomes negative and significant, while cash flow drops out. Finally, when the NSF/Census R&D data are substituted, the debt/equity ratio enters as a significant positive variable. This is counterintuitive, since the cost of debt should rise as the debt/equity ratio rises, thus discouraging new debt. This latter result was found by Guerard and McCabe (1992).

RELATION OF CURRENT RESULTS TO PRIOR RESEARCH

Three new models of statistically significant relationships between financial decisions and the stock market value of a firm have been estimated. They have then been compared with a fourth reference model developed in an earlier study.

1. The compression of the time frame from 1975–1982 to 1978–1982 did not change the structure of the R&D equation or the capital expenditure equation associated with the 303-firm Compustat sample. This means that the interdependence of the R&D, net income, dividends, and capital expenditures decisions were preserved, implying support for the imperfect markets hypothesis.[10]

2. The reduced Compustat data set includes only the firms that are present in both the Compustat (1978–1982) and the NSF/Census (1978–1982) samples (see Guerard, Bean, and Andrews 1987). An important result of this analysis is that the R&D decision is no longer dependent on the net income and dividend decisions, but is driven only by previous R&D. This finding tends to weaken support for the imperfect markets hypothesis. The capital expenditures equation continues to be stable, including a positive association between R&D and capital expenditures, which is consistent with expectations and prior results. The fact that the RCOMP (1978–1982) sample contains only R&D-performing firms, and thus includes a limited range of observed levels of R&D expenditures (lacking a zero point, for example), may explain why R&D expenditures are no longer interdependent with other financial decisions. The dividend equation is consistent with the COMP303 (1978–1982) version. New debt continues to be strongly affected by capital expenditures, but the relationship to dividends changes signs again, thus indicating instability based on sample composition as well as time frame (the preceding item 1).

3. The substitution of the NSF/Census R&D expenditure data for the Compustat R&D data in the reduced sample suggests only minor modifications for the aforementioned findings. R&D expenditures are dependent only on the prior year's R&D. The discussion in item 2 is relevant here. The capital expenditures and dividends equations are unchanged. The new debt equation once again shows some instability. Capital expenditures and dividends are related to new debt, as they were in the RCOMP (1978–1982) series. However, the debt/equity ratio has now entered the equation, but the sign is opposite to expectations. As the debt/equity ratio rises, new debt financing should fall, because the cost of new debt should be rising; thus, the

TABLE 5.7 Stock Price Equation Estimates. Two-Stage Least Squares Regressions

Variables	1978	1979	1980	1981	1982
Constant (t)	−0.0170	−0.0461	−0.0780	0.0060	−0.0134
	(−2.52)	(−4.45)	(−6.00)	(0.40)	(−3.00)
BETA	0.0276	0.0057	0.0268	0.0113	0.0056
	(−1.32)	(0.78)	(3.80)	(1.95)	(1.99)
DIV	0.3898	0.2612	0.4345	0.2508	0.2374
	(4.75)	(1.67)	(2.64)	(1.75)	(3.22)
CE	0.0551	0.2434	0.3310	−0.1675	0.0194
	(1.25)	(3.62)	(4.46)	(−3.85)	(0.77)
RD	0.0431	−0.0248	−0.1490	−0.0059	0.1027
	(0.68)	(−0.28)	(−1.30)	(−0.80)	(2.55)
Book value	0.8360	1.4961	1.4833	0.0026	0.8582
	(18.18)	(19.78)	(17.74)	(0.95)	(21.66)
Patents	0.0547	0.2891	−0.3328	−0.375	0.1438
	(0.62)	(0.89)	(−1.10)	(−1.90)	(1.51)
LPatents	0.0344	−0.1055	0.4237	0.5218	−0.0207
	(0.34)	(−0.54)	(1.33)	(2.29)	(−0.28)
R^2	0.760	0.765	0.745	0.057	0.834

finding is counterintuitive. Moreover, the beta variable, measuring the firm's systematic risk, is positively associated with new debt. A similar relationship was found by Switzer (1984).

The stock price was positively affected by dividends and investment in 1979 and 1980, research and development in 1982, and the book value of common stock, as shown in Table 5.7.[11] However, the financial decision variables, particularly the dividend variable, tend to be positively associated with the firm's stock price. One should notice that the significance levels for the capital expenditures variable tends to decline in the latter years of the study, 1981 and 1982. The book value of the equity variable dominates the stock price (PCS) equation, as is found in Ben-Zion (1984), for all years, with the noted exception of 1981.

EXTENSIONS OF THE SIMULTANEOUS EQUATION APPROACH

After developing the original simultaneous equation model of stock price valuation based on the 303-firm Compustat sample, Guerard and McCabe (1987) and Guerard and Bean (1987) explored several extensions. This section reexamines two of those extensions based on the availability of the NSF/Census data. A multicriteria model was developed from the imperfect markets hypothesis (Guerard and Bean 1987). It was used to determine the research, dividend, and investment allocations, as well as the external financing levels, needed to maximize a firm's common stock price relative to its asset base. Regression coefficients from the three-stage least squares model were used as inputs to a multiple-goal linear programming model that minimizes the firm's underachievement or overachievement (relative to industry averages) of allocations among the financial variables that impact common stock prices. A firm that is interested in minimizing the underachievement of desired research, investment, and dividend expenditures and the overachievement of desired levels of debt in order to maximize its share price, can use the goal programming model to determine the appropriate financial decisions. The objective function of the linear programming model may be written as follows:

$$\text{Min } Z = p_1 d_1^- + p_2 d_2^- + p_3 d_3^- + p_4 d_4^+$$

where

d_1^- = underachievement of desired research expenditures;
d_2^- = underachievement of desired investment expenditures;
d_3^- = underachievement of desired dividend payments;
d_4^+ = overachievement of desired external financing; and
P = the priority goal programming levels.

See Lee (1972) for a description of priority goal programming levels. Guerard and Bean (1987) estimated this model for a firm selected at random from the 303-firm Compustat database for 1982. The sample firm's R&D, capital, and dividend allocations were less than the industry average for firms with its asset base, and its new debt was greater than the industry average, as shown in Table 5.8.

A question that might be raised by this firm's management is whether a change in its financial decisions would improve stockholder wealth and, if

TABLE 5.8 1982 Allocations and Commitments (Million $)

	Sample Firm	Industry Average
R&D	68	148
Capital Expenditures	770	799
Dividends	42	117
New Debt	577	231

TABLE 5.9 Sample Firm—Results of Optimization Procedure Using 1982 Compustat Data (Million $)

	Sample Firm	Optimal Levels	Under/Over
R&D	68	199	$-131.1
Capital Expenditures	770	924	-155
Dividends	42	278	-236
New Debt	577	475	102

so, how the changes should be made.[12] As shown in Table 5.9, these results suggest that the firm, given its asset base, was spending less than optimal levels on R&D, capital replacement/expansion, and dividends, and that it incurred excessive new debt. Under the assumption that its asset base would remain the same, these results imply that the firm could increase its stock price in 1983 by increasing its expenditures on worthy R&D and capital projects, increasing dividends, and reducing issuance of new debt. As pointed out in Guerard and Bean (1987), this approach to maximizing stockholder wealth (that of formulating optimal financial policies) was shown to be more effective than approaches that attempt to maximize return on equity or growth in earnings per share, as discussed by Rappaport (1983). In addition, the approach presented here offers econometric support for the financial modeling approach and extends the theoretical models of Carleton (1970), Hamilton and Moses (1973), and Burton, Damon, and Obel (1979, 1984).

Given these interesting results from the Compustat data, the analysis was replicated using the 1982 NSF/Census data; the results are shown in Table 5.10.

Clearly, the two data sets produce somewhat different results. The NSF/Census data call for smaller adjustments than the Compustat data.

TABLE 5.10 Sample Firm—Results of Optimization Procedure Using 1982 NSF/Census Data (Million $)

	Firm's Actual	Optimal Levels	Under/Over
R&D	68	189	$-120.9
Capital Expenditures	770	882	-112
Dividends	42	139	-97
New Debt	577	540	36

TABLE 5.11 Sample Firm—"Optimal" 1983 Financial Decisions Based on 1982 Actual and Industry Levels (Million $)

Decision Variables	1982 Actual Levels	1982 Industry Levels	Optimal 1983 Compustat	Optimal 1983 NSF/Census
R&D	68	148	199	189
Capital Expenditures	769.5	798.5	924.1	881.5
Dividends	42	117	278	139
New Debt	577	231	475	540

However, the adjustments are all in the same direction relative to the firm's actual allocations, and with respect to the industry averages. Thus, in both cases it appears that the firm's actual financial decisions resulted in depressed stock prices in 1982 relative to its asset base. Given its assets, it was underinvesting in R&D and capital projects, paying lower dividends, and issuing more new debt than the industry average. However, given its asset base, the new debt decision was closer to "optimal" than the other decisions. The most striking contrast between the two cases involves dividend policy. Clearly, in the case of the Compustat data, stock prices in this industry are much more dependent on dividend policy than in the NSF/Census case. The comparisons are summarized in Table 5.11.

SUMMARY AND CONCLUSIONS

An examination of R&D data for matched pairs of firms has shown that Compustat data and NSF/Census data are comparable for a diverse sample of firms. However, on a year-by-year basis, the NSF/Census data appear to be significantly different from the Compustat data in 1979 and, probably, in 1978. In general, it appears that firms overreported R&D expenditures in the Compustat studies relative to NSF/Census studies. This is not surprising, given the exclusion of engineering and technical services, as well as research in the behavioral sciences, from NSF's definitions of R&D. In the analysis of six specific industries, the R&D expenditures data were significantly different in only one, petroleum, where the sample size was relatively small.

When the NSF/Census R&D data were substituted into a series of regression equations used to test for relationships associated with the perfect markets hypothesis, support for the interdependence of R&D and other financial management decisions became progressively weaker as:

- The time frame of the data series was compressed; and
- The diversity of the firms in the sample was reduced.

When the NSF/Census R&D data was substituted for the Compustat data in the more homogeneous sample in the compressed time frame, the regression results were not altered significantly. Thus, it appears that the NSF/Census data were equally useful in testing the perfect markets hypothesis within this sample.

SUGGESTIONS FOR FUTURE RESEARCH

A larger, more diversified sample of firms should be drawn from the NSF/Census data set to reexamine the perfect markets hypothesis. The hypothesis is important to national R&D policy concerns because it addresses the question of wealth creation and its underlying causes. The larger, more diversified sample drawn from the Compustat data provided relatively strong support for the concept of simultaneous determination of the financial decisions within a firm that influence wealth creation. Moreover, the level of R&D funding was an important determinant of wealth creation. A less diversified sample, covering fewer firms and a shorter time frame, provided weaker support for the interdependencies among these decisions, regardless of whether NSF/Census data or Compustat data were used. Since it is well known that technology flows through the economy by interindustry transfer processes (Scherer 1982), it is necessary to include technology-using firms that may not perform R&D, as well as technology-generating (R&D performing) firms, in studies of the role of R&D in wealth creation. If the NSF/Census data set does not include non-R&D-performing firms, it becomes increasingly important to understand the pros and cons of "linking" this data set to others that are more representative of the economy as a whole.

This brief examination of the role of federal R&D expenditures in explaining wealth creation is merely an introduction to an important issue. The addition of federally funded R&D to a firm's own R&D lends support to the perfect markets hypothesis. This suggests that federal support for industrial R&D tends to reduce gaps between social and private returns to R&D (Mansfield 1981, 1984). Much more work is needed on this important issue.[13]

There seems to be little doubt that the firm-aggregate Compustat and NSF/Census data are not the only data necessary to examine the strategic value of industrial R&D. Industrial competitiveness depends on the specific product-market segments toward which industrial R&D is directed, not simply the aggregate level of R&D in firms that are nominally classified by an aggregate SIC code. The large firm's financial decisions, which are seen in the aggregate in Compustat and NSF/Census data, reflect a composite of operating-level decisions associated with specific lines of business. The inclusion of the Federal Trade Commission (FTC) Line-of-Business database (1974–1977) in the analysis would enable an examination of the impact of R&D strategic focus (and other financial decisions) on stock market values. This type of analysis will set the stage for the study of some of the more "micro" aspects of the NSF/Census database, such as the "mix" of R&D activities, its structure by field of science, and employment patterns of science and engineering personnel.

NOTES

1. In view of this difference, the data used for the 1981 regression were reexamined, and no irregularities were found. The decline in significance was partially noteworthy for the R&D variable.
2. Thus, the procedure involved matching firms present in the original 303-firm Compustat database with identical firms in the NSF/Census database on a year-by-year basis: (1) pairwise elimination of cases with missing data; (2) reestimating the original models using the Compustat data; (3) substituting the NSF/Census R&D data for the Compustat R&D data in the relevant firms; and (4) reestimating the original models, once again using the hybrid NSF/Census-Compustat database.
3. This implies that the richer content of the NSF/Census data set regarding the R&D activities of a firm can be brought to bear on questions impossible to answer with the Compustat data alone.
4. Closer examination of the firm-level data might help to explain the reasons for these differences.
5. The fact that the variances were significantly different in 1978 and 1979, and that the means differed in the opposite direction in 1979, raises questions about the way R&D expenditures were reported in these years.
6. It is noteworthy that the differences in variances of the two series become insignificant in 1980–1982 when the sample size drops from 12 to 11, thus suggesting that a single firm could have accounted for the 1975–1979 differences.
7. This result is noted in Guerard, Bean, and Andrews (1987).

8. Thus, the COMP303 data set covers firms with a wider range of R&D expenditures and has a true zero point. Inasmuch as the constant term is not significantly different from zero for any of the samples across the five years, the COMP303 model seems the most plausible. To put it another way, it seems implausible that a firm that did no prior R&D would never spend money on R&D. Clearly, some firms must launch R&D programs even though they previously had none. The COMP303 model can accommodate this event while the others cannot.

9. The change in sign between 1981 and 1982 may reflect the effects of the R&D tax incentives associated with the 1981 tax reforms. This should be examined as a separate issue.

10. The dividends and new debt equations were changed by the reduced time frame. In the dividends equation, new debt was no longer significant, a finding that is consistent with the work of Dhrymes and Kurz (1967) for the 1947–1960 period. New debt, in turn, is influenced by capital expenditures, but not by cash flow. The relationship between dividends and new debt changes from negative to positive, although the relationship is not well behaved. It was positive in three years and negative in two, all five being statistically significant. Thus, it appears that the new debt equation is highly sensitive to the change in time frame. While a positive relationship between new debt and dividends is consistent with the notion that sources and uses of funds should rise and fall together, the relationship is weak for this sample.

11. The results are in reasonable agreement with Ben-Zion (1984), who used ordinary least squares analysis to investigate the relationship between research and development and the firm's market value for 157 firms during the 1969–1977 period. However, patents, stock betas, and lagged patents rarely influence the stock price, counter to Ben-Zion's results. The stock price maximization framework for R&D analysis proposed in Guerard and Bean (1986) is not found to be as significant as reported in the larger sample.

12. Additional assumptions that were needed to estimate the model, such as depreciation schedules, etc., can be found in balance sheets and income statements in Guerard and Bean (1987).

13. We believe that a causal modeling technique (such as LISREL) should be employed to examine the linkages between public and private R&D expenditures and the financial decisions that affect stock market prices.

REFERENCES

Bender, P. S., W. D. Northup, and J. F. Shapiro. "Practical Modeling for Resource Management." *Harvard Business Review* (1981):163–173.

Ben-Zion, U. "The R&D and Investment Decision and Its Relationship to the

Firm's Market Value: Some Preliminary Results." In *R&D, Patents, and Productivity,* edited by Z. Griliches. Chicago: University of Chicago Press, 1984.

Burton, R. M., W. W. Damon, and B. Obel. "Operations Planning and Investment Strategy in the Owner Financed Firm." In *Optimization Models for Strategic Planning,* edited by T. Naylor and C. Thomas. Amsterdam: North-Holland, 1984.

Burton, R. M., W. W. Damon, and B. Obel. "An Organizational Model of Integrated Budgeting for Short-Run Operations and Long-Run Investments." *Journal of Operational Research and Society* 30 (1979):575–585.

Carlton, W. T. "An Analytical Model for Long-Range Financial Planning." *Journal of Finance* 25 (1970):291–315.

Damon, W. W., and R. Schramm. "A Simultaneous Decision Model for Production, Marketing and Finance." *Management Science* 18 (1972):161–172.

Dhrymes, P. J. *Econometrics: Statistical Foundations and Applications.* New York: Springer-Verlag, 1974.

Dhrymes, P., and M. Kurz. "Investment, Dividends, and External Finance Behavior of Firms." In *Determinants of Investment Behavior,* edited by Robert Ferber. New York: Columbia University Press, 1967.

Dhrymes, P. J., and M. Kurz. "On the Dividend Policy of Electric Utilities." *Review of Economics and Statistics* 46 (1964):76–81.

Fama, E. F. "The Empirical Relationship Between the Dividend and Investment Decisions of Firms." *American Economic Review* 64 (1974):304–318.

Fama, E. F., and H. Babiak. "Dividend Policy: An Empirical Analysis." *Journal of the American Statistical Association* 63 (1968):1132–1161.

Grabowski, H. G. "The Determinants of Industrial Research and Development: A Study of the Chemical, Drug, and Petroleum Industries." *Journal of Political Economy* 76 (1968):292–306.

Grabowski, J. G., and D. C. Mueller. "Managerial and Stock-holder Welfare Models of Firm Expenditures." *Review of Economics and Statistics* 54 (1972): 9–24.

Guerard, J. B., Jr., and A. S. Bean. "Identifying and Closing Gaps in Corporate R&D Expenditures. In *Handbook of Technology Management,* edited by D. F. Kocaoglu. New York: John Wiley & Sons, 1987.

Guerard, J. B., Jr., and A. S. Bean. "Goal Setting and Corporate Strategic Planning." Presented at the International Conference on Vector Optimization, Darmstadt, West Germany (1986).

Guerard, J. B., and G. M. McCabe. "The Integration of Research and Development Management into the Firm Decision Process. In *Management of R&D and Engineering,* edited by D. F. Kocaoglu. Amsterdam: North-Holland, 1987.

Guerard, J. B., Jr., and B. K. Stone. "Strategic Planning and the Investment-Financing Behavior of Major Industrial Companies." *Journal of the Operational Research Society* 38 (1987):1039–1050.

Guerard, J. B., Jr., A. S. Bean, and S. Andrews. "R&D Management and Corporate Financial Policy." *Management Science* 33 (1987):1419–1427.

Guerard, J. B., Jr., A. S. Bean, and G. M. McCabe. "Identifying Intertemporal Relationships in Corporate R&D Expenditures." *IEEE Transactions on Engineering Management,* EM-33 (1986):157–161.

Hamilton, W. F., and M. A. Moses. "An Optimization Model for Corporate Financial Planning." *Operations Research* 21 (1973):677–692.

Higgens, R. C. "The Corporate Dividend-Saving Decision." *Journal of Financial and Quantitative Analysis* (1984):1527–1541.

Jalilvand A., and R. S. Harris. "Corporate Behavior in Adjusting to Capital Structure and Dividend Targets: An Econometric Study." *Journal of Finance* 39 (1984):127–145.

Lawrence, K. D., and J. B. Guerard, Jr. "Strategic Planning and the Problem of Capital Budgeting in a Steel Firm: A Multi-Criteria Approach." In *Optimization Models for Strategic Planning,* edited by Thomas Naylor and Celia Thomas. Amsterdam: North-Holland, 1984.

Lee, S. M. *Goal Programming for Decision Analysis.* Philadelphia: Auerbach Publishers, 1972.

Lintner, J. "Distributions of Incomes of Corporations Among Dividends, Retained Earnings and Taxes." *American Economic Review* 46 (1956):97–118.

Mansfield, E., "R&D and Innovation." In *R&D, Patents, and Productivity,* edited by Z. Griliches. Chicago: University of Chicago Press, 1984.

Mansfield, E. "How Economists see R&D." *Harvard Business Review* 59 (1981):98–106.

Mansfield, E., J. Rappaport, A. Romeo, S. Wagner, and G. Beardsley, Social and Private Rates of Return from Industrial Innovations, *Quarterly Journal of Economics* 91 (1977).

Martin, M. J. C. *Managing Technological Innovation and Entrepreneurship.* Reston: Reston Publishing Company, 1984.

McCabe, G. M. "The Empirical Relationship Between Investment and Financing: A New Look." *Journal of Financial and Quantitative Analysis* 14 (1979):119–135.

McDonald, J. G., B. Jacquillat, and M. Nussenbaum. "Dividend, Investment, and Financial Decisions: Empirical Evidence on French Firms." *Journal of Financial and Quantitative Analysis* 10 (1975):741–755.

Meyer, J. R., and E. Kuh. *The Investment Decision.* Cambridge: Harvard University Press, 1975.

Meyers, S. C., and G. A. Pogue. "A Programming Approach to Corporate Financial Management." *Journal of Finance* 29 (1074):579–599.

Miller, M., and F. Modigliani, "Dividend Policy, Growth, and the Valuation of Shares." *Journal of Business* 34 (1961):411–433.

Mueller, D. C., "The Firm Decision Process: An Econometric Investigation." *Quarterly Journal of Economics* 81 (1967):58–87.

National Science Foundation. *A Comparative Analysis of Information on National Industrial R&D Expenditures,* Special Report NSF 85-311 (Washington: NSF, 1985).

Peterson, P., and G. Benesh. "A Reexamination of the Empirical Relationship Between Investment and Financial Decisions." *Journal of Financial and Quantitative Analysis* 18 (1983):439–454.

Rappaport, A. "Corporate Performance Standards and Shareholder Value." *The Journal of Business Strategy* 3 (1983):28–38.

Rosenbloom, R. S., and A. M. Kantrow. "The Nurturing of Corporate Research." *Harvard Business Review* (1982):115–123.

Scherer, F. M. "Inter-Industry Technology Flows in the United States." *Research Policy* 11 (1982).

Scherer, F. M. "Firm Size, Market Structure, Opportunity, and the Output of Patented Inventions." *American Economic Review* (1965): 1104–1113.

Switzer, L. "The Determinants of Industrial R&D: A Funds Flow Simultaneous Equation Approach." *Review of Economics and Statistics* 66 (1984):163–168.

Tinbergen, J. *A Model and its Application to Investment Activity.* Geneva: League of Nations, 1939.

Weston, J. F., and T. E. Copeland. *Managerial Finance.* Chicago: CBS Publishing, 1986.

Zeleny, M. *Multiple Criteria Decision Making.* New York: McGraw-Hill, 1982.

6
MORE ON THE INTERDEPENDENCIES AMONG CORPORATE FINANCIAL POLICIES: THE CASE OF EFFECTIVE DEBT

We have discussed and empirically verified the hypothesis that firms simultaneously determine their research and development, investment, dividend, and new debt policies. In this chapter we introduce the reader to the concept of effective debt, which holds that many firms use cash and/or marketable securities as a store of financing (liquidity). Firms may issue long-term bonds in excess of current need, reduce short-term indebtedness, and put any surplus into cash and/or marketable securities. The effective debt concept represents the net use of debt financing in a given accounting period. The determinants of research and development, dividend, investment, and effective debt decisions of the U.S. firms in the WorldScope database are econometrically estimated during the 1983–1995 period.

The purpose of this chapter is to estimate an econometric model to analyze the interdependencies among the decisions concerning research and development, investment, dividends, and effective debt financing. We extend the simultaneous equations modeling system of a firm's financial decisions that was introduced in Chapter 4 to incorporate the modeling of effective debt. Management decisions on dividends, capital expenditures, and research and development activities are made while minimizing reliance on effective debt funding to generate future profits. Effective debt, defined as the increase in total debt less the increase in cash and marketable securities, specifically addresses the question as to whether economies of scale exist in debt issuance. Management may issue long-term bonds in excess of current needs and allocate the surplus debt into cash and/or marketable securities if economies of scale exist in the debt decision.

A firm has a "pool" of resources, composed of net income, depreciation, and new debt issues, and this pool is reduced by dividend payments, investment in capital projects, and expenditures for research and development activities. We developed and estimated our model having verified the imperfect markets hypothesis concerning financial decisions, as shown in Chapter 4. Financial decisions are interdependent, and simultaneous equations must be used to estimate the equations econometrically.

The goal of this study is to test empirically the independence of financial decisions hypothesis using the Guerard and Stone (1987) framework of effective debt. Guerard and Stone developed and estimated their model using a set of simultaneous equations using the 303-firm sample during the 1978–1995 period, as described in Guerard and McCabe (1992). We update the Guerard and Stone (1987) study and compare its initial 1987 results with those derived from the U.S. firms in the WorldScope database for the 1983–1995 period. We find stronger evidence of the interdependence of financial decisions using the larger U.S. firms in the WorldScope database than we reported in Guerard and Stone (1987) using the Guerard and McCabe and Guerard and Stone 303-firm database for the 1978–1982 period. There is stronger evidence that U.S. financial decisions are interdependent, and there are few significant estimated differences in the use of long-term debt issuance and (net) effective debt issuance equations.

THE MODEL

The model we developed employs investment, dividends, and new capital financing equations to describe the budget constraint facing the manager of a manufacturing firm. The manager may use available funds to undertake capitalized research and development activities (R&D) or new investment (CE), or to pay dividends (DIV) or increase net working capital (LIQ). The sources of funds are represented by net income (NI), depreciation (DEP), and new debt issues (dLTD):

$$RD + CE + dCA - dCL = NI + DEP - DIV + dLTD + NEQ \quad (6.1)$$

where

    dcA = increase in current assets
    dCL = increase in current liabilities
    NEQ = net new-equity issues

Let us define the change in net working capital (NWK) as dCA−dCL. We can rewrite equation (6.1) in terms of effective debt by expanding upon changes in net working capital:

$$dCA = dOCA + dMS + dCASH$$
$$dCL = dOCL + dSTD$$

where

    dOCA = increase in other current assets, current assets less marketable securities and cash
    dMS = increase in marketable securities
  dCASH = increase in cash balances
    dOCL = increase in current liabilities other than short-term debt
    dSTD = increase in short-term debt

$$RD + CE + dOCA - dOCL = $$
$$NI + DEP - DIV + NEQ + dSTD - dMS - dCASH + dLTD \quad (6.2)$$

Guerard and Stone (1987) rewrote equation (6.2) as:

$$RD + CE + dOCA - dOCL = $$
$$NI + DEP - DIV + NEQ + dED \quad (6.3)$$

where dED is the (net) increase in effective debt, the increase in total debt less the increase in cash and marketable securities. In the world of business during the past 30 years, net debt issues have accounted for about 80% to 90% of new capital issues. Here we present the estimation of a simultaneous equations system of the largest capitalized firms during the 1983–1995 period using the WorldScope database and the Compustat database with only minor differences. We find few differences between the estimated effective debt equations in this chapter and the traditional new debt issues equation estimated in Chapter 4.

The following is a summary of the hypothesized equation system:

$$CE = F(\overset{-}{DIV}, \overset{-}{RD}, \overset{+}{dED}, \overset{-}{LIQ}, \overset{+}{NI}, \overset{+}{LSAL}) \quad (6.3)$$
$$DIV = F(\overset{-}{CE}, \overset{-}{RD}, \overset{-}{LIQ}, \overset{+}{dED}, \overset{+}{LDIV}, \overset{+}{NI}) \quad (6.4)$$
$$dED = F(\overset{+}{CE}, \overset{+}{RD}, \overset{+}{DIV}, \overset{-}{NI}, \overset{-}{DEP}, \overset{-}{KD}, \overset{-}{DE}) \quad (6.5)$$
$$RD = F(\overset{+}{LRD}, \overset{-}{CE}, \overset{-}{DIV}, \overset{+}{dED}, \overset{+}{NI}) \quad (6.6)$$

THE DATA

The financial variables are drawn from the WorldScope data files and are divided by total assets to alleviate heteroscedasticity; hence, an "S" precedes standardized variables (for example, effective debt is SED). We also use annual financial data for all U.S. firms during the 1983–1995 period.

ESTIMATED SIMULTANEOUS EQUATION RESULTS

Ordinary least squares (OLS) is used to initially estimate equations (6.3) through (6.6). The simultaneous equation results reported in this study are produced with the use of two-stage (2SLS) and three-stage (3SLS) least squares analysis. Although Dhrymes and Kurz (1967) found that the insignificantly negative association between capital expenditures and dividends in the two-stage least squares regression estimation became a significantly negative association in the three-stage least squares estimations, we found no statistically significant differences in the limited-information (two-stage least squares) and full-information (three-stage least squares) procedures. The two-stage least squares regression equation residuals were not highly correlated, providing the statistical basis for the insignificant coefficient differences in the two- and three-stage least squares estimations. The highest correlations were found in the annual regression residuals between the new debt and capital expenditures equations. We found little differences in the 2SLS and 3SLS results; the OLS, 2SLS, and 3SLS squares regression results are reported in Table 6.1.

The estimated system equations, estimated annually for the U.S. firms on the WorldScope database, are shown in Table 6.1. Here we find stronger evidence of the interdependencies of financial decisions in the larger universe than in the original Guerard and Stone (1987) study. In the estimated capital expenditures investment equation, dividends are an alternative use of funds (large negative and statistically significant coefficient), whereas investments are positively associated with increasing research activities and net effective debt financing at the 10% level of significance. Net income and depreciation have positive and statistically significant coefficients in the investment equation. The change in sales does not positively influence investment, the accelerator position, as was the case in the study presented in Chapter 4, using the WorldScope database and traditionally issued long-term debt. The change in sales is a positive and statistically significant determinant of investment when the 3,000 largest firms in the United States are analyzed, not just the larger firms found on the WorldScope database. In the estimated dividend equation, dividends are negatively associated with capital expenditures and positively associated with R&D and net income. We do not find a positive coefficient on the effective debt variable as one would expect according to the imperfect markets hypothesis. New effective debt financing is significantly associated with higher capital expenditures, dividends, and R&D variables; the larger coefficient is found on the capital investment variable, as was the case in Guerard and Stone (1987). R&D is associated with higher effective debt financing and negatively associated with capital expenditures and dividends. There are significant violations of the independence (perfect markets) hypothesis in the capital expenditures, dividend, effective debt, and research activities equation estimations.

The statistically significant and positive coefficient on the external funds issued variable is convincing in the investment equation and complements the work of McCabe (1979), Peterson and Benesh (1983), and Guerard and McCabe (1992). Dhrymes and Kurz (1967) and Switzer (1984) did not always find a significantly positive relationship between new debt and investments. Mueller (1967) found an inverse relationship between investments and research in his earlier investigation and no relationship between the variables in his later work with Grabowski (1972). Switzer found no significant association between R&D and investment. Decreases in net liquidity are associated with rising investment and dividends. Net income and depreciation positively affect investment and negatively affect new debt financing. Dividends are positively associated with rising net income and last year's dividends, supporting the positions of Lintner (1956), Fama and Babiak (1968), and Switzer (1984).

The rejection of the perfect markets hypothesis is found in (1) the influence of dividends and effective debt financing on the investment decision; (2) evidence that increasing capital expenditures lowers dividends, whereas R&D is positively associated with increasing dividends; and (3) the interdependence between investment and effective debt variables. The empirical evidence concerning U.S. firms in the WorldScope universe confirms the necessity of using simultaneous equations to econometrically model the interdependencies of financial decisions. The evidence is more supportive of the existence of imperfect markets in the larger universe than in the original Guerard and Stone (1987) and Guerard and McCabe (1992) 303-firm sample.

The following is a summary of the 3SLS statistically significant estimated equation system:

$$SCE = F(\overset{-}{SDIV}, \overset{+}{SdED}, \overset{-}{SLIQ}, \overset{+}{SNI}, \overset{+}{SDEP})$$
$$SDIV = F(\overset{-}{SCE}, \overset{+}{SRD}, \overset{+}{SdED}, \overset{-}{SLIQ}, \overset{+}{SLDIV}, \overset{+}{SNI})$$
$$SdED = F(\overset{+}{SCE}, \overset{+}{SRD}, \overset{+}{SDIV}, \overset{-}{SNI}, \overset{-}{SDEP}, \overset{+}{DE})$$
$$SRD = F(\overset{-}{SCE}, \overset{-}{SDIV}, \overset{+}{SdED}, \overset{+}{SRD}, \overset{+}{SNI})$$

Here there is almost complete agreement with the WorldScope traditional long-term debt-issued WorldScope universe results reported in Chapter 4.

TABLE 6.1 The Interdependencies of Financial Decisions: The Investment, Dividend, Effective Debt, and R&D Decisions of U.S. Firms on WorldScope

OLS	SCE	1982		1983		1984	
SDIV	−0.2249	(−6.23)	−0.1956	(−11.27)	−0.2180	(−6.40)	−0.7639
SED	0.1166	(10.63)	0.1939	(14.13)	0.1539	(14.87)	0.2004
SRD	0.1983	(3.73)	0.1265	(3.21)	0.0862	(2.20)	0.1728
SNI	0.1978	(6.55)	0.2705	(12.47)	0.1711	(8.38)	0.7252
SLIQ	−0.0919	(−10.20)	−0.0826	(−12.88)	−0.0615	(−10.39)	−0.0972
SLSAL	−0.0032	(−0.57)	−0.0130	(−3.78)	0.0053	(1.24)	−0.0590
SDEP	1.6677	(24.56)	1.3219	(29.13)	1.2435	(27.16)	2.0147

		1989		1990		1991	
SDIV	−0.1064	(−8.84)	−0.3522	(−18.03)	−0.4748	(−22.36)	−0.7344
SED	0.1138	(11.62)	0.1635	(13.03)	0.2172	(15.19)	0.3050
SRD	0.0061	(0.21)	0.2756	(8.52)	0.3659	(10.59)	0.3495
SNI	0.1228	(10.28)	0.4416	(33.82)	0.5693	(46.33)	0.8311
SLIQ	−0.0353	(−6.58)	−0.1364	(−22.32)	−0.1774	(−28.70)	−0.1838
SLSAL	0.0036	(1.70)	−0.0033	(−0.74)	−0.0837	(−28.91)	−0.0856
SDEP	1.0996	(29.70)	0.6210	(16.08)	0.5504	(11.23)	0.5235

OLS	SDIV	1982		1983		1984	
SLDIV	1.4361	(146.68)	0.5633	(287.01)	0.2995	(19.13)	1.3383
SCE	−0.0208	(−4.86)	0.0099	(1.88)	−0.0665	(−4.27)	−0.0613
SED	0.0117	(5.70)	−0.0064	(−1.97)	−0.0025	(−0.34)	0.0096
SRD	0.0341	(3.51)	−0.0257	(−2.70)	0.0054	(0.19)	0.0379
SNI	0.0644	(12.44)	0.0499	(10.54)	0.3014	(26.56)	0.0551
SLIQ	−0.0128	(−7.56)	−0.0008	(−0.51)	−0.0309	(−7.19)	−0.0054

		1989		1990		1991	
SLDIV	0.6966	(55.02)	0.4436	(37.92)	0.6997	(42.06)	0.6190
SCE	−0.1109	(−5.39)	−0.1119	(−6.88)	−0.1025	(−7.34)	−0.1736
SED	0.1467	(13.22)	0.1137	(11.35)	0.0150	(1.30)	0.0481
SRD	0.0696	(2.15)	0.0757	(2.88)	0.0481	(1.76)	0.0793
SNI	0.2223	(16.75)	0.2109	(17.74)	0.1285	(16.40)	0.1871
SLIQ	−0.0297	(−4.80)	−0.0249	(−4.67)	−0.0186	(−3.36)	−0.0305

OLS	SED	1982		1983		1984	
SDIV	0.2023	(2.17)	0.1251	(4.03)	−0.0463	(−0.55)	0.1077
SCE	0.6721	(9.91)	0.6357	(12.30)	0.8484	(13.66)	0.2642
SRD	0.4318	(3.29)	0.2645	(3.34)	0.4522	(4.67)	0.1161
SNI	−0.2045	(−2.64)	−0.2009	(−4.52)	−0.1405	(−2.75)	−0.1571
DE	0.0419	(1.68)	0.0143	(0.99)	0.0905	(5.65)	−0.0062
SDEP	−1.2231	(−5.77)	−1.0104	(−8.48)	−1.9410	(−14.15)	−0.8110

ESTIMATED SIMULTANEOUS EQUATION RESULTS

1985		1986		1987		1988	
(−29.45)	−0.7039	(−36.01)	−0.7570	(−38.76)	−0.1175	(−5.80)	
(10.97)	0.3550	(22.70)	0.2140	(20.80)	0.1067	(10.70)	
(3.15)	0.1240	(2.07)	0.1800	(3.22)	0.0178	(0.47)	
(82.08)	0.7163	(81.60)	0.7623	(93.13)	0.1160	(8.29)	
(−10.80)	−0.1391	(−13.90)	−0.1143	(−12.39)	−0.0279	(−4.18)	
(−10.96)	−0.0650	(−9.37)	−0.0241	(−4.78)	0.0032	(1.16)	
(31.24)	2.0002	(29.38)	1.5361	(29.32)	0.5984	(16.94)	

1992		1993		1994		1995	
(−33.80)	−0.5881	(−23.32)	−0.6049	(−25.11)	−0.5172	(−25.04)	
(17.18)	0.0889	(4.57)	0.2091	(11.74)	0.2757	(14.21)	
(7.94)	0.4078	(8.57)	0.3485	(7.47)	0.5619	(14.77)	
(110.00)	0.6648	(65.31)	0.7304	(86.96)	0.5741	(54.37)	
(−23.74)	−0.1838	(−18.11)	−0.1846	(−19.35)	−0.1448	(−16.38)	
(−12.97)	−0.0457	(−5.83)	−0.0662	(−8.85)	−0.1199	(−18.32)	
(17.10)	0.5727	(13.06)	1.2036	(24.33)	0.9781	(18.03)	

1985		1986		1987		1988	
(99.47)	0.9516	(123.43)	0.4990	(33.43)	0.7245	(60.76)	
(−10.87)	−0.0249	(−4.16)	−0.2901	(−23.18)	−0.0292	(−2.11)	
(1.80)	0.0170	(3.26)	0.0768	(10.68)	0.0338	(5.17)	
(2.33)	−0.0057	(−0.31)	0.1343	(3.31)	−0.0252	(−1.02)	
(10.16)	0.0273	(4.89)	0.3011	(26.19)	0.0537	(5.71)	
(−1.98)	−0.0037	(−1.14)	−0.0353	(−5.12)	−0.0036	(−0.82)	

1992		1993		1994		1995	
(56.46)	0.9431	(135.17)	0.7876	(82.35)	0.8342	(141.16)	
(−19.17)	−0.0386	(−7.95)	−0.0332	(−4.66)	−0.0275	(−4.95)	
(5.32)	−0.0055	(−1.15)	0.0538	(8.07)	0.0091	(1.51)	
(3.46)	0.0067	(0.56)	0.0070	(0.38)	0.0182	(1.51)	
(23.02)	0.0353	(8.67)	0.0408	(6.42)	0.0356	(8.83)	
(−7.07)	0.0019	(0.71)	−0.0045	(−1.12)	−0.0070	(−2.45)	

1985		1986		1987		1988	
(2.38)	0.4600	(12.93)	0.7435	(13.46)	0.2389	(5.32)	
(7.85)	0.6100	(20.30)	1.0414	(24.05)	0.5086	(10.27)	
(1.51)	0.0816	(0.95)	0.5049	(4.22)	0.0971	(1.16)	
(−5.36)	−0.4886	(−18.66)	−0.7503	(−18.64)	−0.2314	(−7.29)	
(−0.45)	0.0166	(1.11)	−0.2046	(−11.26)	0.0309	(2.15)	
(−6.89)	−1.2610	(−10.45)	−1.6680	(−12.29)	−0.5983	(−7.12)	

Continued on following page

130 MORE ON THE INTERDEPENDENCIES AMONG CORPORATE FINANCIAL POLICIES

TABLE 6.1 (*Continued*)

	1989		1990		1991		
SDIV	0.2059	(7.97)	0.3153	(9.43)	0.0109	(0.39)	0.2037
SCE	0.5126	(11.65)	0.4065	(13.10)	0.2511	(10.91)	0.2372
SRD	0.2222	(3.74)	0.1238	(2.30)	0.3237	(6.58)	0.0730
SNI	−0.1987	(−7.48)	−0.2182	(−8.12)	−0.0578	(−3.85)	−0.2390
DE	−0.0295	(−2.73)	−0.0361	(−3.44)	−0.0009	(−0.10)	0.0545
SDEP	−0.8706	(−9.48)	−0.4048	(−6.15)	−0.5338	(−8.09)	−0.3738

OLS   SRD

	1982		1983		1984		
SLRD	1.2283	(51.84)	1.1325	(83.03)	1.1156	(99.81)	1.2058
SCE	0.0118	(1.59)	0.0111	(1.80)	0.0081	(1.49)	−0.0056
SDIV	0.1147	(10.45)	−0.0233	(−5.81)	−0.0524	(−6.25)	−0.0128
SED	0.0203	(5.56)	0.0254	(6.42)	0.0163	(6.37)	0.0112
SNI	−0.1063	(−11.99)	0.0351	(6.26)	0.0366	(7.76)	0.0071

	1989		1990		1991		
SLRD	1.0783	(124.67)	1.0722	(106.13)	1.0658	(138.66)	1.0141
SCE	0.0119	(2.33)	0.0114	(2.21)	−0.0049	(−1.51)	−0.0037
SDIV	−0.0157	(−4.70)	0.0033	(0.59)	−0.0112	(−2.75)	−0.0059
SED	0.0154	(5.49)	0.0043	(1.19)	0.0240	(7.95)	0.0094
SNI	0.0187	(5.69)	−0.0094	(−2.21)	0.0027	(1.27)	0.0048

2SLS   SCE

	1982		1983		1984		
SDIV	−0.3658	(−3.89)	−0.4962	(−9.23)	−0.6535	(−4.10)	−0.6001
SED	0.9382	(4.58)	1.0754	(8.88)	0.8423	(7.47)	2.6728
SRD	0.0468	(0.30)	0.1257	(1.40)	−0.0856	(−0.93)	0.0568
SNI	0.3730	(4.41)	0.5902	(9.54)	0.4750	(5.74)	0.6950
SLIQ	−0.1288	(−5.33)	−0.1250	(−8.57)	−0.1456	(−7.94)	−0.0486
SLSAL	−0.0517	(−2.87)	−0.0915	(−7.23)	−0.0671	(−4.66)	−0.2159
SDEP	1.6523	(10.16)	1.4032	(14.80)	1.7313	(13.98)	2.4592

	1989		1990		1991		
SDIV	−0.0829	(−5.02)	−0.6543	(−7.26)	−0.4574	(−4.25)	−0.9176
SED	0.2818	(5.20)	1.9175	(8.21)	2.1821	(11.74)	1.7752
SRD	−0.0457	(−1.37)	0.0442	(0.38)	−0.2353	(−1.86)	0.1061
SNI	0.1135	(7.38)	0.5909	(12.15)	0.5622	(11.74)	0.8551
SLIQ	−0.0312	(−5.28)	−0.0877	(−4.32)	−0.1405	(−7.35)	−0.1810
SLSAL	0.0016	(0.64)	−0.1097	(−5.80)	−0.0967	(−9.90)	−0.1977
SDEP	1.1668	(26.67)	0.9950	(7.62)	1.6178	(9.33)	0.9047

2SLS   SDIV

	1982		1983		1984		
SLDIV	1.4127	(85.86)	0.5643	(272.86)	0.2991	(18.69)	1.3210
SCE	−0.0662	(−6.34)	0.0175	(2.37)	−0.0424	(−1.89)	−0.0747

1992		1993		1994		1995
(7.10)	−0.0086	(−0.30)	0.1660	(5.38)	0.1129	(5.16)
(12.09)	0.0858	(4.29)	0.1912	(8.97)	0.2190	(12.45)
(1.46)	0.0332	(0.67)	0.1379	(2.56)	0.0426	(1.08)
(−12.17)	−0.0074	(−0.42)	−0.1119	(−5.78)	−0.1363	(−10.03)
(6.56)	−0.0224	(−2.53)	−0.0068	(−0.73)	−0.0240	(−3.06)
(−10.63)	−0.3618	(−7.74)	−0.6515	(−10.36)	−0.5193	(−9.06)

1985		1986		1987		1988
(101.47)	1.1082	(100.05)	1.0010	(104.36)	1.0142	(76.44)
(−1.70)	−0.0077	(−2.52)	0.0098	(3.11)	−0.0052	(−0.77)
(−2.26)	−0.0101	(−2.41)	0.0091	(2.09)	−0.0256	(−3.98)
(3.42)	0.0130	(4.51)	0.0025	(1.42)	0.0206	(6.45)
(2.14)	0.0085	(2.81)	−0.0099	(−3.19)	0.0221	(5.01)

1992		1993		1994		1995
(137.65)	1.0232	(149.50)	0.9207	(135.79)	0.9429	(94.56)
(−1.33)	−0.0024	(−0.93)	−0.0014	(−0.52)	0.1182	(29.83)
(−1.41)	−0.0050	(−1.29)	−0.0039	(−0.94)	−0.0325	(−6.00)
(3.27)	0.0048	(1.79)	0.0018	(0.69)	−0.0008	(−0.16)
(1.76)	0.0040	(1.76)	0.0022	(0.88)	0.0400	(11.96)

1985		1986		1987		1988
(−5.69)	−0.7675	(−26.53)	−1.0316	(−39.88)	−0.2903	(−4.90)
(6.52)	0.9871	(21.97)	0.3054	(18.45)	0.8382	(5.13)
(0.26)	0.2635	(2.81)	0.1488	(2.28)	0.2171	(2.22)
(20.94)	0.7674	(59.12)	0.8155	(87.83)	0.2841	(6.15)
(−1.42)	−0.1327	(−9.12)	−0.1083	(−10.76)	−0.0840	(−4.64)
(−6.69)	−0.0903	(−8.95)	−0.0084	(−1.36)	−0.0373	(−3.58)
(9.83)	1.7658	(17.76)	1.3900	(24.05)	0.7840	(9.82)

1992		1993		1994		1995
(−17.77)	−0.6259	(−23.10)	−0.7466	(−12.34)	−0.6206	(−10.62)
(16.78)	0.3003	(3.63)	1.9475	(11.29)	2.6493	(12.10)
(1.12)	0.4221	(8.17)	−0.0561	(−0.47)	0.2972	(2.50)
(54.89)	0.6681	(62.48)	0.7206	(37.38)	0.7038	(22.61)
(−11.73)	−0.1801	(−17.00)	−0.1294	(−5.90)	−0.0484	(−1.91)
(−13.18)	−0.0569	(−6.36)	−0.2777	(−10.68)	−0.2016	(−10.61)
(−13.31)	0.6269	(12.18)	1.8497	(14.35)	1.6907	(10.45)

1985		1986		1987		1988
(91.50)	0.9416	(103.30)	0.4863	(31.07)	0.7285	(58.74)
(−11.42)	-0.0359	(−4.47)	−0.3080	(−21.50)	−0.0008	(−0.03)

Continued on following page

132 MORE ON THE INTERDEPENDENCIES AMONG CORPORATE FINANCIAL POLICIES

TABLE 6.1 (*Continued*)

SED	0.1003	(5.12)	−0.0177	(−2.26)	−0.0343	(−1.55)	0.0204
SRD	0.0075	(0.39)	−0.0145	(−1.36)	0.0457	(1.49)	0.0449
SNI	0.0809	(9.01)	0.0472	(9.40)	0.2988	(25.83)	0.0672
SLIQ	−0.0194	(−6.26)	−0.0004	(−0.25)	−0.0283	(−5.63)	−0.0080
	1989		1990		1991		
SLDIV	0.7035	(54.15)	0.4457	(36.17)	0.6860	(28.35)	0.6220
SCE	0.0130	(0.40)	−0.0924	(−4.23)	0.0568	(2.01)	−0.1684
SED	0.0711	(1.38)	0.0262	(0.56)	−0.5641	(−7.99)	0.0526
SRD	0.0847	(2.38)	0.1020	(3.42)	0.1275	(2.96)	0.0905
SNI	0.2091	(15.07)	0.2051	(15.40)	0.1070	(9.00)	0.1829
SLIQ	−0.0275	(−4.30)	−0.0254	(−4.45)	−0.0041	(−0.49)	−0.0302
2SLS	SED						
	1982		1983		1984		
SDIV	0.1253	(1.28)	0.1298	(3.96)	−0.4407	(−2.64)	−0.0350
SCE	0.2842	(1.97)	0.6100	(6.38)	0.5065	(4.18)	0.0872
SRD	0.2408	(1.50)	0.1373	(1.58)	0.3361	(3.18)	0.0827
SDEP	−0.5386	(−1.74)	−0.9623	(−5.91)	−1.5041	(−8.13)	−0.4109
SNI	−0.1515	(−1.86)	−0.2076	(−4.41)	0.0498	(0.60)	−0.0232
DE	0.0447	(1.72)	0.0207	(1.40)	0.0848	(4.94)	−0.0160
	1989		1990		1991		
SDIV	−0.0508	(−1.47)	−0.0001	(−0.00)	−0.1405	(−3.66)	0.1648
SCE	0.3486	(3.47)	0.2592	(6.43)	0.1928	(7.61)	0.1814
SRD	0.2110	(3.14)	0.2043	(3.34)	0.2572	(4.88)	0.0749
SDEP	−0.6449	(−4.82)	−0.3157	(−4.58)	−0.5061	(−7.58)	−0.3509
SNI	0.0033	(0.10)	−0.0344	(−0.93)	−0.0133	(−0.80)	−0.1919
DE	−0.0578	(−4.91)	−0.0503	(−4.47)	0.0079	(0.83)	0.0563
2SLS	SRD						
	1982		1983		1984		
SLRD	1.2000	(27.98)	1.1331	(82.61)	1.1138	(95.47)	1.2104
SCE	−0.0246	(−1.29)	0.0029	(0.35)	−0.0002	(−0.02)	−0.0159
SDIV	0.0911	(4.47)	−0.0231	(−5.61)	−0.0814	(−4.78)	−0.0223
SED	0.2068	(5.91)	0.0359	(3.86)	0.0214	(2.96)	0.0539
SNI	−0.0792	(−4.76)	0.0352	(6.10)	0.0491	(6.33)	0.0168
	1989		1990		1991		
SLRD	1.0795	(122.63)	1.0694	(93.06)	1.0522	(110.56)	1.0130
SCE	0.0043	(0.54)	−0.0048	(−0.62)	−0.0272	(−5.19)	−0.0110
SDIV	−0.0182	(−4.34)	0.0125	(1.25)	0.0002	(0.03)	−0.0131
SED	0.0367	(2.88)	0.0900	(5.02)	0.1201	(7.61)	0.0457
SNI	0.0219	(5.74)	−0.0083	(−1.32)	0.0034	(1.20)	0.0112

(1.05)	0.0363	(2.92)	0.1001	(10.25)	0.0089	(0.32)
(2.55)	0.0042	(0.20)	0.1318	(3.00)	−0.0168	(−0.54)
(10.81)	0.0376	(5.01)	0.3185	(24.46)	0.0484	(4.56)
(−2.80)	−0.0060	(−1.74)	−0.0366	(−5.16)	−0.0020	(−0.38)
1992		1993		1994		1995
(55.01)	0.9400	(129.00)	0.7879	(80.65)	0.8506	(109.69)
(−16.56)	−0.0419	(−7.76)	−0.0312	(−4.07)	0.0001	(0.01)
(2.35)	−0.0605	(−3.67)	0.0170	(0.99)	−0.2261	(−7.78)
(3.68)	0.0078	(0.60)	0.0092	(0.46)	0.0087	(0.49)
(20.52)	0.0389	(8.72)	0.0400	(5.93)	0.0128	(2.22)
(−6.83)	0.0004	(0.14)	−0.0041	(−1.00)	−0.0088	(−2.33)
1985		1986		1987		1988
(−0.64)	0.4560	(11.53)	1.4064	(15.73)	0.1361	(2.39)
(1.90)	0.5807	(16.32)	1.4242	(22.63)	0.2232	(1.81)
(1.00)	−0.0481	(−0.52)	0.4496	(3.34)	−0.1275	(−1.27)
(−3.00)	−1.1567	(−8.97)	−2.1278	(−13.76)	−0.4110	(−3.88)
(−0.61)	−0.4766	(−15.93)	−1.1379	(−19.17)	−0.1875	(−5.26)
(−1.13)	0.0234	(1.53)	−0.1436	(−7.15)	0.0504	(3.19)
1992		1993		1994		1995
(4.43)	−0.0441	(−1.43)	0.0110	(0.30)	0.0593	(2.54)
(8.03)	0.0207	(0.96)	0.0956	(4.04)	0.1448	(7.38)
(1.40)	0.0496	(0.94)	0.1980	(3.42)	0.0107	(0.24)
(−9.89)	−0.3305	(−7.01)	−0.5631	(−8.75)	−0.4232	(−7.20)
(−8.36)	0.0347	(1.88)	−0.0233	(−1.08)	−0.0918	(−6.35)
(6.71)	−0.0186	(−2.09)	−0.0100	(−1.08)	−0.0174	(−2.18)
1985		1986		1987		1988
(96.22)	1.1092	(99.26)	1.0014	(104.09)	1.0388	(62.96)
(−3.86)	−0.0124	(−3.12)	0.0055	(1.26)	−0.0242	(−1.59)
(−3.28)	−0.0151	(−3.01)	0.0038	(0.56)	−0.0372	(−3.83)
(4.31)	0.0135	(2.04)	0.0013	(0.51)	0.1179	(8.11)
(4.08)	0.0129	(3.27)	−0.0060	(−1.36)	0.0366	(5.97)
1992		1993		1994		1995
(132.80)	1.0291	(137.21)	0.9205	(133.44)	0.9359	(85.55)
(−3.19)	−0.0048	(−1.59)	−0.0036	(−1.26)	0.1295	(26.47)
(−2.34)	−0.0052	(−1.15)	−0.0058	(−1.18)	−0.0272	(−4.30)
(6.29)	0.0619	(6.30)	0.0232	(3.35)	−0.1064	(−5.39)
(3.31)	0.0042	(1.57)	0.0037	(1.36)	0.0305	(7.46)

Continued on following page

TABLE 6.1 (Continued)

3SLS	SCE						
		1982		1983		1984	
SDIV	−0.3947	(−4.75)	−0.3322	(−7.00)	0.3070	(2.26)	−0.5431
SED	1.2849	(15.13)	1.4101	(13.63)	0.6962	(10.11)	3.3587
SRD	−0.2445	(−1.75)	−0.0841	(−0.97)	−0.3074	(−3.61)	−0.0907
SNI	0.3833	(5.58)	0.4374	(7.63)	−0.0539	(−0.82)	0.6751
SLIQ	−0.0726	(−6.24)	−0.0632	(−4.91)	−0.0187	(−1.58)	−0.0471
SLSAL	−0.0260	(−4.08)	−0.0425	(−4.55)	0.0133	(1.85)	−0.0749
SDEP	0.6945	(5.72)	1.4600	(15.52)	1.6766	(15.84)	2.4467
		1989		1990		1991	
SDIV	−0.0698	(−4.28)	−0.4895	(−5.56)	0.5473	(7.09)	−0.9211
SED	0.4426	(8.90)	2.4429	(16.92)	2.6589	(29.25)	1.9194
SRD	−0.0791	(−2.40)	−0.1644	(−1.55)	−0.4263	(−3.93)	−0.1312
SNI	0.1103	(7.26)	0.5021	(10.69)	0.1250	(4.11)	0.8664
SLIQ	−0.0224	(−4.12)	−0.0447	(−4.68)	−0.0463	(−5.87)	−0.0650
SLSAL	0.0006	(0.29)	−0.0330	(−4.03)	−0.0172	(−4.05)	−0.0357
SDEP	1.2201	(28.47)	0.8573	(7.45)	0.3969	(3.04)	0.8180

3SLS	SDIV						
		1982		1983		1984	
SLDIV	1.3756	(90.66)	0.5647	(273.30)	0.2635	(16.81)	1.3167
SCE	−0.1279	(−15.18)	0.0191	(2.59)	−0.0666	(−2.98)	−0.0778
SED	0.1879	(14.67)	−0.0259	(−3.32)	−0.0568	(−2.61)	0.0275
SRD	−0.0170	(−0.95)	−0.0140	(−1.32)	0.0779	(2.55)	0.0468
SNI	0.1031	(12.72)	0.0461	(9.19)	0.3172	(27.76)	0.0699
SLIQ	−0.0216	(−10.44)	−0.0002	(−0.13)	−0.0363	(−7.45)	−0.0090
		1989		1990		1991	
SLDIV	0.7043	(54.31)	0.4332	(35.61)	0.5380	(24.79)	0.5925
SCE	0.0540	(1.66)	−0.0912	(−4.21)	0.2522	(10.70)	−0.2032
SED	0.0296	(0.58)	−0.0373	(−0.81)	−1.0635	(−19.36)	0.0936
SRD	0.0867	(2.44)	0.1176	(3.96)	0.2156	(5.27)	0.0976
SNI	0.2057	(14.85)	0.2109	(15.94)	0.0921	(8.13)	0.2126
SLIQ	−0.0302	(−4.76)	−0.0273	(−4.85)	−0.0041	(−0.73)	−0.0334

3SLS	SED						
		1982		1983		1984	
SDIV	0.2807	(3.15)	0.1160	(3.91)	−0.5892	(−3.73)	0.1458
SCE	0.6575	(10.03)	0.5775	(6.26)	1.1232	(14.39)	0.2823
SRD	0.2942	(1.96)	0.1531	(1.81)	0.4662	(4.58)	0.0438
SDEP	−0.1186	(−0.87)	−0.8873	(−5.74)	−2.1082	(−14.62)	−0.7483
SNI	−0.2654	(−3.68)	−0.1864	(−4.46)	0.1318	(1.76)	−0.1962
DE	0.0341	(3.77)	0.0098	(1.32)	0.0168	(2.07)	0.0186

1985		1986		1987		1988	
(−5.39)	−0.7667	(−26.70)	−1.1902	(−48.19)	−0.2190	(−4.34)	
(14.30)	1.1274	(40.95)	0.4326	(36.52)	0.9555	(11.25)	
(−0.45)	0.0215	(0.24)	0.1027	(1.63)	0.0809	(0.96)	
(21.18)	0.7769	(63.36)	0.9053	(108.12)	0.2369	(6.99)	
(−3.32)	−0.0411	(−5.51)	−0.0375	(−5.08)	−0.0350	(−3.59)	
(−4.09)	−0.0221	(−5.09)	−0.0093	(−2.67)	−0.0078	(−1.51)	
(10.77)	1.9055	(20.00)	0.8745	(18.47)	0.5829	(8.66)	
1992		1993		1994		1995	
(−17.96)	−0.6223	(−22.98)	−0.6668	(−11.09)	−0.4794	(−8.51)	
(30.80)	0.4537	(5.62)	2.3188	(24.99)	2.9650	(32.97)	
(−1.45)	0.4212	(8.17)	−0.2060	(−1.90)	0.5908	(5.38)	
(57.96)	0.6637	(62.20)	0.6901	(37.07)	0.6142	(23.12)	
(−7.68)	−0.1761	(−16.92)	−0.0651	(−5.42)	−0.0123	(−1.40)	
(−4.71)	−0.0535	(−6.16)	−0.0992	(−7.26)	−0.0333	(−4.58)	
(13.60)	0.6303	(12.36)	2.0417	(18.08)	1.1831	(9.29)	

1985		1986		1987		1988	
(91.32)	0.9225	(102.03)	0.2763	(20.52)	0.7285	(58.75)	
(−11.92)	−0.0555	(−6.99)	−0.4846	(−39.19)	−0.0015	(−0.06)	
(1.42)	0.0640	(5.21)	0.2078	(23.01)	0.0007	(0.03)	
(2.66)	0.0106	(0.52)	0.1613	(3.72)	−0.0157	(−0.51)	
(11.27)	0.0568	(7.66)	0.4922	(43.56)	0.0478	(4.51)	
(−3.15)	−0.0062	(−1.81)	−0.0317	(−4.95)	−0.0025	(−0.49)	
1992		1993		1994		1995	
(53.04)	0.9326	(128.16)	0.7856	(80.42)	0.8740	(114.93)	
(−20.39)	−0.0513	(−9.56)	−0.0347	(−4.55)	0.0618	(8.61)	
(4.19)	−0.1150	(−7.20)	−0.0094	(−0.55)	−0.4101	(−16.80)	
(3.98)	0.0140	(1.07)	0.0181	(0.91)	−0.0511	(−3.00)	
(24.32)	0.0465	(10.47)	0.0428	(6.34)	−0.0234	(−4.35)	
(−7.57)	−0.0035	(−1.30)	−0.0076	(−1.87)	-0.0096	(−3.15)	

1985		1986		1987		1988	
(3.22)	0.6611	(19.77)	2.2811	(30.72)	0.1901	(3.41)	
(10.05)	0.8611	(36.79)	1.9673	(40.42)	0.8712	(13.34)	
(0.55)	0.0450	(0.51)	0.0538	(0.41)	−0.0581	(−0.61)	
(−7.09)	−1.7503	(−15.88)	−2.0956	(−15.86)	−0.4758	(−6.06)	
(−8.37)	−0.6763	(−32.95)	−1.7635	(−39.46)	−0.2243	(−6.66)	
(3.12)	0.0157	(2.23)	−0.0100	(−0.69)	0.0291	(3.44)	

Continued on following page

TABLE 6.1 (Continued)

	1989		1990		1991		
SDIV	0.0340	(1.00)	0.1790	(3.65)	−0.3197	(−8.90)	0.4274
SCE	1.0739	(11.90)	0.3764	(11.97)	0.3501	(19.38)	0.4609
SRD	0.1978	(2.97)	0.0693	(1.20)	0.1664	(3.34)	0.1083
SDEP	−1.4693	(−11.93)	−0.3218	(−5.35)	0.0368	(0.77)	−0.3984
SNI	−0.1050	(−3.31)	−0.2066	(−7.44)	0.0012	(0.08)	−0.4162
DE	−0.0092	(−0.84)	0.0241	(4.12)	0.0039	(1.50)	0.0165

3SLS	SRD						
	1982		1983		1984		
SLRD	1.1222	(27.43)	1.1316	(82.50)	1.1123	(95.38)	1.2095
SCE	−0.1438	(−9.09)	0.0011	(0.14)	−0.0029	(−0.37)	−0.0208
SDIV	0.0412	(2.05)	−0.0238	(−5.76)	−0.0820	(−4.82)	−0.0253
SED	0.3737	(18.96)	0.0475	(5.12)	0.0249	(3.45)	0.0906
SNI	−0.0330	(−2.03)	0.0363	(6.29)	0.0495	(6.39)	0.0213

	1989		1990		1991		
SLRD	1.0798	(122.68)	1.0635	(93.00)	1.0323	(110.17)	1.0119
SCE	−0.0035	(−0.44)	−0.0332	(−4.60)	−0.0639	(−13.50)	−0.0263
SDIV	−0.0186	(−4.41)	0.0043	(0.44)	0.0337	(4.88)	−0.0291
SED	0.0479	(3.80)	0.1707	(10.48)	0.2241	(16.44)	0.0724
SNI	0.0230	(6.02)	0.0031	(0.51)	0.0032	(1.15)	0.0252

CONCLUSIONS

An econometric model has been developed to determine interdependence between a firm's research, dividends, investment, and effective debt. The interdependence of decisions in these areas is of considerable importance to the manager who is interested in integrating them with research and development decisions. We find significant relationships between research, capital expenditure, dividend, and effective debt decisions in the United States. However, more work is necessary in examining the interdependence of financial decisions and how such decisions impact corporate expenditures and stockholder wealth.

REFERENCES

Ben-Zion, U. "The R&D and Investment Decision and Its Relationship to the Firm's Market Value: Some Preliminary Results." In *R&D, Patents, and Productivity,* edited by Z. Griliches. Chicago: University of Chicago Press, 1984.

Damon, W. W., and R. Schramm. "A Simultaneous Decision Model for Production, Marketing and Finance." *Management Sci.* 18 (1972):161–172.

1992		1993		1994		1995	
(13.03)	0.0671	(2.18)	0.2661	(7.94)	0.1512	(6.92)	
(32.04)	0.1895	(8.97)	0.4071	(28.24)	0.3302	(33.21)	
(2.10)	−0.0255	(−0.49)	0.1067	(1.89)	−0.2222	(−5.28)	
(−11.74)	−0.4523	(−10.56)	−0.9089	(−15.36)	−0.3487	(−7.53)	
(−26.11)	−0.1054	(−5.80)	−0.2888	(−19.98)	−0.1996	(−18.18)	
(3.88)	0.0230	(2.92)	0.0175	(3.54)	0.0044	(1.57)	
1985		1986		1987		1988	
(96.18)	1.1102	(99.35)	1.0025	(104.21)	1.0433	(63.32)	
(−5.16)	−0.0141	(−3.57)	0.0015	(0.34)	−0.0820	(−5.82)	
(−3.76)	−0.0168	(−3.36)	−0.0016	(−0.23)	−0.0488	(−5.03)	
(7.61)	0.0154	(2.32)	0.0030	(1.20)	0.1960	(15.81)	
(5.26)	0.0147	(3.71)	−0.0020	(−0.44)	0.0503	(8.32)	
1992		1993		1994		1995	
(132.74)	1.0335	(137.83)	0.9227	(133.88)	0.9074	(83.90)	
(−7.86)	−0.0152	(−5.00)	−0.0123	(−4.34)	0.1572	(34.15)	
(−5.26)	−0.0101	(−2.24)	−0.0126	(−2.58)	−0.0121	(−1.93)	
(10.21)	0.1029	(11.20)	0.0392	(5.75)	−0.1896	(−10.30)	
(7.60)	0.0102	(3.84)	0.0107	(3.96)	0.0109	(2.77)	

Dhrymes, P. J. *Econometrics: Statistical Foundations and Applications,* New York: Springer-Verlag, 1974.

Dhrymes, P. J., and M. Kurz. "Investment, Dividends, and External Finance Behavior of Firms." In *Determinants of Investment Behavior,* edited by Robert Ferber. New York: Columbia University Press, 1967.

Dhrymes, P. J., and M. Kurz. "On the Dividend Policy of Electric Utilities." *Rev. Economics and Statist.* 46 (1964):76–81.

Fama, E. F. "The Empirical Relationship Between the Dividend and Investment Decisions of Firms." *Amer. Economic Rev.* 63 (1974):304–318.

Grabowski, H. G. "The Determinants of Industrial Research and Development: A Study of the Chemical, Drug and Petroleum Industries." *Journal of Political Economy* 76 (1968):292–306.

Grabowski, H. G., and D. C. Mueller. "Managerial and Stockholder Welfare Models of Firm Expenditures." *Rev. Economics and Static.* 54 (1972):9–24.

Guerard, J. B., Jr., and G. M. McCabe. "The Integration of Research and Development Management into the Firm Decision Process." In *Management of R&D and Engineering,* edited by D. F. Kocaoglu. Amsterdam: North-Holland, 1992.

Guerard, J. B., Jr., and B. K. Stone. "Strategic Planning and the Investment-Financing Behaviour of Major Industrial Companies." *Journal of the Operational Research Society* 38 (1987):1039–1050.

Guerard, J. B., Jr., A. S. Bean, and S. Andrews. "R&D Management and Corporate Financial Policy." *Management Science* 33 (1987):1419–1427.

Hambrick, D. C., I. C. MacMillan, and R. R. Barbosa. "Changes in Product R&D Budgets." *Management Sci.* 29 (1983):757–769.

Higgins, R. C. "The Corporate Dividend-Saving Decision." *J. Financial and Quantitative Anal.* 7 (1972):1527–1541.

Jalilvand, A., and R. S. Harris. "Corporate Behavior in Adjusting to Capital Structure and Dividend Targets: An Econometric Study." *J Finance* 39 (1984):127–145.

Lintner, J. "Distributions of Incomes of Corporations Among Dividends, Retained Earnings and Taxes." *Amer. Economic. Rev.* 46 (1979):119–135.

Mansfield,, E. "R&D and Innovation." In *R&D, Patents, and Productivity*, edited by Z. Griliches. Chicago: University of Chicago Press, 1984.

Mansfield, E. "Composition of R&D Expenditures: Relationship to Size of Firm, Concentration, and Innovative Output." *Review of Economics and Statistics* 63 (1981a):610–615.

Mansfield, E. "How Economists See R&D." *Harvard Business Review* 59 (1981b):98–106.

Mansfield, E. "Size of Firm, Market Structure and Innovation." *J. Political Economy* 71 (1963):556–576.

McCabe, G. M. "The Empirical Relationship Between Investment and Financing: A New Look." *J. Financial and Quantitative Anal.* 14 (1979):119–135.

McDonald, J. G., B. Jacquillat, and M. Nussenbaum. "Dividend, Investment, and Financial Decisions: Empirical Evidence on French Firms." *J. Financial and Quantitative Anal.* 10 (1975):741–755.

Meyer, J. R., and E. Kuh. *The Investment Decision.* Cambridge: Harvard University Press, 1957.

Miller, M., and F. Modigliani. "Dividend Policy, Growth, and the Valuation of Shares." *J. Business.* 34 (1961):411–433.

Mueller, D. C. "The Firm Decision Process: An Econometric Investigation." *Quart. J. Economics* 81 (1967):58–87.

Peterson, P., and G. Benesh. "A Reexamination of the Empirical Relationship Between Investment and Financial Decisions." *J. Financial and Quantitative Anal.* 18 (1983):439–454.

Scherer, F. M. "Firm Size, Market Structure, Opportunity, and the Output of Patented Inventions." *Amer. Economic Rev.* 55 (1965):1104–1113.

Switzer, L. "The Determinants of Industrial R&D: A Funds Flow Simultaneous Equation Approach." *Rev, Economics and Statist.* 66 (1984):163–168.

Tinbergen, J. *A Method and Its Application to Investment Activity.* Geneva: League of Nations, 1939.

7
HISTORICAL DATA AND ANALYSTS' FORECASTS IN THE CREATION OF EFFICIENT PORTFOLIOS

This chapter demonstrates that statistically significant stock selection models can be developed using historical fundamental data and earnings breadth, forecasts, and revisions. Earnings forecasting creates portfolios that dominate the historical model analysis. We find that composite models using fundamental and earnings forecasting data substantially outperformed the market during the 1982–1994 period. Weighted regression is used in this study to create Japanese and U.S. portfolios.

Here we address several issues in financial modeling: (1) the development and estimation of composite models using fundamental data, such as earnings, book value, cash flow, and sales, found in the Compustat database and concensus analysts forecasting data found on the I/B/E/S database; (2) the question of weighting variables by regression or equal weighting of factors in estimating backtest models; (3) the estimation of an multifactor risk model; and (4) the construction of mean-variance efficient portfolios using the composite models scores that historically produced statistically significant excess returns. In this study we use I/B/E/S earnings forecasts for the 1982–1994 period. We will show that the use of outlier-adjusted regression techniques yield statistically significant composite models that produce significant excess returns. These models are used as inputs to a mean-variance optimization process, generating statistically significant backtested excess returns annually in the United States during the 1982–1994 period, very similar to the modeling and optimization results reported for U.S. and Japanese equities in Guerard, Takano, and Yamane (1993) for the 1976–1990 and 1974–1990 periods, respectively. We find that the index-tracking optimization techniques of Guerard, Takano, and Yamane (1993), which were extremely useful, can be replicated by merely increasing the number of securities in an efficient portfolio. A manager may often be more concerned with underperforming an index than simply maximizing the portfolio geometric mean. A four-quarter outlier-adjusted regression model is estimated and shown to be the superior model (of our tested group) in the U.S. market. It is interesting to note that the real-time results have substantiated the portfolio construction process.[1] We model the 3,000 largest firms in the United States in June 1982 and expand the sample to include the largest 3,000 securities every June during the 1983–1994 period.

In this study we use traditionally analyzed fundamental variables. The earnings-to-price (EP), book-to-price (BP), cash flow-to-price (CP), and sales-to-price (SP) ratios are used in this analysis, just as they were in Chan, Hamao, and Lakonishok (1991), Fama and French (1992, 1995), Guerard (1990), Guerard and Stone (1992), Guerard, Takano, and Yamane (1993), Jacobs and Levy (1988), and Ziemba (1990, 1992). The dividend yield (DY) and net current asset value (NCAV) are also used in this chapter. The net current asset value is defined as current assets less all liabilities of the firm. The NCAV was advocated as a fundamental variable in Graham and Dodd (1962) and Vu (1990). We also use the current fundamental ratio relative to its five-year mean (the "relative" variables, denoted by "R" preceding the fundamental ratio; that is, REP is the relative EP). A consensus analyst growth variable, PRGR, is constructed using the mean FY1 (fiscal year 1) and FY2 (fiscal year 2) forecasted earnings-to-price, monthly revisions of FY1 and FY2, and the forecast breadths (number of upward revisions less number of downward revisions, divided by total forecasts). Much of the development of this variable is found in Guerard, Gultekin, and Stone (1996), who found that (1) I/B/E/S breadth dominated forecasts and revisions; (2) FY1 data was more significant than FY2 data in generating total returns; (3) quarterly models dominated monthly models; and (4) the composite model information coefficients (ICs) produced statistically significant portfolio excess returns.

Here we reexamine and expand the Guerard, Gultekin, and Stone (1996) analysis by estimating several forms of composite models integrating fundamental data and earnings expectational data.

$$TR_T = a_0 + a_1 EP_t + a_2 BP_t + a_3 CP_t + a_4 SP_t + a_5 REP_t + a_6 RBP_t + a_7 RCP_t + a_8 RSP_t + e_t \quad (7.1)$$

where

TR$_T$ = total returns in the subsequent period from the Center for Research in Security Prices (CRSP) file
EP, BP, CP, SP, REP, RBP, RCP, RSP, PRGR = previously defined fundamental financial variables
$a_0, a_1, \ldots, a_9$ = regression parameters
e = randomly distributed error term

$$\text{TR}_T = a_0 + a_1\text{EP}_t + a_2\text{BP}_t + a_3\text{CP}_t + a_4\text{SP}_t + a_5\text{REP}_t + a_6\text{RBP}_t + a_7\text{RCP}_t + a_8\text{RSP}_t + a_9\text{PRGR}_t + e_t \quad (7.2)$$

$$\text{TR}_T = a_0 + a_1\text{EP}_t + a_2\text{BP}_t + a_3\text{CP}_t + a_4\text{SP}_t + a_5\text{DY}_t + a_6\text{NCAV}_t + a_7\text{REP}_t + a_8\text{RBP}_t + a_9\text{RCP}_t + a_{10}\text{RSP}_t + a_{11}\text{RDY}_t + a_{12}\text{RNCAV}_t + a_{13}\text{PRGR}_t + e_t \quad (7.3)$$

One can use ordinary least squares (OLS), an outlier adjustment, a robust (ROB) regression procedure proposed by Beaton and Tukey (1974), and latent root regression (LRR) for decomposing the correlation matrix into its nonpredictive near-singularities (addressing multicolinearity). Latent root regression on the robust-weighted data (referred to as weighted latent root regression (WLRR)) was used in Guerard, Takano, and Yamane (1993). Guerard and Stone (1992) found that the WLRR technique produced the lowest out-of-sample forecasting errors in analyzing composite earnings models using time series models and consensus I/B/E/S forecasts. Guerard, Takano, and Yamane (1993) and Bloch et al. (1993) found that the WLRR technique produced higher geometric means in the first-section Tokyo Stock Exchange (TSE) securities than alternative formulations in Japan and the United States. We use the Beaton-Tukey (1974) robust procedure in this study because the regression condition numbers are in the 5 to 15 range, substantially less than the value of 30.0, advanced by Belsley, Kuh, and Welsch (1980) as being indicative of severe multicolinearity. The average observation has a weight of 0.92, quite consistent with the Guerard and Stone (1992) estimation of U.S. returns during the 1981–1985 period. The OLS regression results of equation (7.1) are shown in Table 7.1, and the ROB regression results of equation (7.1) are shown in Table 7.2.

The regression results may be summarized as follows:

1. The robust (ROB) regression results, shown in Table 7.2, dominate the ordinary least squares (OLS) in terms of producing higher adjusted goodness-of-fit measures (R^2 and F-statistics).
2. The EP, SP, REP, RBP, and RSP variables are consistently positive and statistically significant in equation (7.1).

The robust regression results of equation (7.2) are shown in Table 7.3 and indicate the importance of the proprietary growth variable, PRGR. The I/B/E/S variable complements the EP, SP, REP, RBP, and RSP variables, supporting the complete regression results of Guerard, Gultekin, and Stone (1996) modeling results. If one uses robust regression techniques to estimate equation (7.3), shown in Table 7.4, one finds no support for the addition of the DY and NCAV or RNCAV variables, but substantial evidence to incorporate the RDY variable. The quarterly regression results support the use of the earnings-to-price and sales-to-price variables as well as the relative EP, BP, SP, and DY variables and the proprietary growth variable. We will use the proprietary growth variable in Chapters 8 and 9.

Guerard, Gultekin, and Stone (1996) introduced a 60-month excess return beta and found that the beta is negative as often as it is positive; however, if one examines the periods when the subsequent 3-months total returns are negative, the beta coefficient is negative and statistically significant in 18 or 20 quarters, as was found in Grundy and Malkiel (1996). Moreover, if the coming 3-month subsequent total returns are positive, the beta coefficient is positive and statistically significant in 18 of 21 quarters.

One may question whether the I/B/E/S variable works better for smaller stocks than for larger stocks. We ran equation (7.3) using three samples for the 1984–1994 period: (1) the largest 1,000 market-capitalized securities in each quarter; (2) market-capitalized securities with ranks 1,001–1,999 (the "mid-cap" portfolio concept); and (3) securities with market capitalizations exceeding 2,000 (the smallest stocks). We found little differences in the average PRGR coefficients for the three samples, as reported in Guerard, Gultekin, and Stone (1996). The middle market-capitalized model has the highest growth coefficient (0.176), versus 0.128 and 0.123 for the largest and smallest securities, respectively. There does not appear to be an extreme small-stock bias in our analysis, a result consistent with our socially responsible investing results shown in Chapter 9 (see Table 7.5).

COMPOSITE MODEL ESTIMATIONS

In our initial analysis, we create quarterly fundamental ratios by using annual earnings, book value, cash flow and sales divided by quarterly prices. In our composite modeling process, we use the quarterly regressions using the previously discussed ROB procedure to determine which variables were significantly associated with total returns in the previous quarter. We create composite weights by (1) identifying those variables that have coefficients in accord with economic theory (i.e., positive); (2) using the variables in the next quarterly composite model that were statistically significant at the 10% level during the latest quarterly cross-sectional regression; and (3) averaging the weights over the last four quarters (leading Guerard and Takano (1991) to refer to the quarterly models with the "Q_p4" designation). The justification for the four-quarter weighting scheme can be found in Bloch, Guerard, Markowitz, Todd and Xu (1993). We use the forecasted I/B/E/S data in creating the final composite model weights and rankings. The four-quarter weights are shown in Figures 7.1 through 7.8 for the public model. It is interesting to note that the book-to-price ratio, the fundamental variable found by Chan (1991) to be most statistically associated with security returns in Japan, and by Fama and French (1992, 1995) to be the most important variable in the United States, has an average of 0.022. We find that the SP and RSP variables are more important than the BP variable. The PRGR variable weights average approximately 33%.

The regression-weighted composite model is statistically significant at identifying undervalued securities, as it produces an information coefficient (IC), measuring the association between the ranked composite model score

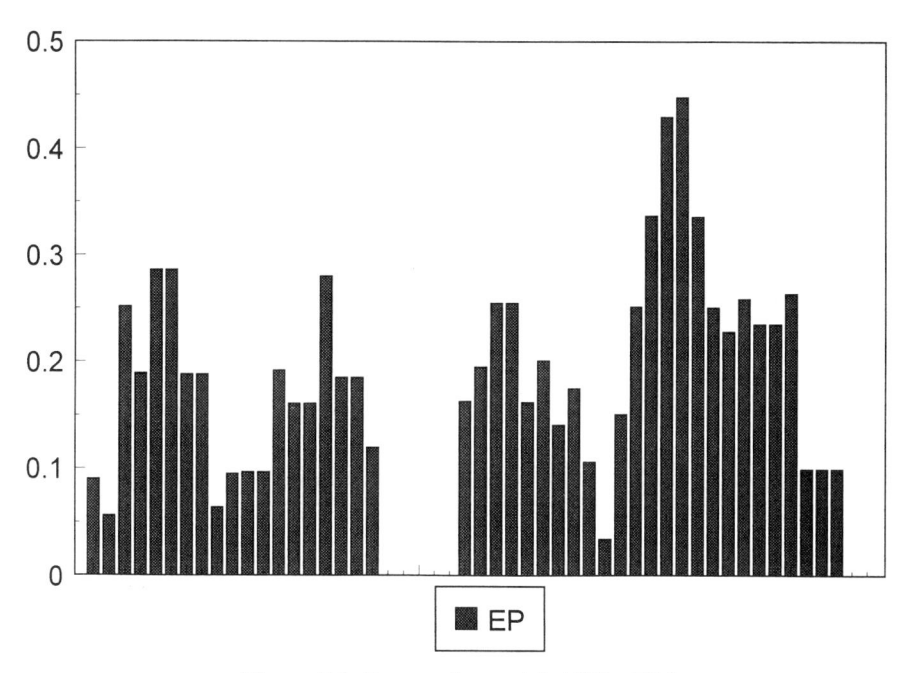

Figure 7.1 Composite model, 1982–1994.

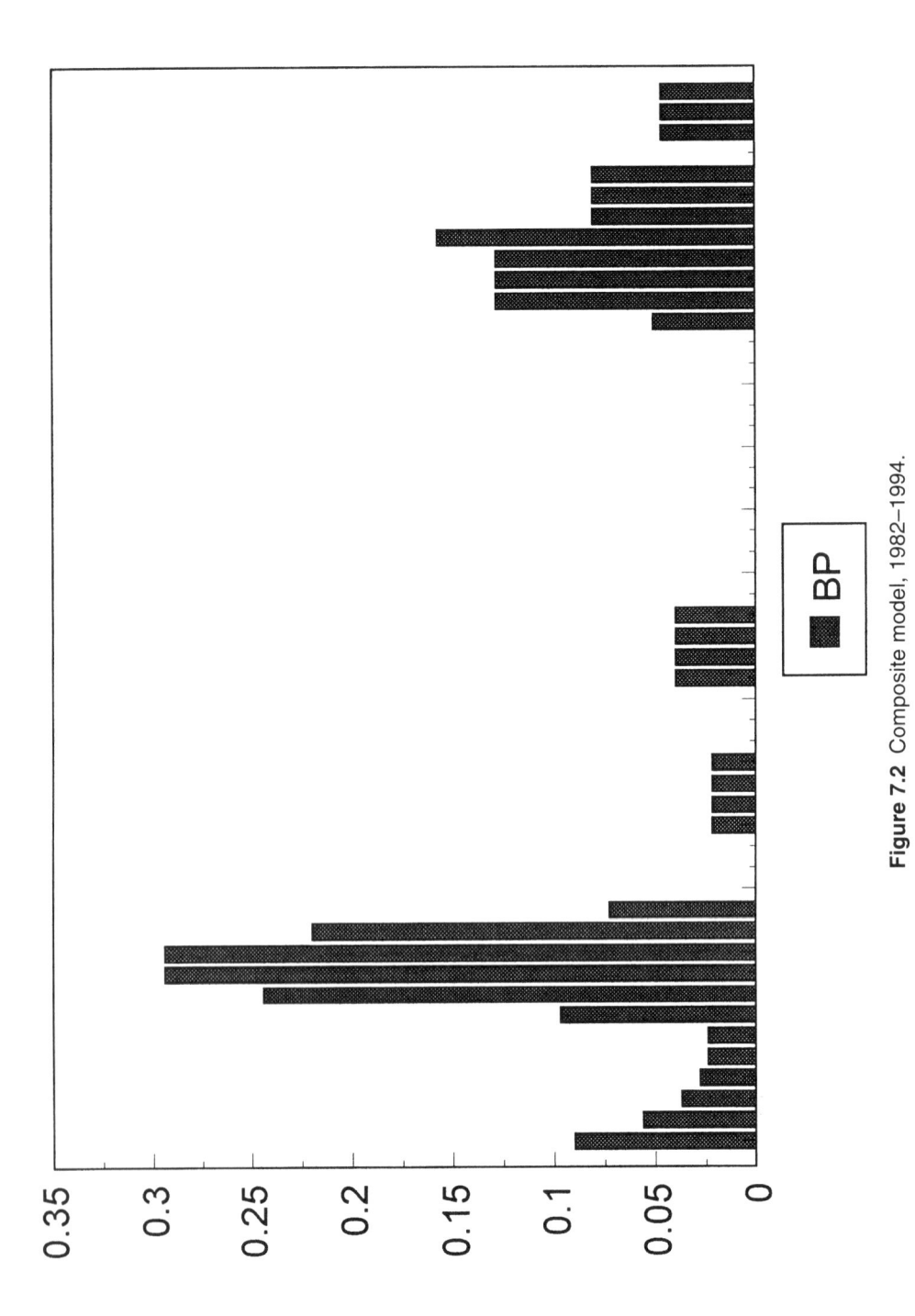

Figure 7.2 Composite model, 1982–1994.

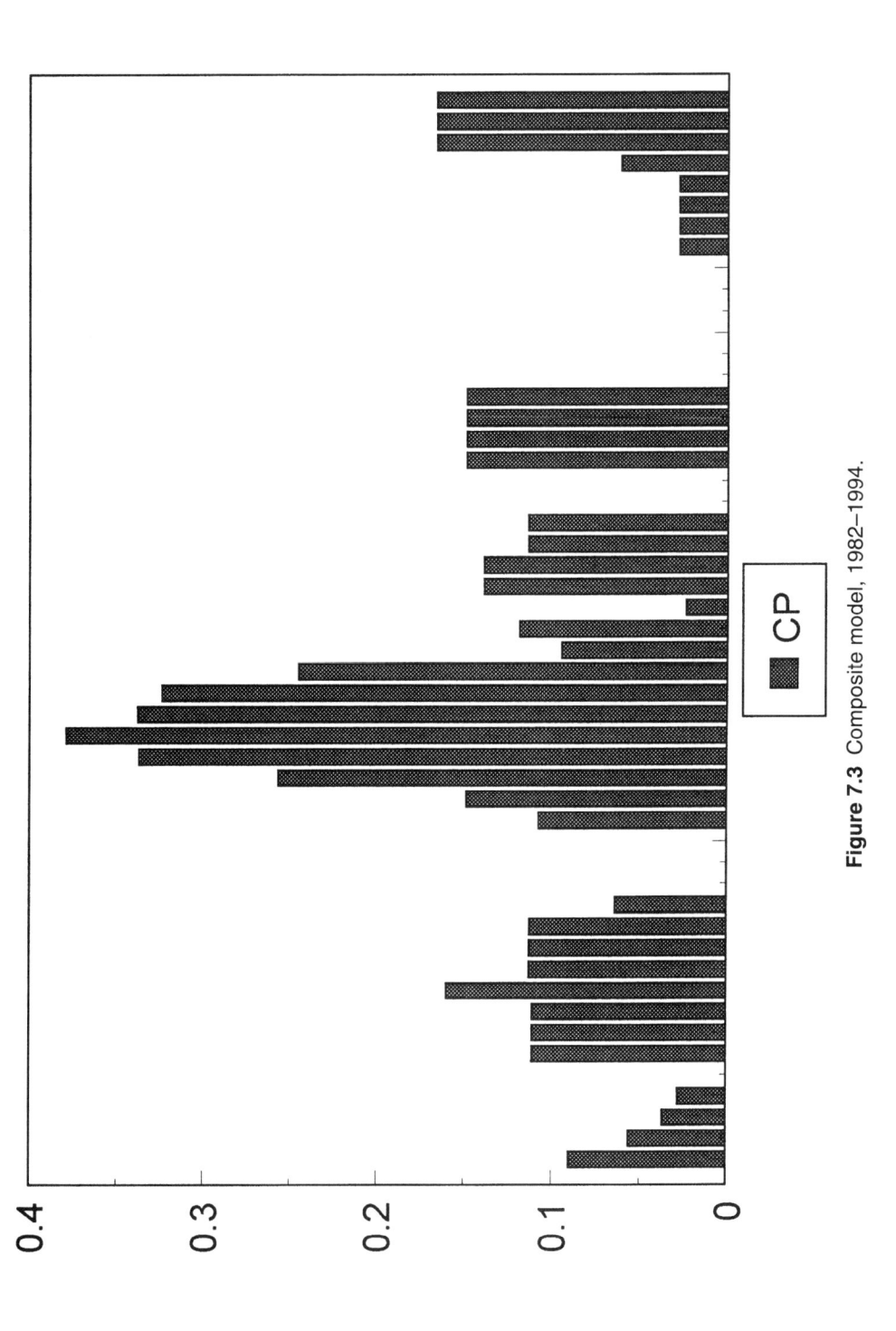

Figure 7.3 Composite model, 1982–1994.

146 HISTORICAL DATA AND ANALYSTS' FORECASTS

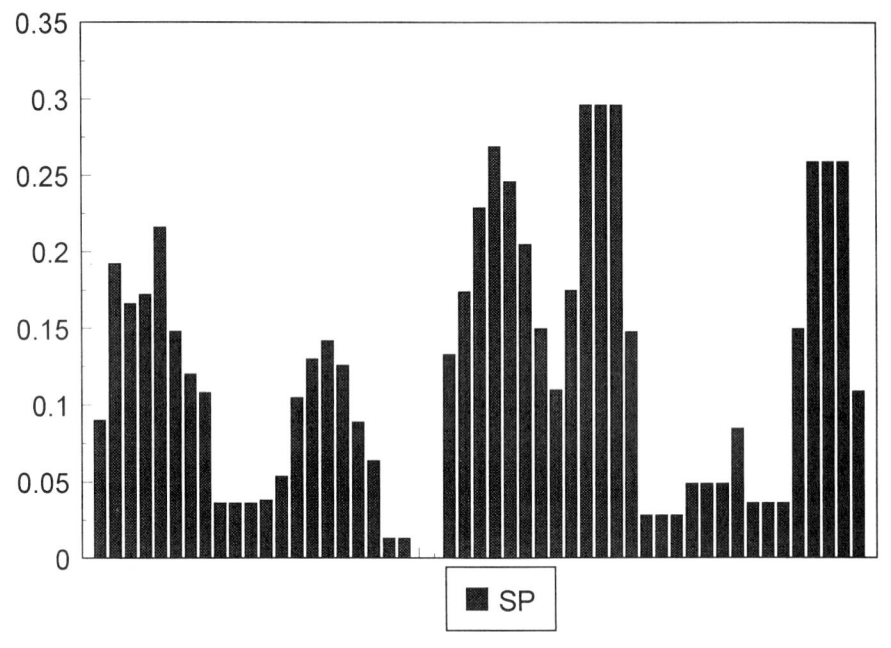

Figure 7.4 Composite model, 1982–1994.

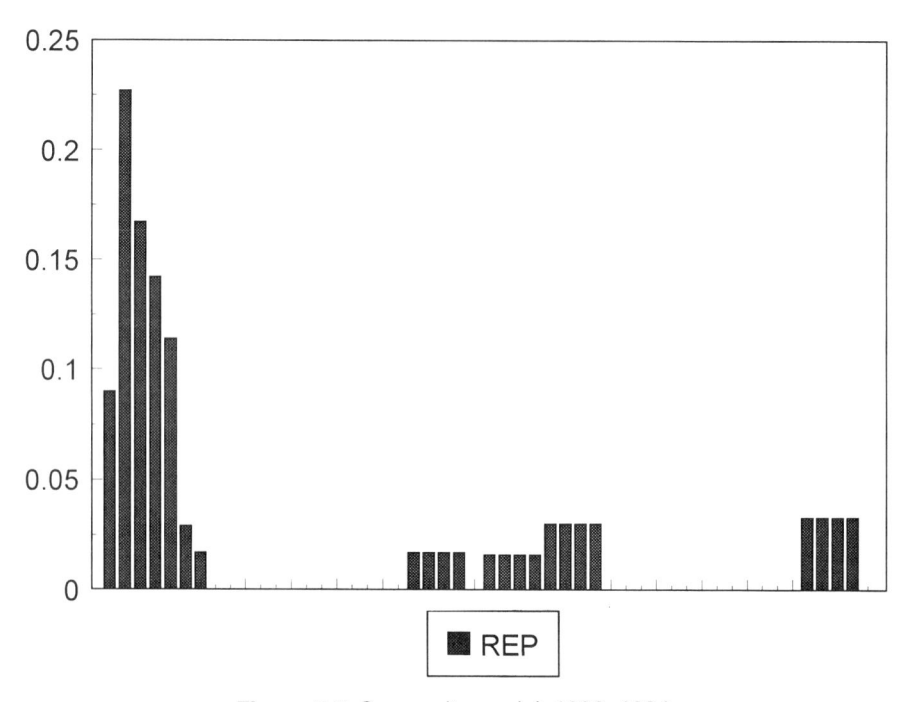

Figure 7.5 Composite model, 1982–1994.

COMPOSITE MODEL ESTIMATIONS **147**

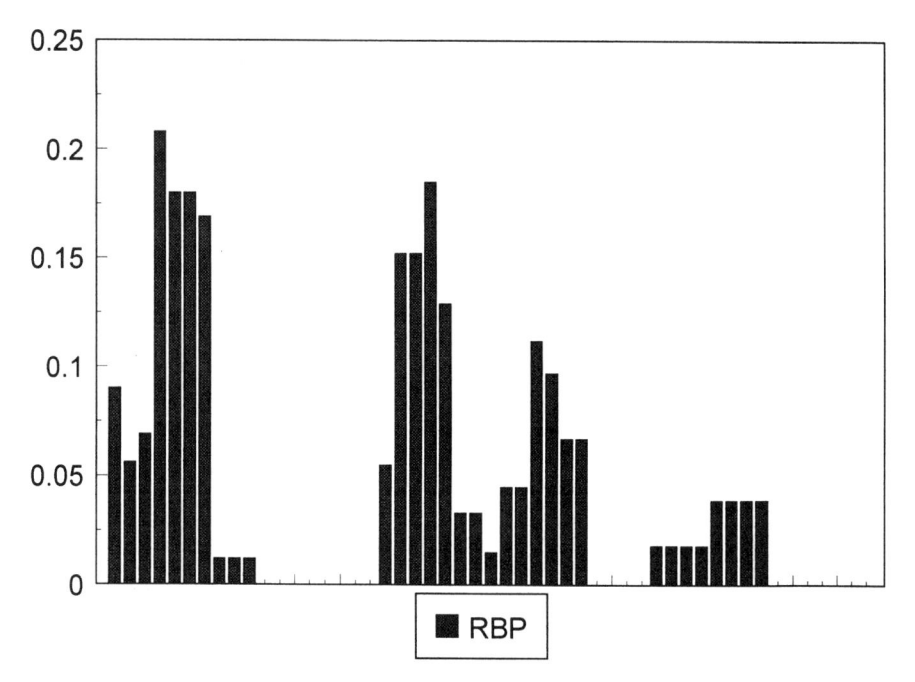

Figure 7.6 Composite model, 1982–1994.

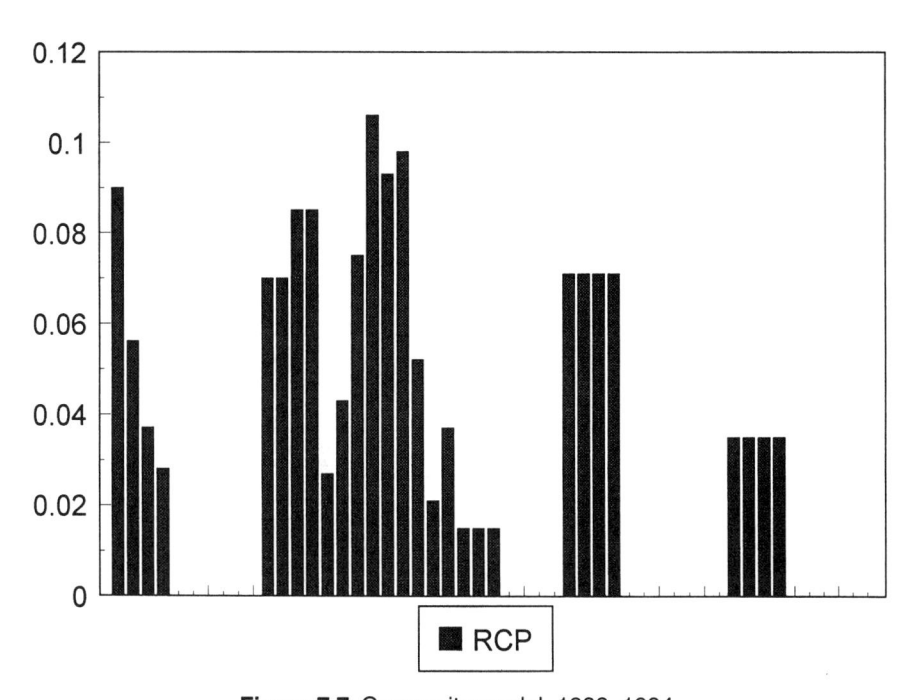

Figure 7.7 Composite model, 1982–1994.

Figure 7.8 Composite model, 1982–1994.

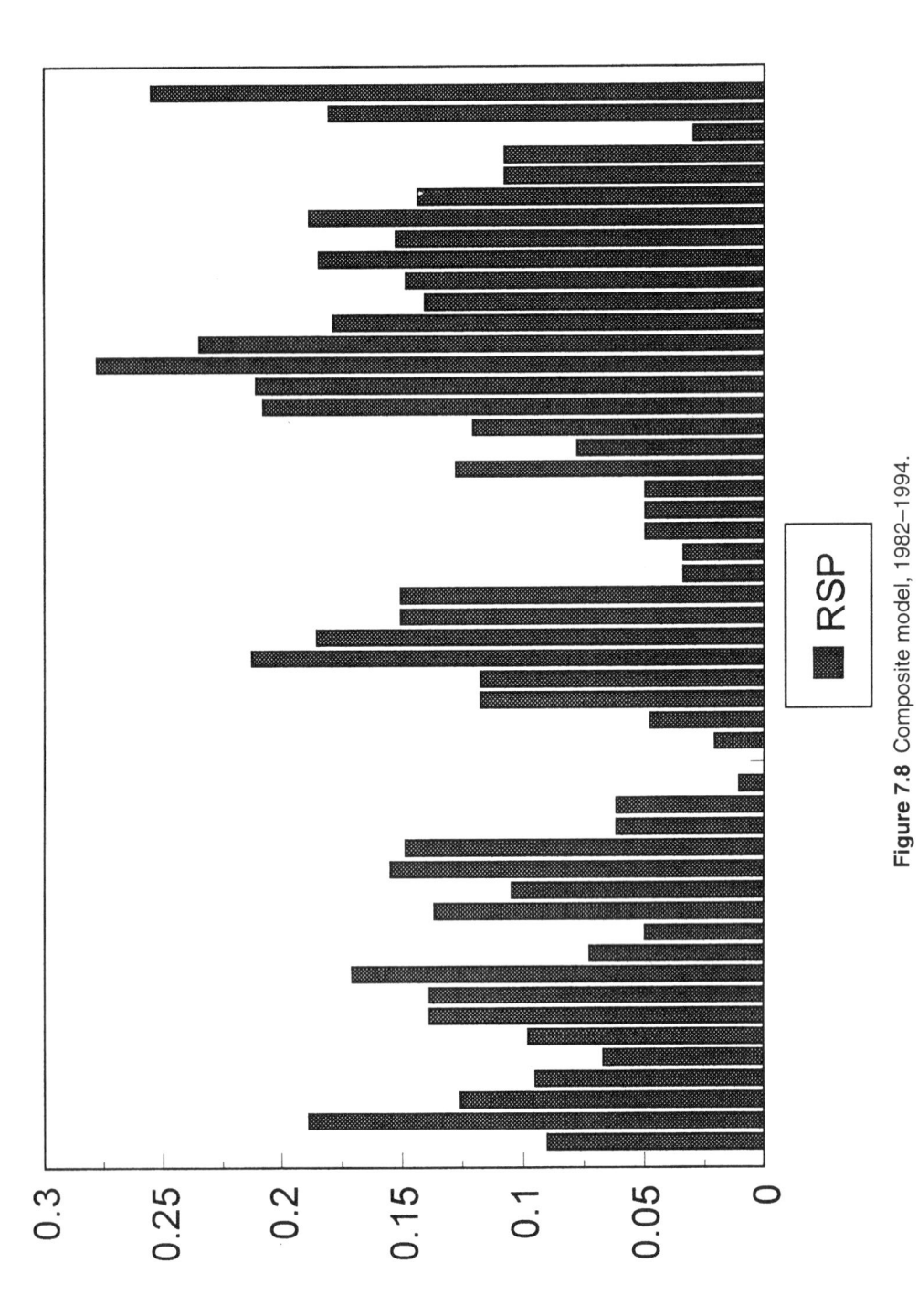

and ranked subsequent total returns of 0.08 with an average *t*-value of 4.71. The equally weighted composite model produced an IC of 0.05 and an average *t*-value of 3.89.[2] The robust regression technique creates a more effective ranking procedure that should produce higher excess returns when one simulates the strategy.

ARBITRAGE PRICING MODEL THEORY AND ESTIMATION

The first part of this section develops the formulas for assessing conditional expected returns, assuming that the return-generating process for securities is jointly normal, stationary, and independent over time and that the parameters of the joint distribution are known. The second part incorporates factor models into the formulas and shows that the effect of factor models is to place restrictions on the estimated covariance matrix. To introduce some notation that draws heavily from Blume, Gultekin, and Gultekin (1990), let r_i be the return on asset i less its unconditional expectation; s_{ii}, is the variance of the return on asset i, and s_{ij} is the covariance between the returns of asset i and asset j; there are N assets.

If one assumes normality, the expectation of rj conditional on the returns of the remaining (N − 1) assets is a linear function of these remaining returns. Specifically, if R^i is the vector of returns with the return of asset i deleted, the conditional expected return is given by:

$$E[r_i | R^i] = \Sigma \, w_k \, r_k \qquad (7.4)$$

where w_k are weights appropriate to asset k, and S is the summation function where k is not equal to i. From normal theory, the weights themselves are given by:

$$W^i = (\Sigma^i)^{-1} C^i \qquad (7.5)$$

If an investor's loss function is quadratic, the natural measure of loss is the mean-squared error. The W^i is a column vector of the (N − 1) weights, C^i is a column vector of the covariances of the returns of asset i with respect to each of the other (N − 1) assets, and Σ^i is a square matrix with dimension (N − 1) obtained by deleting the ith row and ith column of the full covariance matrix of all N securities. The weights, given by equation (7.5), have the important property that they minimize the variance of r_i conditional on R^i. This is not a surprising result, since these weights are nothing more than the expected value of the estimated coefficients of a regression of r_i on the returns of the remaining (N − 1) assets. The essence of least-squares regression is to minimize mean-squared errors. Thus, the process of estimating a least-squares regression can be viewed as consisting of two steps. First, estimate the covariance matrix of the dependent and independent variables. Second, use this estimated matrix to estimate the regression coefficients, which can then be used as the weights in equation (7.5). Viewing a regression this way helps to clarify the role of factor models in forming conditional expectations.

Using a factor model to assess conditional expected returns is similar to a regression but with an important exception: factor models place restrictions on the structure of the covariance matrix of returns, whereas the usual least-squares regression places no restrictions on this matrix. To develop these restrictions, consider the factor model:

$$r_{it} = \sum_{k=1}^{K} a_{ik} k_{ft} + n_{it} \tag{7.6}$$

where K is the number of factors, a_{ik} is the so-called factor loading of asset i on factor k, k_{ft} is the score or value of factor k during interval t, and n_{it} is a mean-zero independent disturbance. The expected value of the factor score is zero, and the covariances between the different factors are zero. Within the estimation period and within the class of linear estimators, estimates of the conditional expected returns for asset i that place no restrictions on the estimated covariance matrix will mathematically produce the minimum mean-squared errors. However, outside the estimation period, there is no guarantee that such an unrestricted estimate of the covariance matrix will yield the minimum mean-squared errors, or even the minimum expected mean-squared errors. If the restrictions that factor models impose on the covariance matrix are valid, it is possible that calculating conditional expected returns using a covariance matrix estimated with restrictions will yield lesser mean-squared errors in the prediction period than using an unrestricted estimate.

Nonstationarities complicate the story. Without restrictions, an estimate of the covariance matrix may "discover" nonexistent relations among the returns. With restrictions, an estimate of the covariance matrix may be less prone to discover nonexistent relations. In turn, it is possible that restrictions, even if not perfectly true, may improve the accuracy of conditional expectations in out-of-the-estimation period analysis. Validating various models with different data from those used in estimating the models provides some insight into these two issues: restrictions on the estimated covariance matrix and nonstationarities.

Blume, Gultekin and Gultekin (1990) used monthly returns of 57 sets of size-ranked portfolios of New York Stock Exchange (NYSE) stocks constructed from the Center for Research in Security Prices (CRSP) file to estimate their multifactor or Arbitrage Pricing Theory (APT) model. The first set consists of all securities in the CRSP files with complete data for the six years 1926 through 1931, five years for the estimation period and the sixth year for the forecast period. These securities were ranked by their market value as of December 1930 and then partitioned into 20 size-ranked portfolios with as close to an equal number of securities as possible. The experiment uses the portfolio returns from 1926 through 1930 to estimate the parameters of a particular model, and the 12 monthly returns in 1931 to validate the model. This process was repeated year by year to 1986. The estimation period for the last set is 1982–1986, and the validation year is 1987. The validation year will be used to identify a specific set, so that the first set is the 1931 set and the last set is the 1987 set. The total number of securities used in the analysis starts at 361 for the 1931 set, increases to 763 for the 1949 set, and gradually reaches 980 for the 1987 set. An analysis of the basic data discloses dramatic changes in the variability of the returns of the portfolios over time. The variability is greatest in the 1930s, as was found by Morgan and Morgan (1987), but even in the later years the variability does change somewhat from one year to the next.

Blume, Gultekin, and Gultekin (1990) (hereafter, BGG) found that the smaller portfolios display greater variability in returns than the larger portfolios. These changes in variability make summary measures of mean-squared errors misleading without some adjustment for these changes. BGG used the maximum likelihood method to estimate the factor models; the usual way to assess the number of required factors is to rerun the procedure, successively increasing the number of factors until the chi-squared test for the goodness of fit indicates that the number of factors is sufficient. To use this criterion, one must specify the level of significance, 5% in their test. The level of significance is important since there is a direct relation between the level of significance and the number of significant factors. However, there is no direct relation between this arbitrary level of significance and the criterion of minimizing the mean-squared errors in the forecast period. The number of required factors varies over time. More factors are required at the beginning and the end of the 1930–1986 period than in the midpart. Further analysis of the required number of factors reveals a positive relation between the number of factors and the variability of returns during the estimation period. Interestingly, there is little change over time in the relative size of the portfolio consisting of the largest stocks, even though the market value of all of the portfolios increased almost tenfold from 1930 through 1986. During periods of relatively low volatility, most of the volatility is firm-specific and it is difficult to identify the common factors. In more volatile times, the common factors are relatively more important than the firm-specific factors, making it easier to identify them. As mentioned before, if the process generating returns were stationary over time and correctly specified, one could rely on the sample statistics within the estimation period. However, if the process is nonstationary with respect to the specified process, validating the model with data different from those used in estimating the model provides insight into the usefulness of the model. Like the variability of the monthly returns, average mean-squared errors vary substantially for each 12-month predictive period as a function of both time and the number of factors. In view of this substantial variation, any summary measure of these mean-squared errors over time or portfolio size would be misleading without some form of scaling or normalization. The scale factor used in this study is the mean-squared error associated with a naive forecast.

The *naive forecast* is the average return for each portfolio in the estimation period, that is, an estimate of the unconditional expectation. An analysis of the scaled mean-squared errors shows that this normalization removes a large portion of the time trends in the annual mean-squared errors over time for a given portfolio size. However, substantial differences still remain among the size-ranked portfolios. As one moves from a one-factor to a two-factor model, BGG found that the mean-squared errors dropped dramatically for both large and small portfolios, while there was little change for the midsize. As one moves to the three- or, possibly, four-factor model, the mean-squared errors for the large and small portfolios drop further, though only slightly. In addition, the minimum mean-squared error for the midsize portfolios tends to occur with fewer than five factors. The mean-squared errors in the forecast period for the factor models selected by the chi-squared criterion are slightly greater than the mean-squared errors associated with the best-performing factor model in the forecast period for each portfolio size. The behavior of the mean-squared errors as a function of the number of factors leads to the conjecture that the arbitrary selection of two or three factors for midsize portfolios, and three or four factors for the largest and smallest portfolios, leads to lesser mean-squared errors than using the standard chi-squared test. The chi-squared criterion yields little difference in the mean-squared errors among different levels of commonly used significance, because any criterion that points to two to five factors will lead to similar mean-squared errors. With a significance level of 5%, the median number of factors over the 57 estimation periods is four; with a significance level of 10%, the median number is also four; and with a significance level of 20%, the median number is three. In contrast to the predictions in the validation period, BGG found that the average mean-squared errors decrease monotonically for each portfolio as the number of factors increases from one to five. On the surface, this result suggests that the greater the number of factors the better. However, the validation of the models with additional data shows that there is little difference between models with anywhere from two to five factors.[3] This difference in results between the estimation period and the validation period shows the value of validating a statistical model with data outside the estimation period.

BGG also found that the factor models produced smaller mean square forecasting errors than the market model or an APT model using the Chen, Roll, and Ross (1986) prespecified macroeconomics variables. The factor models generally produced lower mean squared forecasting errors for the equally weighted NYSE Index as opposed to the value-weighted NYSE index.

SIMULATING THE COMPOSITE MODEL STRATEGY

In this section we use the BGG maximum likelihood estimated k-factor models using five years of monthly data on 30 market-capitalization-ranked stock portfolios. The factors are reestimated quarterly during the 1983–1994 period. Portfolios of 50 stocks are created using the ROB regression-weighted expected returns model in equation (7.1). The 50-stock portfolios substantially outperform the Standard & Poors (S&P) 500 during the December 1983–June 1994 period, after subtracting 100 basis points (each way) for transactions costs and market impact. A $1.00 investment in the S&P 500 Index in December 1983 grows to $3.89 by the end of June 1994, whereas the investment increases to $8.71 through our composite model strategy. The portfolios' average 52 stocks have an average annual tracking of 3.55%, incur 17.4% quarterly turnover, and have an average quarterly return of 6.18%. The average quarterly excess return is 2.60%. An equally weighted universe return of the top 3,000 securities as of June of each year reaches a value of $3.35. The composite model strategy outperforms the equally weighted universe index in 60% of the quarters and the S&P 500 Index in 55% of the quarters, a result consistent with Guerard, Takano, and Yamane (1993). Moreover, we find that if one uses the I/B/E/S variable as a constraint, rather than the regression-based expected return model, one may obtain a better portfolio excess return of approximately 420 basis points (see Blin, Bender, and Guerard 1996).[4] The portfolios significantly outperform the CRSP equal-weighted and value-weighted indices.

TABLE 7.1 The Determinants of Security Returns, 1982–1994: OLS

1982	Q1		Q2		Q3		Q4	
EP	0.058	(0.79)	0.353	(6.06)	−0.179	(−2.76)	−0.034	(−0.62)
BP	−0.126	(−3.41)	0.024	(0.63)	−0.173	(−4.47)	0.063	(1.69)
CP	−0.077	(−1.21)	−0.383	(−6.81)	−0.015	(−0.25)	−0.176	(−3.20)
SP	0.171	(4.33)	0.126	(3.37)	0.210	(5.53)	0.284	(7.61)
REP	0.135	(1.13)	−0.101	(−1.15)	0.120	(1.47)	0.024	(0.30)
RBP	−0.050	(−0.74)	0.286	(3.88)	0.597	(7.15)	0.241	(2.56)
RCP	0.076	(1.02)	−0.135	(−1.24)	−0.010	(−0.10)	0.029	(0.34)
RSP	0.054	(1.26)	−0.096	(−2.48)	−0.163	(−4.06)	−0.109	(−2.63)
ADJ-R^2	0.010		0.024		0.040		0.058	
F-STAT	3.786		8.160		12.861		18.713	
1983	Q1		Q2		Q3		Q4	
EP	−0.169	(−6.21)	0.480	(6.41)	0.246	(3.45)	−0.301	(−3.44)
BP	−0.117	(−6.32)	0.285	(9.41)	0.167	(5.42)	0.201	(6.09)
CP	0.076	(2.67)	−0.208	(−3.10)	−0.109	(−1.79)	0.270	(4.50)
SP	0.068	(3.76)	0.010	(0.28)	0.141	(3.87)	−0.074	(−2.08)
REP	0.081	(1.66)	−0.024	(−0.43)	−0.003	(−0.05)	−0.004	(−0.07)
RBP	0.095	(3.54)	−0.191	(−3.37)	−0.266	(−4.26)	−0.244	(−3.52)
RCP	−0.111	(−0.79)	−0.080	(−0.91)	−0.159	(−1.86)	−0.202	(−2.73)
RSP	0.010	(0.39)	0.049	(1.24)	0.007	(0.17)	0.072	(1.53)
ADJ-R^2	0.049		0.074		0.049		0.045	
F-STAT	16.860		25.621		17.062		15.417	
1984	Q1		Q2		Q3		Q4	
EP	0.319	(4.15)	0.544	(6.92)	0.578	(6.18)	−0.106	(−1.22)
BP	−0.130	(−4.22)	0.040	(1.33)	0.041	(1.28)	−0.092	(−2.44)
CP	−0.110	(−1.98)	−0.174	(−2.68)	−0.032	(−0.47)	0.044	(0.60)
SP	0.070	(2.01)	0.120	(3.93)	0.088	(2.84)	0.108	(3.49)
REP	0.017	(0.21)	0.021	(0.28)	0.000	(0.00)	0.121	(1.54)
RBP	0.043	(0.26)	−0.163	(−0.85)	−0.499	(−2.73)	0.042	(0.22)
RCP	0.019	(0.37)	0.035	(0.73)	0.103	(2.02)	−0.101	(−2.02)
RSP	0.077	(2.39)	0.000	(0.01)	−0.139	(−4.98)	0.126	(3.86)
ADJ-R^2	0.014		0.032		0.070		0.021	
F-STAT	6.745		14.146		30.792		9.624	
1985	Q1		Q2		Q3		Q4	
EP	0.422	(7.10)	0.209	(2.67)	1.052	(10.95)	0.712	(7.40)
BP	−0.042	(−1.59)	−0.045	(−1.25)	−0.083	(−1.98)	0.053	(0.95)
CP	−0.120	(−2.47)	−0.055	(−0.83)	−0.569	(−7.16)	−0.581	(−7.33)
SP	0.030	(1.38)	0.074	(2.25)	0.150	(4.80)	0.227	(6.75)
REP	0.038	(1.24)	0.072	(1.39)	0.026	(0.45)	−0.088	(−1.67)
RBP	−0.173	(−4.83)	0.095	(1.49)	−0.048	(−0.83)	−0.175	(−2.93)
RCP	0.105	(1.79)	0.062	(0.75)	0.267	(3.74)	−0.008	(−0.11)
RSP	0.049	(1.80)	−0.083	(−2.45)	0.123	(3.56)	0.023	(0.54)
ADJ-R^2	0.051		0.007		0.051		0.032	
F-STAT	22.881		3.804		22.734		14.341	

TABLE 7.1 (Continued)

1986	Q1		Q2		Q3		Q4	
EP	−0.040	(−0.62)	−0.133	(−1.29)	0.234	(2.54)	−0.474	(−7.53)
BP	−0.272	(−5.74)	0.200	(4.53)	−0.117	(−3.15)	−0.052	(−1.54)
CP	0.036	(0.63)	0.231	(2.47)	−0.162	(−1.88)	0.315	(5.54)
SP	−0.010	(−0.30)	−0.044	(−1.40)	0.091	(3.16)	−0.015	(−0.56)
REP	−0.154	(−1.67)	−0.056	(−1.31)	0.081	(1.84)	−0.072	(−1.80)
RBP	0.074	(1.33)	−0.106	(−2.59)	0.113	(1.70)	0.135	(3.39)
RCP	0.002	(0.02)	0.212	(2.23)	0.132	(1.58)	0.047	(0.62)
RSP	0.002	(0.07)	−0.072	(−2.25)	−0.014	(−0.44)	0.087	(2.46)
ADJ-R^2	0.017		0.020		0.007		0.069	
F-STAT	8.157		9.097		3.837		31.132	

1987	Q1		Q2		Q3		Q4	
EP	−0.639	(−8.34)	0.009	(0.13)	0.258	(4.69)	−0.144	(−1.88)
BP	0.011	(0.32)	−0.034	(−1.34)	0.010	(0.39)	0.001	(0.04)
CP	0.516	(7.47)	0.005	(0.09)	−0.031	(−0.61)	0.068	(0.95)
SP	−0.042	(−1.38)	0.011	(0.44)	−0.169	(−6.66)	0.130	(4.48)
REP	0.005	(0.10)	0.026	(0.48)	−0.051	(−1.03)	−0.027	(−0.47)
RBP	0.097	(1.18)	0.041	(0.81)	−0.228	(−4.90)	0.188	(2.44)
RCP	0.063	(0.85)	0.043	(0.75)	0.075	(1.56)	−0.025	(−0.50)
RSP	0.068	(2.44)	0.098	(3.41)	−0.149	(−5.85)	0.109	(3.50)
ADJ-R^2	0.046		0.013		0.159		0.062	
F-STAT	22.758		6.892		85.957		30.334	

1988	Q1		Q2		Q3		Q4	
EP	0.447	(5.71)	0.030	(0.93)	0.214	(3.32)	−0.223	(−3.62)
BP	−0.084	(−2.47)	−0.052	(−3.25)	−0.046	(−1.45)	−0.059	(−2.10)
CP	−0.287	(−3.86)	0.102	(3.23)	−0.007	(−0.12)	0.244	(4.20)
SP	0.069	(2.25)	0.039	(2.70)	0.019	(0.67)	0.154	(6.04)
REP	0.057	(1.02)	0.063	(1.60)	−0.059	(−0.86)	0.074	(1.15)
RBP	−0.014	(−0.36)	0.014	(0.74)	0.091	(2.31)	0.015	(0.39)
RCP	0.088	(0.53)	−0.140	(−1.76)	−0.269	(−1.70)	−0.239	(−1.58)
RSP	0.060	(2.00)	0.010	(0.80)	−0.054	(−2.05)	0.014	(0.56)
ADJ-R^2	0.012		0.014		0.013		0.028	
F-STAT	6.647		7.322		6.817		14.369	

1989	Q1		Q2		Q3		Q4	
EP	0.094	(1.07)	0.414	(3.83)	0.181	(1.54)	−0.030	(−0.29)
BP	−0.099	(−2.70)	−0.098	(−3.15)	−0.164	(−3.93)	−0.026	(−0.78)
CP	0.008	(0.09)	−0.324	(−3.12)	0.028	(0.26)	0.177	(1.75)
SP	−0.018	(−0.57)	−0.089	(−3.32)	−0.171	(−6.15)	0.003	(0.10)
REP	0.047	(0.18)	0.464	(1.64)	0.308	(1.16)	0.307	(1.39)
RBP	−0.027	(−0.63)	0.080	(1.93)	0.050	(1.03)	−0.118	(−3.01)
RCP	0.028	(0.29)	−0.194	(−2.20)	0.006	(0.07)	0.096	(1.19)
RSP	0.128	(4.78)	0.151	(6.75)	0.086	(3.72)	0.058	(2.38)
ADJ-R^2	0.009		0.029		0.027		0.004	
F-STAT	5.269		14.792		13.662		2.871	

Continued on following page

TABLE 7.1 (Continued)

1990	Q1		Q2		Q3		Q4	
EP	0.227	(2.01)	−0.086	(−0.61)	0.434	(5.45)	0.076	(0.92)
BP	−0.305	(−5.92)	0.035	(0.91)	−0.332	(−7.41)	0.023	(0.56)
CP	−0.250	(−2.22)	0.098	(0.83)	−0.125	(−1.62)	−0.268	(−3.36)
SP	−0.166	(−5.98)	−0.200	(−6.59)	−0.199	(−6.86)	0.089	(3.23)
REP	−0.033	(−0.64)	0.028	(0.59)	0.015	(0.34)	0.012	(0.34)
RBP	−0.040	(−1.03)	−0.042	(−1.12)	0.058	(1.65)	0.063	(1.84)
RCP	0.039	(1.10)	0.026	(0.59)	0.038	(0.97)	0.038	(1.07)
RSP	0.247	(9.28)	−0.080	(−2.76)	0.197	(7.04)	0.219	(7.68)
ADJ-R^2	0.049		0.034		0.074		0.105	
F-STAT	24.526		17.284		37.598		54.336	

1991	Q1		Q2		Q3		Q4	
EP	0.289	(4.16)	0.527	(4.86)	0.452	(3.31)	0.169	(1.93)
BP	−0.056	(−1.51)	−0.145	(−4.41)	−0.231	(−4.13)	0.223	(4.27)
CP	−0.263	(−4.05)	−0.462	(−4.29)	−0.078	(−0.55)	−0.454	(−5.45)
SP	−0.002	(−0.08)	−0.074	(−2.39)	−0.126	(−4.46)	0.263	(9.91)
REP	0.248	(0.91)	0.002	(0.01)	0.211	(1.51)	−0.048	(−0.39)
RBP	0.013	(0.37)	0.119	(3.68)	0.034	(1.09)	0.039	(1.20)
RCP	0.033	(0.58)	−0.039	(−0.92)	−0.042	(−1.17)	−0.018	(−0.59)
RSP	0.121	(4.42)	0.101	(3.63)	0.053	(1.87)	−0.014	(−0.48)
ADJ-R^2	0.016		0.032		0.034		0.094	
F-STAT	7.986		15.540		16.262		46.463	

1992	Q1		Q2		Q3		Q4	
EP	0.223	(2.80)	0.116	(1.73)	0.334	(4.87)	−0.291	(−4.07)
BP	0.145	(3.33)	0.052	(1.29)	−0.017	(−0.43)	0.334	(8.39)
CP	−0.118	(−1.52)	0.058	(1.15)	−0.258	(−4.93)	0.194	(3.27)
SP	−0.073	(−2.47)	−0.106	(−3.72)	0.083	(2.73)	0.070	(2.25)
REP	0.039	(0.64)	−0.005	(−0.06)	0.057	(0.78)	−0.083	(−1.32)
RBP	0.158	(3.38)	0.006	(0.13)	0.117	(2.54)	−0.132	(−2.76)
RCP	0.030	(0.61)	0.033	(1.00)	−0.002	(−0.07)	−0.030	(−0.83)
RSP	0.001	(0.03)	0.045	(1.87)	0.116	(4.51)	−0.098	(−3.57)
ADJ-R^2	0.016		0.011		0.028		0.045	
F-STAT	8.372		5.960		13.928		22.252	

1993	Q1		Q2		Q3		Q4	
EP	−0.227	(−3.54)	0.043	(0.85)	0.042	(0.70)	−0.131	(−1.14)
BP	−0.023	(−0.59)	−0.012	(−0.54)	−0.077	(−3.12)	0.016	(0.58)
CP	0.015	(0.28)	−0.037	(−0.82)	0.003	(0.05)	0.176	(1.60)
SP	−0.041	(−1.47)	−0.026	(−1.14)	0.121	(4.62)	0.174	(6.70)
REP	−0.017	(−0.28)	−0.030	(−0.81)	−0.021	(−0.47)	0.061	(1.42)
RBP	0.244	(1.64)	0.067	(0.58)	−0.058	(−0.44)	0.010	(0.07)
RCP	0.009	(0.20)	−0.086	(−2.13)	−0.003	(−0.07)	−0.148	(−3.10)
RSP	0.074	(4.07)	0.040	(2.44)	0.126	(6.55)	−0.052	(−2.24)
ADJ-R^2	0.014		0.002		0.021		0.021	
F-STAT	8.026		1.818		11.431		11.361	

TABLE 7.1 (Continued)

1994	Q1		Q2		Q3		Q4	
EP	−0.094	(−1.50)	−0.033	(−0.54)	0.158	(1.46)	0.200	(1.55)
BP	0.096	(3.01)	−0.068	(−3.06)	−0.083	(−3.73)	0.005	(0.19)
CP	0.131	(2.33)	0.024	(0.41)	−0.046	(−0.44)	−0.152	(−1.18)
SP	−0.083	(−3.18)	0.014	(0.58)	−0.019	(−0.77)	−0.017	(−0.66)
REP	0.040	(0.65)	−0.071	(−1.20)	−0.107	(−1.79)	0.023	(0.36)
RBP	0.060	(1.08)	0.114	(2.28)	0.109	(1.93)	−0.012	(−0.19)
RCP	0.014	(0.26)	0.062	(0.70)	0.005	(0.06)	0.014	(0.15)
RSP	−0.015	(−0.61)	0.087	(2.90)	−0.097	(−3.27)	0.139	(4.80)
ADJ-R^2	0.004		0.010		0.013		0.011	
F-STAT	3.505		6.304		8.282		7.094	

TABLE 7.2 The Determinants of Security Returns, 1982–1994: ROB

1982	Q1		Q2		Q3		Q4	
EP	0.198	(3.89)	0.690	(7.05)	0.225	(2.38)	0.254	(2.31)
BP	−0.011	(−0.64)	−0.024	(−1.12)	−0.148	(−7.73)	0.043	(2.74)
CP	−0.139	(−3.07)	−0.499	(−5.38)	−0.273	(−3.07)	−0.645	(−5.90)
SP	0.062	(3.86)	0.068	(3.99)	0.121	(7.51)	0.111	(8.20)
REP	−0.202	(−3.79)	0.022	(0.34)	0.026	(0.45)	−0.111	(−2.13)
RBP	−0.398	(−1.43)	0.425	(1.15)	3.031	(7.52)	0.350	(1.41)
RCP	0.056	(1.65)	−0.034	(−0.43)	−0.026	(−0.36)	0.018	(0.36)
RSP	0.074	(6.22)	0.089	(7.11)	−0.080	(−6.26)	−0.054	(−4.77)
ADJ-R^2	0.016		0.023		0.030		0.050	
F-STAT	10.467		14.759		18.421		30.918	

1983	Q1		Q2		Q3		Q4	
EP	−0.133	(−1.92)	0.239	(6.26)	0.382	(6.98)	−0.052	(−1.01)
BP	−0.035	(−2.37)	0.169	(12.69)	0.195	(10.90)	0.200	(10.79)
CP	0.007	(0.11)	−0.100	(−2.80)	−0.206	(−4.01)	0.084	(1.76)
SP	0.010	(0.68)	0.044	(3.71)	0.105	(6.31)	0.031	(1.97)
REP	−0.070	(−1.33)	0.002	(0.07)	0.087	(1.65)	−0.015	(−0.38)
RBP	0.080	(0.17)	−0.044	(−0.29)	−0.086	(−0.41)	−0.015	(−0.08)
RCP	0.042	(1.06)	0.048	(1.50)	0.115	(2.56)	−0.008	(−0.18)
RSP	0.077	(7.46)	0.071	(8.25)	−0.070	(−5.67)	0.040	(2.99)
ADJ-R^2	0.017		0.104		0.074		0.058	
F-STAT	11.035		66.622		46.979		35.867	

Continued on following page

TABLE 7.2 (Continued)

1984	Q1		Q2		Q3		Q4	
EP	0.356	(6.60)	0.487	(6.00)	0.746	(9.50)	0.006	(0.08)
BP	0.040	(1.76)	0.050	(2.22)	0.095	(4.72)	−0.060	(−2.40)
CP	−0.143	(−2.88)	−0.159	(−1.91)	−0.434	(−5.53)	−0.119	(−1.66)
SP	0.066	(3.63)	0.110	(6.51)	0.076	(5.36)	0.059	(3.93)
REP	0.034	(0.94)	0.068	(1.58)	0.126	(3.49)	0.052	(1.37)
RBP	−0.052	(−1.49)	0.049	(1.14)	−0.186	(−5.01)	0.104	(2.43)
RCP	0.040	(0.68)	−0.201	(−1.83)	−0.029	(−0.35)	0.126	(1.60)
RSP	0.013	(0.85)	−0.015	(−1.08)	−0.094	(−7.93)	0.022	(1.42)
ADJ-R^2	0.020		0.045		0.089		0.008	
F-STAT	12.601		28.205		57.367		5.683	
1985	Q1		Q2		Q3		Q4	
EP	0.317	(5.35)	0.428	(6.27)	0.829	(11.31)	0.315	(5.04)
BP	−0.051	(−2.43)	0.064	(2.80)	−0.079	(−3.18)	−0.090	(−3.71)
CP	−0.088	(−1.56)	−0.168	(−2.63)	−0.555	(−8.23)	−0.339	(−5.86)
SP	0.059	(3.77)	0.059	(3.64)	0.079	(4.64)	0.138	(8.68)
REP	−0.075	(−0.98)	−0.089	(−1.34)	0.127	(1.78)	−0.053	(−0.79)
RBP	0.070	(0.64)	0.147	(1.12)	0.374	(2.71)	0.205	(1.40)
RCP	0.130	(1.69)	−0.328	(−3.38)	−0.091	(−0.89)	0.024	(0.26)
RSP	−0.026	(−2.00)	0.037	(2.90)	−0.042	(−3.15)	−0.081	(−5.79)
ADJ-R^2	0.025		0.028		0.060		0.031	
F-STAT	16.769		18.379		39.467		20.412	
1986	Q1		Q2		Q3		Q4	
EP	0.100	(1.85)	0.086	(1.60)	0.218	(4.58)	−0.654	(−14.61)
BP	−0.183	(−7.32)	0.169	(7.70)	−0.037	(−1.74)	−0.157	(−5.15)
CP	0.048	(−0.99)	0.047	(0.96)	−0.041	(−0.93)	0.466	(12.09)
SP	−0.058	(3.39)	0.045	(2.91)	0.066	(4.53)	−0.030	(−2.35)
REP	−0.162	(−2.86)	0.010	(0.24)	0.022	(0.48)	−0.007	(−0.17)
RBP	0.151	(3.22)	0.063	(0.64)	0.013	(0.11)	0.096	(1.51)
RCP	0.143	(−2.31)	0.128	(2.04)	−0.093	(−1.59)	−0.170	(−2.47)
RSP	−0.011	(−0.67)	0.007	(0.56)	0.052	(−4.28)	0.056	(4.30)
ADJ-R^2	0.015		0.032		0.024		0.066	
F-STAT	10.364		20.930		15.886		43.053	
1987	Q1		Q2		Q3		Q4	
EP	−0.127	(−2.86)	−0.173	(−1.76)	0.035	(0.34)	−0.034	(−0.42)
BP	0.028	(1.11)	−0.049	(−1.53)	0.199	(4.08)	−0.011	(−0.43)
CP	0.116	(2.87)	0.229	(2.65)	0.402	(4.83)	0.037	(0.65)
SP	−0.021	(−1.49)	0.006	(0.35)	−0.096	(−5.19)	0.075	(4.71)
REP	0.004	(0.14)	0.104	(1.91)	0.017	(0.23)	0.018	(−0.25)
RBP	0.038	(2.14)	0.065	(4.07)	−0.059	(−3.23)	0.014	(0.78)
RCP	−0.045	(−0.56)	−0.305	(−2.88)	0.257	(2.24)	−0.005	(−0.17)
RSP	0.076	(5.32)	0.033	(2.51)	−0.099	(−6.46)	0.100	(6.52)
ADJ-R^2	0.028		0.013		0.075		0.032	
F-STAT	18.284		8.831		49.963		20.817	

TABLE 7.2 (Continued)

1988	Q1		Q2		Q3		Q4	
EP	0.048	(0.57)	0.068	(2.42)	0.111	(2.22)	−0.244	(−4.49)
BP	−0.117	(−5.29)	−0.003	(−0.22)	−0.025	(−0.94)	−0.039	(−1.67)
CP	0.070	(1.42)	0.060	(2.21)	0.153	(3.26)	0.228	(4.39)
SP	−0.001	(−0.04)	0.052	(5.53)	0.003	(0.20)	0.042	(2.96)
REP	−0.027	(−0.74)	0.013	(0.52)	−0.024	(−0.56)	−0.110	(−2.91)
RBP	0.077	(2.28)	0.055	(2.89)	0.103	(3.13)	−0.029	(−1.17)
RCP	0.152	(1.79)	0.017	(−0.55)	−0.001	(−0.03)	0.028	(0.55)
RSP	0.091	(6.70)	−0.027	(3.64)	−0.086	(−6.52)	−0.015	(−1.11)
ADJ-R^2	0.020		0.031		0.042		0.009	
F-STAT	14.056		21.058		28.934		6.735	

1989	Q1		Q2		Q3		Q4	
EP	0.038	(1.28)	0.071	(1.49)	0.069	(1.15)	−0.079	(−1.50)
BP	−0.013	(−1.31)	−0.073	(−2.67)	−0.013	(−0.42)	0.111	(3.30)
CP	0.000	(0.01)	0.035	(0.75)	0.061	(1.05)	0.042	(0.83)
SP	0.006	(0.86)	−0.100	(−6.89)	−0.177	(−11.47)	−0.004	(−0.29)
REP	0.067	(1.52)	−0.109	(−0.86)	0.098	(0.89)	0.098	(1.02)
RBP	0.062	(2.46)	0.074	(1.72)	−0.070	(−1.57)	−0.100	(−2.93)
RCP	−0.000	(−0.01)	0.177	(2.76)	0.087	(1.55)	0.021	(0.39)
RSP	0.037	(6.66)	0.028	(2.25)	−0.027	(−2.02)	0.022	(1.38)
ADJ-R^2	0.016		0.025		0.052		0.004	
F-STAT	10.755		16.584		34.761		3.332	

1990	Q1		Q2		Q3		Q4	
EP	0.111	(3.68)	−0.151	(−2.48)	0.333	(4.24)	0.026	(0.49)
BP	−0.118	(−6.33)	0.149	(5.76)	−0.199	(−7.79)	−0.028	(−1.17)
CP	−0.083	(−2.85)	0.258	(4.29)	−0.120	(−1.57)	−0.127	(−2.53)
SP	−0.073	(−7.73)	−0.130	(−7.28)	−0.131	(−7.38)	0.043	(2.95)
REP	−0.013	(−0.72)	0.010	(0.28)	0.094	(2.28)	0.048	(1.22)
RBP	0.057	(2.38)	−0.023	(−0.47)	0.160	(4.17)	0.103	(3.37)
RCP	−0.006	(−0.28)	−0.012	(−0.30)	0.120	(2.57)	0.051	(1.47)
RSP	0.100	(11.13)	0.077	(5.44)	0.041	(2.69)	0.179	(11.77)
ADJ-R^2	0.050		0.022		0.071		0.094	
F-STAT	32.580		14.430		46.883		62.962	

1991	Q1		Q2		Q3		Q4	
EP	0.219	(4.40)	0.309	(5.73)	0.208	(4.21)	0.058	(1.46)
BP	−0.034	(−1.84)	−0.151	(−5.82)	−0.117	(−4.70)	0.073	(4.36)
CP	−0.190	(−3.81)	−0.264	(−5.11)	−0.037	(−0.80)	−0.151	(−3.89)
SP	−0.054	(−4.53)	−0.057	(−3.70)	−0.031	(−2.13)	0.101	(10.57)
REP	0.093	(1.15)	0.144	(1.21)	0.064	(0.58)	−0.079	(−1.12)
RBP	−0.037	(0.93)	0.117	(5.26)	0.013	(0.34)	0.128	(5.22)
RCP	−0.068	(−0.71)	−0.055	(−1.31)	−0.019	(−0.45)	−0.030	(−1.26)
RSP	0.158	(13.42)	0.058	(4.45)	−0.010	(−0.78)	0.040	(4.29)
ADJ-R^2	0.045		0.037		0.036		0.119	
F-STAT	29.172		24.028		23.345		80.390	

Continued on following page

TABLE 7.2 (Continued)

1992	Q1		Q2		Q3		Q4	
EP	0.218	(3.47)	0.249	(3.69)	0.286	(4.29)	−0.097	(−1.44)
BP	0.034	(1.11)	0.034	(0.86)	−0.135	(−2.79)	0.318	(9.86)
CP	−0.065	(−1.10)	−0.004	(−0.08)	−0.093	(−1.48)	0.039	(0.61)
SP	−0.060	(−3.86)	−0.051	(−2.69)	0.078	(4.15)	0.065	(3.74)
REP	0.029	(0.67)	−0.100	(−1.39)	0.098	(1.31)	0.036	(0.53)
RBP	0.112	(5.18)	−0.014	(−0.46)	0.072	(2.44)	−0.185	(−5.92)
RCP	0.015	(0.47)	0.042	(0.98)	−0.126	(−2.87)	0.049	(1.32)
RSP	0.120	(8.56)	0.092	(6.22)	0.017	(1.00)	0.046	(2.89)
ADJ-R^2	0.062		0.024		0.012		0.044	
F-STAT	39.974		15.682		7.961		28.328	
1993	Q1		Q2		Q3		Q4	
EP	−0.116	(−2.59)	0.142	(1.99)	0.005	(0.25)	−0.029	(−0.77)
BP	−0.006	(−0.18)	0.022	(0.92)	0.004	(0.56)	0.131	(6.96)
CP	0.083	(1.99)	−0.026	(−0.38)	−0.002	(−0.09)	0.025	(0.67)
SP	−0.030	(−2.25)	0.006	(0.40)	0.028	(5.22)	0.139	(9.34)
REP	−0.050	(−1.85)	−0.018	(−0.59)	0.009	(0.89)	0.051	(1.23)
RBP	0.015	(0.39)	−0.005	(−0.10)	0.024	(1.22)	−0.017	(−0.39)
RCP	−0.029	(−0.92)	−0.126	(−2.12)	−0.009	(−0.63)	0.033	(0.86)
RSP	0.063	(6.04)	0.005	(0.43)	0.013	(3.44)	−0.002	(−0.19)
ADJ-R^2	0.012		0.002		0.013		0.052	
F-STAT	8.279		1.914		9.364		35.083	
1994	Q1		Q2		Q3		Q4	
EP	−0.024	(−0.51)	0.219	(4.32)	0.112	(1.90)	−0.101	(−1.60)
BP	0.136	(3.97)	−0.044	(−1.57)	−0.060	(−3.24)	0.029	(1.48)
CP	0.180	(3.87)	−0.169	(−3.51)	0.027	(0.48)	0.093	(1.53)
SP	−0.015	(−1.18)	0.015	(1.00)	0.028	(1.76)	−0.054	(−3.59)
REP	−0.074	(−1.20)	−0.039	(−0.74)	−0.046	(−0.88)	0.005	(0.10)
RBP	−0.018	(−0.75)	0.163	(4.91)	−0.030	(−0.81)	0.077	(1.71)
RCP	−0.054	(−1.22)	0.035	(0.81)	0.036	(0.78)	−0.020	(−0.43)
RSP	0.069	(6.45)	0.083	(6.91)	−0.036	(−2.83)	0.064	(4.45)
ADJ-R^2	0.024		0.025		0.012		0.010	
F-STAT	18.924		19.808		9.658		7.972	
1995	Q1		Q2		Q3		Q4	
EP	0.309	(3.01)	0.213	(2.96)	0.355	(7.30)	−0.307	(−7.14)
BP	−0.035	(−1.52)	−0.041	(−1.67)	0.042	(2.58)	0.037	(2.73)
CP	−0.229	(−2.29)	−0.371	(−5.74)	−0.001	(−0.02)	0.079	(1.89)
SP	−0.045	(−3.23)	−0.035	(−2.59)	−0.185	(−10.69)	0.019	(1.62)
REP	−0.001	(−0.03)	0.034	(0.61)	0.126	(2.01)	0.023	(0.56)
RBP	0.053	(1.08)	0.056	(1.46)	−0.036	(−0.76)	0.005	(0.18)
RCP	−0.075	(−2.07)	−0.099	(−3.23)	0.022	(0.56)	−0.029	(−1.14)
RSP	0.113	(10.10)	0.048	(4.76)	−0.059	(−4.98)	0.037	(3.71)
ADJ-R^2	0.021		0.020		0.079		0.035	
F-STAT	17.908		16.542		68.338		29.146	

TABLE 7.3 The Determinants of Security Returns, 1982–1994: ROB

1982	Q1		Q2		Q3		Q4	
EP	0.049	(0.75)	0.328	(6.40)	−0.195	(−3.41)	−0.028	(−0.55)
BP	−0.123	(−3.73)	0.006	(0.17)	−0.148	(−4.40)	0.082	(2.52)
CP	−0.097	(−1.70)	−0.348	(−7.05)	−0.029	(−0.55)	−0.205	(−4.20)
SP	0.167	(4.72)	0.142	(4.30)	0.211	(6.34)	0.294	(9.00)
REP	0.059	(0.54)	−0.111	(−1.45)	0.115	(1.60)	0.018	(0.26)
RBP	0.006	(0.10)	0.374	(5.54)	0.695	(8.98)	0.376	(4.29)
RCP	0.064	(0.94)	−0.114	(−1.12)	0.032	(0.36)	0.107	(1.38)
RSP	0.028	(0.70)	−0.127	(−3.60)	−0.187	(−5.08)	−0.149	(−3.99)
PRGR	0.022	(0.37)	0.196	(4.32)	0.218	(4.38)	0.095	(2.10)
ADJ-R^2	0.011		0.040		0.060		0.080	
F-STAT	3.812		11.447		17.061		23.114	

1983	Q1		Q2		Q3		Q4	
EP	−0.167	(−7.47)	0.423	(6.35)	0.216	(3.29)	−0.294	(−3.74)
BP	−0.110	(−7.29)	0.283	(10.87)	0.176	(6.57)	0.188	(6.58)
CP	0.072	(3.08)	−0.163	(−2.77)	−0.091	(−1.65)	0.289	(5.45)
SP	0.052	(3.45)	−0.022	(−0.70)	0.121	(3.78)	−0.092	(−2.98)
REP	0.072	(1.75)	−0.024	(−0.48)	0.028	(0.56)	0.011	(0.20)
RBP	0.084	(3.78)	−0.186	(−3.72)	−0.269	(−4.99)	−0.259	(−4.23)
RCP	−0.116	(−0.98)	−0.136	(−1.72)	−0.137	(−1.78)	−0.246	(−3.59)
RSP	0.023	(1.09)	0.045	(1.28)	0.010	(0.27)	0.082	(1.98)
PRGR	0.022	(0.94)	0.193	(4.86)	0.103	(2.35)	−0.074	(−1.23)
ADJ-R^2	0.066		0.103		0.068		0.056	
F-STAT	20.467		32.781		20.970		17.281	

1984	Q1		Q2		Q3		Q4	
EP	0.358	(5.34)	0.609	(8.91)	0.621	(7.15)	−0.156	(−2.04)
BP	−0.134	(−5.03)	0.066	(2.61)	0.099	(3.53)	−0.096	(−2.91)
CP	−0.111	(−2.32)	−0.196	(−3.54)	−0.052	(−0.85)	0.050	(0.78)
SP	0.081	(2.66)	0.112	(4.33)	0.065	(2.39)	0.118	(4.51)
REP	0.041	(0.58)	0.039	(0.59)	0.011	(0.16)	0.138	(2.00)
RBP	0.202	(1.43)	−0.247	(−1.46)	−0.927	(−5.03)	0.134	(0.83)
RCP	0.063	(1.42)	0.085	(1.93)	0.117	(2.63)	−0.138	(−3.17)
RSP	0.051	(1.81)	−0.000	(−0.00)	−0.097	(−3.66)	0.152	(5.50)
PRGR	0.273	(5.13)	0.246	(6.50)	0.396	(8.02)	0.245	(4.75)
ADJ-R^2	0.031		0.069		0.130		0.044	
F-STAT	12.074		27.144		53.208		17.024	

Continued on following page

TABLE 7.3 (Continued)

1985	Q1		Q2		Q3		Q4	
EP	0.424	(8.26)	0.165	(2.44)	1.207	(13.59)	0.759	(8.08)
BP	−0.024	(−1.09)	−0.055	(−1.77)	−0.059	(−1.62)	0.049	(0.99)
CP	−0.115	(−2.76)	−0.027	(−0.48)	−0.661	(−9.20)	−0.597	(−7.72)
SP	0.034	(1.88)	0.073	(2.65)	0.178	(6.64)	0.238	(8.13)
REP	0.017	(0.65)	0.018	(0.39)	−0.048	(−0.92)	−0.098	(−2.06)
RBP	−0.132	(−4.32)	0.096	(1.79)	−0.048	(−0.92)	−0.142	(−2.74)
RCP	0.093	(1.85)	0.033	(0.41)	0.205	(3.23)	−0.058	(−0.97)
RSP	0.028	(1.26)	−0.083	(−2.87)	0.164	(5.26)	0.031	(0.83)
PRGR	0.279	(8.70)	0.132	(3.65)	0.319	(6.42)	0.463	(7.37)
ADJ-R^2	0.097		0.014		0.098		0.057	
F-STAT	39.594		6.152		39.938		22.622	
1986	Q1		Q2		Q3		Q4	
EP	−0.015	(−0.27)	−0.104	(−1.13)	0.183	(2.30)	−0.600	(−10.74)
BP	−0.282	(−6.90)	0.252	(6.38)	−0.116	(−3.62)	−0.074	(−2.57)
CP	0.044	(0.88)	0.246	(2.95)	−0.107	(−1.43)	0.427	(8.56)
SP	−0.003	(−0.09)	−0.060	(−2.16)	0.076	(3.11)	0.001	(0.03)
REP	−0.112	(−1.45)	−0.076	(−2.05)	0.072	(1.92)	−0.087	(−2.59)
RBP	0.116	(2.27)	−0.178	(−4.89)	0.111	(1.95)	0.165	(4.80)
RCP	0.016	(0.17)	0.265	(2.92)	0.072	(0.98)	0.086	(1.31)
RSP	−0.019	(−0.56)	−0.059	(−2.10)	−0.013	(−0.48)	0.066	(2.12)
PRGR	0.282	(5.73)	0.126	(3.34)	0.258	(4.93)	0.164	(4.31)
ADJ-R^2	0.036		0.047		0.016		0.109	
F-STAT	14.573		18.660		6.889		44.865	
1987	Q1		Q2		Q3		Q4	
EP	−0.718	(−10.47)	−0.000	(−0.00)	0.324	(6.24)	−0.157	(−2.43)
BP	−0.027	(−0.95)	−0.027	(−1.25)	0.022	(1.00)	0.021	(0.88)
CP	0.590	(9.53)	−0.015	(−0.30)	−0.079	(−1.60)	0.092	(1.51)
SP	−0.023	(−0.88)	0.011	(0.50)	−0.177	(−7.78)	0.155	(6.43)
REP	−0.011	(−0.26)	0.005	(0.12)	−0.041	(−0.92)	−0.038	(−0.76)
RBP	0.158	(2.26)	0.041	(0.95)	−0.249	(−5.79)	0.233	(3.58)
RCP	0.015	(0.25)	0.031	(0.60)	0.042	(0.97)	−0.053	(−1.20)
RSP	0.048	(2.02)	0.094	(3.79)	−0.162	(−6.94)	0.114	(4.42)
PRGR	0.146	(3.14)	0.213	(6.65)	0.273	(7.21)	0.121	(2.87)
ADJ-R^2	0.060		0.026		0.233		0.112	
F-STAT	26.628		11.651		121.701		50.971	

TABLE 7.3 (Continued)

1988	Q1		Q2		Q3		Q4	
EP	0.342	(5.14)	0.084	(3.03)	0.123	(2.24)	−0.265	(−4.82)
BP	−0.094	(−3.30)	−0.043	(−3.45)	−0.086	(−3.23)	−0.059	(−2.50)
CP	−0.208	(−3.33)	0.036	(1.33)	0.069	(1.38)	0.251	(4.81)
SP	0.067	(2.61)	0.046	(4.08)	0.028	(1.19)	0.148	(6.99)
REP	0.075	(1.62)	0.060	(1.91)	−0.030	(−0.54)	0.109	(1.94)
RBP	0.027	(0.79)	0.011	(0.78)	0.148	(4.50)	0.020	(0.62)
RCP	0.140	(0.86)	−0.147	(−2.41)	−0.263	(−2.00)	−0.227	(−1.78)
RSP	0.040	(1.59)	−0.003	(−0.34)	−0.088	(−4.07)	0.018	(0.85)
PRGR	0.512	(10.12)	0.029	(2.02)	0.556	(11.35)	0.223	(4.96)
ADJ-R^2	0.045		0.017		0.056		0.045	
F-STAT	20.316		8.012		25.471		20.220	

1989	Q1		Q2		Q3		Q4	
EP	0.063	(0.79)	0.440	(4.77)	0.119	(1.09)	−0.034	(−0.35)
BP	−0.107	(−3.34)	−0.094	(−3.49)	−0.165	(−4.37)	−0.033	(−1.18)
CP	0.070	(0.95)	−0.299	(−3.38)	0.180	(1.74)	0.165	(1.76)
SP	−0.031	(−1.18)	−0.088	(−3.86)	−0.195	(−8.17)	−0.002	(−0.07)
REP	0.023	(0.11)	0.524	(2.30)	0.263	(1.18)	0.116	(0.64)
RBP	−0.007	(−0.17)	0.087	(2.49)	0.028	(0.67)	−0.115	(−3.57)
RCP	0.048	(0.58)	−0.202	(−2.73)	−0.038	(−0.50)	0.127	(1.91)
RSP	0.115	(4.92)	0.143	(7.59)	0.080	(4.03)	0.037	(1.82)
PRGR	0.328	(5.85)	0.343	(8.11)	0.405	(6.74)	0.322	(6.64)
ADJ-R^2	0.021		0.058		0.058		0.018	
F-STAT	9.604		25.804		25.759		8.359	

1990	Q1		Q2		Q3		Q4	
EP	0.234	(2.33)	−0.153	(−1.13)	0.366	(5.27)	0.189	(2.64)
BP	−0.389	(−8.40)	0.054	(1.54)	−0.377	(−9.42)	−0.002	(−0.06)
CP	−0.248	(−2.48)	0.148	(1.32)	−0.085	(−1.28)	−0.371	(−5.46)
SP	−0.127	(−5.28)	−0.212	(−7.65)	−0.202	(−8.04)	0.105	(4.56)
REP	0.015	(0.34)	0.011	(0.25)	0.027	(0.71)	0.037	(1.22)
RBP	0.016	(0.47)	−0.053	(−1.58)	0.093	(3.04)	0.086	(2.96)
RCP	0.027	(0.89)	0.045	(1.12)	0.045	(1.30)	0.026	(0.86)
RSP	0.219	(9.58)	−0.124	(−4.73)	0.176	(7.35)	0.226	(9.50)
PRGR	0.268	(4.82)	0.095	(2.60)	0.607	(12.36)	−0.241	(−5.01)
ADJ-R^2	0.069		0.062		0.140		0.169	
F-STAT	31.064		27.695		66.906		82.771	

Continued on following page

164 HISTORICAL DATA AND ANALYSTS' FORECASTS

TABLE 7.3 (Continued)

1991	Q1		Q2		Q3		Q4	
EP	0.323	(5.12)	0.338	(3.43)	0.380	(2.93)	0.141	(1.90)
BP	−0.067	(−2.02)	−0.140	(−4.95)	−0.225	(−4.50)	0.350	(7.64)
CP	−0.280	(−4.73)	−0.306	(−3.16)	−0.084	(−0.60)	−0.420	(−5.91)
SP	−0.021	(−0.86)	−0.114	(−4.24)	−0.134	(−5.51)	0.262	(11.78)
REP	0.282	(1.12)	0.097	(0.83)	0.226	(1.83)	−0.015	(−0.14)
RBP	0.043	(1.41)	0.120	(4.37)	0.037	(1.35)	0.024	(0.88)
RCP	0.054	(1.08)	−0.085	(−2.29)	−0.020	(−0.62)	0.018	(0.69)
RSP	0.104	(4.48)	0.100	(4.24)	0.059	(2.41)	−0.014	(−0.62)
PRGR	0.233	(5.39)	0.249	(8.39)	0.329	(7.47)	0.121	(2.45)
ADJ-R^2	0.031		0.062		0.059		0.151	
F-STAT	13.311		26.733		25.542		70.291	
1992	Q1		Q2		Q3		Q4	
EP	0.316	(3.86)	0.143	(2.30)	0.296	(4.58)	−0.556	(−8.40)
BP	0.187	(4.76)	−0.015	(−0.44)	−0.042	(−1.21)	0.349	(10.00)
CP	−0.154	(−1.95)	0.029	(0.63)	−0.273	(−5.68)	0.353	(6.51)
SP	−0.093	(−3.56)	−0.102	(−4.21)	0.083	(3.18)	0.025	(0.95)
REP	0.051	(0.93)	0.042	(0.61)	0.043	(0.68)	−0.038	(−0.70)
RBP	0.183	(4.47)	0.017	(0.44)	0.163	(4.09)	−0.076	(−1.84)
RCP	0.036	(0.82)	0.047	(1.64)	−0.000	(−0.01)	0.004	(0.14)
RSP	0.008	(0.33)	0.053	(2.60)	0.126	(5.65)	−0.100	(−4.28)
PRGR	0.444	(8.58)	0.296	(8.11)	0.311	(7.22)	0.611	(12.03)
ADJ-R^2	0.049		0.038		0.054		0.104	
F-STAT	21.397		16.608		24.029		47.484	
1993	Q1		Q2		Q3		Q4	
EP	−0.248	(−4.29)	0.023	(0.54)	0.023	(0.43)	−0.113	(−1.15)
BP	−0.019	(−0.56)	0.003	(0.15)	−0.101	(−4.70)	−0.006	(−0.25)
CP	0.036	(0.77)	−0.018	(−0.46)	0.008	(0.17)	0.156	(1.66)
SP	−0.054	(−2.27)	−0.027	(−1.38)	0.128	(5.58)	0.181	(8.11)
REP	0.014	(0.28)	−0.012	(−0.36)	−0.027	(−0.70)	0.045	(1.21)
RBP	0.214	(1.72)	0.079	(0.84)	−0.026	(−0.23)	0.033	(0.29)
RCP	0.006	(0.16)	−0.070	(−2.05)	−0.007	(−0.18)	−0.146	(−3.54)
RSP	0.075	(4.74)	0.021	(1.52)	0.128	(7.57)	−0.058	(−2.89)
PRGR	0.203	(4.37)	0.149	(6.31)	0.095	(2.71)	0.070	(1.61)
ADJ-R^2	0.019		0.011		0.028		0.027	
F-STAT	9.497		5.787		13.617		12.990	
1994	Q1		Q2		Q3		Q4	
EP	−0.091	(−1.64)	−0.090	(−1.72)	−0.211	(−2.10)	−0.140	(−0.86)
BP	0.104	(3.75)	−0.045	(−2.41)	−0.063	(−3.40)	0.022	(0.97)
CP	0.094	(1.89)	0.062	(1.23)	0.151	(1.58)	0.213	(1.30)
SP	−0.107	(−4.66)	0.003	(0.15)	−0.081	(−3.81)	−0.038	(−1.72)
REP	0.041	(0.75)	−0.090	(−1.76)	−0.081	(−1.65)	0.021	(0.39)

TABLE 7.3 (Continued)

1994	Q1		Q2		Q3		Q4	
REP	0.041	(0.75)	−0.090	(−1.76)	−0.081	(−1.65)	0.021	(0.39)
RBP	0.081	(1.67)	0.098	(2.37)	0.065	(1.41)	−0.005	(−0.10)
RCP	0.050	(1.11)	0.042	(0.55)	−0.030	(−0.41)	−0.008	(−0.11)
RSP	−0.013	(−0.60)	0.079	(3.19)	−0.091	(−3.76)	0.120	(4.94)
PRGR	0.383	(10.51)	0.161	(6.95)	0.497	(14.71)	0.245	(6.04)
ADJ-R^2	0.030		0.022		0.067		0.021	
F-STAT	16.200		11.951		35.943		11.609	

TABLE 7.4 The Determinants of Security Returns, 1982–1994: ROB

1982	Q1		Q2		Q3		Q4	
EP	0.075	(1.08)	0.202	(3.70)	−0.127	(−2.10)	0.093	(1.75)
BP	−0.113	(−2.73)	−0.077	(−1.98)	−0.033	(−0.82)	0.214	(5.52)
CP	−0.137	(−2.30)	−0.274	(−5.29)	−0.041	(−0.74)	−0.252	(−4.94)
SP	0.169	(4.66)	0.161	(4.81)	0.176	(5.24)	0.253	(7.66)
DY	0.027	(0.94)	0.053	(2.09)	−0.149	(−6.06)	−0.131	(−5.44)
NCAV	0.075	(2.56)	−0.102	(−3.40)	−0.002	(−0.06)	0.068	(2.60)
REP	0.039	(0.35)	−0.095	(−1.24)	0.111	(1.54)	−0.010	(−0.14)
RBP	−0.057	(−0.86)	0.463	(6.42)	0.578	(7.05)	0.169	(1.83)
RCP	0.083	(1.21)	−0.104	(−1.03)	0.013	(0.15)	0.053	(0.69)
RSP	0.031	(0.77)	−0.221	(−5.83)	−0.140	(−3.55)	−0.029	(−0.73)
RDY	0.048	(1.81)	0.127	(4.16)	0.036	(1.31)	−0.032	(−1.21)
RCAV	0.070	(1.01)	−0.021	(−0.30)	−0.079	(−1.25)	−0.187	(−3.03)
PRGR	0.031	(0.52)	0.191	(4.21)	0.256	(5.11)	0.149	(3.21)
ADJ-R^2	0.016		0.056		0.073		0.098	
F-STAT	3.801		11.487		14.913		20.213	

1983	Q1		Q2		Q3		Q4	
EP	−0.104	(−4.45)	0.207	(2.82)	0.098	(1.38)	−0.391	(−4.81)
BP	−0.006	(−0.34)	0.140	(4.16)	0.104	(3.09)	0.135	(3.92)
CP	0.051	(2.10)	−0.061	(−0.96)	−0.073	(−1.26)	0.310	(5.74)
SP	0.025	(1.66)	0.008	(0.24)	0.128	(3.99)	−0.093	(−3.01)
DY	−0.144	(−11.96)	0.132	(5.12)	0.069	(2.72)	0.019	(0.76)
NCAV	0.011	(0.88)	−0.078	(−2.93)	0.012	(0.46)	−0.047	(−1.83)
REP	0.047	(1.15)	−0.040	(−0.81)	0.026	(0.52)	−0.012	(−0.22)
RBP	0.019	(0.82)	−0.114	(−2.19)	−0.254	(−4.46)	−0.233	(−3.67)
RCP	−0.133	(−1.19)	−0.096	(−1.21)	−0.115	(−1.50)	−0.234	(−3.42)
RSP	0.089	(4.06)	−0.049	(−1.33)	−0.044	(−1.17)	0.027	(0.63)
RDY	0.016	(1.37)	0.116	(3.28)	0.137	(5.38)	0.127	(4.86)
RCAV	−0.060	(−2.44)	0.078	(0.77)	−0.013	(−0.18)	−0.009	(−0.12)
PRGR	0.073	(3.06)	0.183	(4.62)	0.103	(2.37)	−0.063	(−1.05)
ADJ-R^2	0.125		0.122		0.087		0.068	
F-STAT	28.228		27.550		19.229		14.963	

Continued on following page

TABLE 7.4 (Continued)

1984	Q1		Q2		Q3		Q4	
EP	0.294	(4.28)	0.341	(4.91)	0.361	(4.32)	−0.131	(−1.70)
BP	−0.204	(−6.48)	−0.101	(−3.36)	−0.157	(−5.01)	−0.041	(−1.12)
CP	−0.129	(−2.67)	−0.067	(−1.20)	0.015	(0.25)	0.078	(1.22)
SP	0.102	(3.38)	0.111	(4.33)	0.102	(3.86)	0.104	(3.96)
DY	0.120	(4.92)	0.065	(2.80)	0.224	(9.64)	−0.118	(−5.09)
NCAV	0.080	(3.38)	−0.214	(−8.29)	−0.205	(−8.25)	−0.065	(−2.59)
REP	0.041	(0.58)	0.034	(0.51)	−0.027	(−0.43)	0.139	(2.02)
RBP	0.190	(1.34)	−0.092	(−0.54)	−0.214	(−1.26)	−0.051	(−0.31)
RCP	0.070	(1.58)	0.077	(1.79)	0.129	(2.98)	−0.140	(−3.24)
RSP	0.037	(1.32)	−0.022	(−0.92)	−0.149	(−6.07)	0.166	(5.89)
RDY	0.037	(1.77)	0.118	(5.43)	0.013	(0.62)	0.025	(1.15)
RCAV	0.021	(0.27)	−0.002	(−0.03)	−0.066	(−1.29)	0.008	(0.16)
PRGR	0.243	(4.56)	0.241	(6.46)	0.299	(6.19)	0.270	(5.25)
ADJ-R^2	0.047		0.106		0.183		0.053	
F-STAT	12.988		29.649		55.432		14.541	

1985	Q1		Q2		Q3		Q4	
EP	0.298	(5.85)	0.165	(2.43)	1.177	(13.20)	0.759	(8.00)
BP	−0.137	(−5.66)	−0.007	(−0.20)	−0.090	(−2.24)	0.013	(0.23)
CP	−0.062	(−1.50)	−0.057	(−0.99)	−0.633	(−8.84)	−0.587	(−7.58)
SP	0.037	(2.05)	0.066	(2.37)	0.169	(6.27)	0.227	(7.73)
DY	0.052	(3.28)	−0.015	(−0.67)	−0.043	(−1.84)	−0.033	(−1.30)
NCAV	−0.141	(−8.73)	0.109	(4.92)	−0.107	(−4.51)	−0.092	(−4.08)
REP	0.020	(0.81)	0.018	(0.38)	−0.042	(−0.81)	−0.095	(−2.01)
RBP	−0.116	(−3.77)	0.064	(1.16)	−0.071	(−1.32)	−0.159	(−3.00)
RCP	0.062	(1.25)	0.041	(0.52)	0.219	(3.48)	−0.037	(−0.63)
RSP	0.018	(0.81)	−0.081	(−2.80)	0.166	(5.35)	0.037	(0.99)
RDY	0.057	(4.31)	0.060	(2.67)	0.073	(3.33)	0.068	(3.15)
RCAV	0.009	(0.27)	−0.151	(−1.99)	0.100	(1.32)	0.042	(0.54)
PRGR	0.226	(7.16)	0.141	(3.91)	0.294	(5.94)	0.441	(7.03)
ADJ-R^2	0.138		0.024		0.104		0.063	
F-STAT	40.740		7.048		29.880		17.783	

TABLE 7.4 (Continued)

1986	Q1		Q2		Q3		Q4	
EP	−0.031	(−0.56)	−0.295	(−3.31)	0.160	(1.97)	−0.578	(−10.32)
BP	−0.347	(−7.93)	0.066	(1.57)	−0.071	(−2.04)	0.020	(0.63)
CP	0.044	(0.88)	0.302	(3.75)	−0.132	(−1.74)	0.435	(8.80)
SP	−0.005	(−0.20)	−0.047	(−1.72)	0.075	(3.07)	−0.017	(−0.74)
DY	0.041	(1.22)	0.257	(12.13)	0.027	(1.23)	−0.156	(−7.09)
NCAV	−0.085	(−3.80)	0.076	(2.88)	0.181	(8.04)	0.043	(2.04)
REP	−0.112	(−1.45)	−0.080	(−2.21)	0.088	(2.36)	−0.084	(−2.52)
RBP	0.127	(2.44)	−0.121	(−3.36)	0.053	(0.93)	0.122	(3.51)
RCP	0.012	(0.14)	0.265	(2.96)	0.098	(1.35)	0.092	(1.41)
RSP	−0.017	(−0.51)	−0.100	(−3.65)	−0.013	(−0.47)	0.074	(2.40)
RDY	0.015	(0.77)	0.028	(1.27)	0.062	(3.04)	0.107	(5.94)
RCAV	0.011	(0.34)	−0.033	(−0.91)	−0.045	(−1.36)	−0.043	(−1.47)
PRGR	0.273	(5.54)	0.119	(3.25)	0.258	(4.98)	0.178	(4.72)
ADJ-R^2	0.042		0.106		0.039		0.126	
F-STAT	11.926		30.563		11.072		36.973	

1987	Q1		Q2		Q3		Q4	
EP	−0.739	(−10.68)	0.006	(0.11)	0.322	(6.18)	−0.143	(−2.22)
BP	0.024	(0.76)	0.038	(1.56)	−0.071	(−2.83)	0.115	(4.29)
CP	0.595	(9.61)	−0.018	(−0.35)	−0.115	(−2.36)	0.071	(1.18)
SP	−0.037	(−1.44)	−0.011	(−0.49)	−0.126	(−5.59)	0.126	(5.21)
DY	−0.100	(−3.78)	−0.133	(−5.49)	0.281	(12.76)	−0.182	(−6.94)
NCAV	0.066	(2.72)	0.063	(2.57)	0.098	(3.62)	0.144	(3.81)
REP	−0.008	(−0.20)	0.005	(0.11)	−0.045	(−1.03)	−0.059	(−1.19)
RBP	0.125	(1.78)	−0.003	(−0.06)	−0.182	(−4.30)	0.165	(2.51)
RCP	0.010	(0.16)	0.038	(0.75)	0.014	(0.33)	−0.046	(−1.06)
RSP	0.034	(1.42)	0.099	(3.98)	−0.201	(−8.77)	0.111	(4.26)
RDY	0.102	(6.20)	0.077	(3.94)	−0.028	(−1.66)	0.100	(5.43)
RCAV	0.027	(0.67)	0.021	(0.49)	0.027	(0.72)	−0.021	(−0.49)
PRGR	0.148	(3.23)	0.212	(6.67)	0.242	(6.55)	0.122	(2.91)
ADJ-R^2	0.071		0.035		0.271		0.127	
F-STAT	22.121		11.048		103.630		41.118	

Continued on following page

TABLE 7.4 (Continued)

1988	Q1		Q2		Q3		Q4	
EP	0.325	(4.86)	0.077	(2.77)	0.144	(2.59)	−0.273	(−4.93)
BP	−0.084	(−2.71)	−0.057	(−4.14)	−0.026	(−0.89)	−0.053	(−2.08)
CP	−0.194	(−3.09)	0.042	(1.55)	0.030	(0.60)	0.287	(5.47)
SP	0.053	(2.02)	0.047	(4.16)	0.026	(1.09)	0.134	(6.34)
DY	−0.072	(−2.49)	0.011	(0.61)	−0.047	(−1.32)	−0.140	(−4.00)
NCAV	−0.044	(−1.11)	−0.028	(−1.58)	0.150	(5.19)	−0.111	(−3.50)
REP	0.076	(1.63)	0.061	(1.94)	−0.036	(−0.65)	0.107	(1.91)
RBP	0.017	(0.49)	0.011	(0.75)	0.128	(3.87)	0.000	(0.01)
RCP	0.137	(0.84)	−0.130	(−2.13)	−0.326	(−2.48)	−0.205	(−1.60)
RSP	0.050	(1.96)	−0.006	(−0.62)	−0.090	(−4.06)	0.042	(1.95)
RDY	0.037	(2.31)	0.018	(2.06)	0.025	(1.36)	0.007	(0.40)
RCAV	−0.120	(−1.57)	0.045	(1.56)	0.070	(1.25)	−0.111	(−2.17)
PRGR	0.513	(10.08)	0.030	(2.03)	0.559	(11.38)	0.226	(5.05)
ADJ-R^2	0.047		0.020		0.064		0.054	
F-STAT	14.979		6.645		20.258		17.036	
1989	Q1		Q2		Q3		Q4	
EP	0.042	(0.52)	0.419	(4.49)	0.169	(1.53)	0.056	(0.62)
BP	−0.129	(−3.75)	−0.103	(−3.58)	−0.081	(−2.05)	0.057	(1.92)
CP	0.079	(1.07)	−0.243	(−2.72)	0.068	(0.65)	0.090	(1.02)
SP	−0.030	(−1.13)	−0.101	(−4.44)	−0.182	(−7.63)	−0.004	(−0.17)
DY	0.005	(0.12)	−0.280	(−5.08)	0.254	(3.93)	−0.048	(−1.04)
NCAV	−0.064	(−1.65)	−0.156	(−3.90)	0.244	(7.85)	0.275	(6.96)
REP	0.026	(0.13)	0.525	(2.32)	0.283	(1.29)	0.139	(0.78)
RBP	−0.003	(−0.07)	0.030	(0.83)	0.104	(2.47)	−0.099	(−3.01)
RCP	0.049	(0.59)	−0.198	(−2.69)	−0.032	(−0.42)	0.129	(1.96)
RSP	0.112	(4.69)	0.170	(8.74)	0.042	(2.09)	0.021	(0.99)
RDY	0.024	(1.39)	0.042	(2.15)	−0.018	(−0.89)	−0.033	(−1.85)
RCAV	0.019	(0.26)	−0.002	(−0.04)	−0.012	(−0.28)	0.062	(1.71)
PRGR	0.328	(5.84)	0.341	(8.07)	0.449	(7.49)	0.311	(6.51)
ADJ-R^2	0.022		0.066		0.074		0.036	
F-STAT	7.156		20.809		23.356		11.330	

TABLE 7.4 (Continued)

1990	Q1		Q2		Q3		Q4	
EP	0.388	(3.55)	−0.153	(−1.14)	0.342	(4.86)	0.189	(2.62)
BP	−0.260	(−5.30)	0.054	(1.45)	−0.373	(−8.63)	0.024	(0.64)
CP	−0.348	(−3.22)	0.056	(0.51)	−0.076	(−1.13)	−0.365	(−5.37)
SP	−0.139	(−5.72)	−0.185	(−6.66)	−0.202	(−8.02)	0.095	(4.11)
DY	−0.111	(−2.10)	0.299	(6.18)	−0.057	(−1.32)	−0.149	(−3.27)
NCAV	0.152	(3.60)	0.181	(5.02)	−0.001	(−0.02)	−0.015	(−0.62)
REP	0.010	(0.22)	−0.002	(−0.04)	0.028	(0.73)	0.041	(1.33)
RBP	0.047	(1.38)	−0.001	(−0.03)	0.079	(2.56)	0.069	(2.34)
RCP	0.038	(1.24)	0.039	(0.98)	0.038	(1.10)	0.025	(0.83)
RSP	0.221	(9.33)	−0.179	(−6.66)	0.173	(7.08)	0.241	(9.94)
RDY	−0.096	(−5.60)	0.040	(1.80)	0.062	(3.33)	0.014	(0.85)
RCAV	0.067	(1.41)	−0.097	(−1.65)	0.026	(0.44)	0.009	(0.17)
PRGR	0.286	(5.20)	0.091	(2.50)	0.608	(12.35)	−0.246	(−5.11)
ADJ-R^2	0.087		0.080		0.141		0.170	
F-STAT	27.885		25.323		46.913		58.252	
1991	Q1		Q2		Q3		Q4	
EP	0.269	(4.27)	0.401	(4.06)	0.407	(3.11)	0.217	(2.90)
BP	−0.046	(−1.33)	−0.181	(−6.05)	−0.235	(−4.50)	0.360	(7.47)
CP	−0.254	(−4.32)	−0.265	(−2.74)	−0.076	(−0.55)	−0.444	(−6.31)
SP	−0.019	(−0.80)	−0.127	(−4.75)	−0.140	(−5.73)	0.248	(11.15)
DY	−0.255	(−5.38)	−0.054	(−2.15)	−0.051	(−1.96)	−0.171	(−6.67)
NCAV	0.017	(0.47)	−0.216	(−6.93)	−0.070	(−2.13)	−0.136	(−4.42)
REP	0.301	(1.21)	0.124	(1.08)	0.248	(2.01)	−0.003	(−0.03)
RBP	0.018	(0.59)	0.113	(4.08)	0.023	(0.86)	0.004	(0.15)
RCP	0.056	(1.12)	−0.080	(−2.19)	−0.018	(−0.56)	0.016	(0.62)
RSP	0.111	(4.71)	0.116	(4.84)	0.063	(2.56)	0.012	(0.51)
RDY	0.164	(7.35)	−0.014	(−0.69)	0.041	(2.18)	0.046	(2.68)
RCAV	0.022	(0.52)	0.012	(0.20)	0.087	(1.70)	0.082	(1.87)
PRGR	0.237	(5.52)	0.235	(7.92)	0.321	(7.26)	0.119	(2.42)
ADJ-R^2	0.046		0.075		0.061		0.161	
F-STAT	13.905		23.030		18.734		53.001	

Continued on following page

TABLE 7.4 (Continued)

1992	Q1		Q2		Q3		Q4	
EP	0.232	(2.92)	0.132	(2.12)	0.330	(5.27)	−0.578	(−8.92)
BP	0.000	(0.01)	0.016	(0.45)	0.035	(0.99)	0.236	(6.73)
CP	−0.084	(−1.11)	0.028	(0.60)	−0.238	(−5.08)	0.357	(6.72)
SP	−0.095	(−3.75)	−0.100	(−4.07)	0.052	(2.02)	0.018	(0.68)
DY	0.073	(1.84)	−0.054	(−1.56)	−0.321	(−9.90)	0.109	(3.22)
NCAV	−0.426	(−9.61)	0.080	(2.75)	−0.078	(−2.52)	−0.274	(−8.00)
REP	0.022	(0.43)	0.046	(0.67)	0.047	(0.75)	−0.056	(−1.04)
RBP	0.089	(2.22)	0.017	(0.44)	0.144	(3.66)	−0.081	(−1.99)
RCP	0.033	(0.80)	0.052	(1.79)	−0.008	(−0.28)	−0.012	(−0.38)
RSP	0.004	(0.14)	0.048	(2.32)	0.163	(7.39)	−0.118	(−5.07)
RDY	0.210	(11.77)	0.034	(1.63)	−0.060	(−2.94)	0.115	(5.98)
RCAV	0.003	(0.05)	−0.046	(−0.68)	0.071	(1.03)	−0.018	(−0.27)
PRGR	0.322	(6.44)	0.305	(8.28)	0.298	(7.07)	0.574	(11.50)
ADJ-R^2	0.133		0.040		0.098		0.141	
F-STAT	43.348		12.493		30.989		46.220	
1993	Q1		Q2		Q3		Q4	
EP	−0.235	(−4.09)	0.028	(0.67)	0.005	(0.09)	−0.158	(−1.59)
BP	0.011	(0.33)	−0.019	(−0.98)	−0.046	(−2.05)	0.029	(1.16)
CP	0.035	(0.74)	−0.008	(−0.21)	0.023	(0.47)	0.156	(1.67)
SP	−0.051	(−2.12)	−0.036	(−1.85)	0.121	(5.30)	0.175	(7.86)
DY	−0.047	(−1.33)	−0.091	(−4.35)	−0.107	(−3.92)	−0.128	(−3.92)
NCAV	0.063	(2.15)	−0.155	(−6.35)	0.128	(5.51)	0.043	(1.78)
REP	0.023	(0.46)	−0.015	(−0.48)	−0.028	(−0.73)	0.043	(1.15)
RBP	0.217	(1.74)	0.063	(0.67)	−0.049	(−0.44)	0.036	(0.32)
RCP	−0.005	(−0.14)	−0.088	(−2.58)	0.021	(0.51)	−0.140	(−3.40)
RSP	0.074	(4.55)	0.044	(3.05)	0.110	(6.31)	−0.058	(−2.77)
RDY	−0.015	(−0.97)	0.035	(2.19)	0.037	(1.93)	0.031	(1.62)
RCAV	0.068	(1.48)	−0.023	(−0.50)	0.006	(0.11)	−0.072	(−1.49)
PRGR	0.214	(4.60)	0.137	(5.84)	0.128	(3.64)	0.087	(2.00)
ADJ-R^2	0.022		0.023		0.039		0.031	
F-STAT	7.627		8.265		13.109		10.734	

TABLE 7.4 (Continued)

1994	Q1		Q2		Q3		Q4	
EP	−0.121	(−2.18)	−0.065	(−1.24)	−0.249	(−2.55)	−0.079	(−0.49)
BP	0.059	(2.09)	0.000	(0.00)	−0.024	(−1.23)	−0.003	(−0.11)
CP	0.110	(2.20)	0.066	(1.33)	0.179	(1.92)	0.170	(1.06)
SP	−0.122	(−5.35)	−0.018	(−0.88)	−0.103	(−4.87)	−0.031	(−1.37)
DY	−0.135	(−4.32)	−0.112	(−4.14)	−0.019	(−0.69)	0.014	(0.45)
NCAV	−0.235	(−8.71)	0.152	(6.46)	0.160	(7.72)	−0.072	(−3.75)
REP	0.024	(0.44)	−0.054	(−1.07)	−0.066	(−1.36)	0.014	(0.27)
RBP	0.053	(1.09)	0.081	(1.95)	0.064	(1.38)	−0.020	(−0.38)
RCP	0.040	(0.90)	0.074	(0.99)	0.007	(0.10)	−0.027	(−0.35)
RSP	−0.008	(−0.37)	0.166	(5.99)	−0.049	(−1.82)	0.127	(4.74)
RDY	0.131	(7.70)	−0.078	(−3.69)	−0.003	(−0.16)	−0.054	(−2.66)
RCAV	0.098	(1.64)	0.016	(0.47)	0.015	(0.48)	0.034	(1.02)
PRGR	0.319	(8.84)	0.199	(8.53)	0.559	(16.13)	0.224	(5.37)
ADJ-R^2	0.060		0.044		0.080		0.025	
F-STAT	22.598		16.387		30.209		9.523	

TABLE 7.5 The Determinants of Security Returns, 1982–1994, Firm Size and the Impact of Fundamental Data and Earnings Forecasting

R38403

	OLS CAPL*			ROBUST CAPL	
EP8403	0.3891	(2.14)	EP8403	0.4604	(2.97)
BP8403	−0.3423	(−6.93)	BP8403	−0.3031	(−7.10)
CP8403	0.0388	(0.38)	CP8403	−0.0090	(−0.10)
SP8403	0.1742	(2.30)	SP8403	0.1247	(1.88)
REP8403	0.0272	(0.08)	REP8403	0.0723	(0.25)
RBP8403	−0.0108	(−0.13)	RBP8403	−0.0519	(−0.70)
RCP8403	0.3227	(2.64)	RCP8403	0.3308	(3.23)
RSP8403	−0.0268	(−0.35)	RSP8403	−0.0204	(−0.31)
PRGR8403	0.1594	(2.48)	PRGR8403	0.1459	(2.64)
F Value	10.0100		F Value	12.5620	
R-square	0.0900		R-square	0.1126	

R38406

	OLS CAPL			ROBUST CAPL	
EP8406	0.4854	(6.28)	EP8406	0.5535	(8.65)
BP8406	0.0729	(1.77)	BP8406	0.0767	(2.29)
CP8406	−0.2385	(−3.39)	CP8406	−0.2364	(−4.16)
SP8406	0.1921	(2.95)	SP8406	0.1914	(3.65)
REP8406	−0.0435	(−0.28)	REP8406	−0.0537	(−0.37)
RBP8406	−0.1577	(−3.28)	RBP8406	−0.1063	(−2.63)
RCP8406	−0.0138	(−0.08)	RCP8406	−0.0012	(−0.01)
RSP8406	0.1476	(3.01)	RSP8406	0.1245	(3.06)
PRGR8406	0.0542	(0.76)	PRGR8406	0.0817	(1.43)
F Value	8.7560		F Value	15.6410	
R-square	0.0762		R-square	0.1349	

R38409

	OLS CAPL			ROBUST CAPL	
EP8409	0.6109	(7.71)	EP8409	0.5877	(8.56)
BP8409	0.1019	(2.50)	BP8409	0.1315	(3.88)
CP8409	−0.1816	(−2.63)	CP8409	−0.2050	(−3.55)
SP8409	0.0271	(0.43)	SP8409	0.0126	(0.24)
REP8409	0.2362	(1.87)	REP8409	0.1912	(1.76)
RBP8409	−0.3671	(−7.39)	RBP8409	−0.3954	(−9.07)
RCP8409	−0.1154	(−0.88)	RCP8409	−0.1259	(−0.85)
RSP8409	0.1553	(3.08)	RSP8409	0.1847	(4.13)
PRGR8409	0.0974	(1.76)	PRGR8409	0.1175	(2.54)
F Value	21.1010		F Value	29.3930	
R-square	0.1783		R-square	0.2345	

*CAPL = Large-Capitalized Firms; CAPM = Medium-Capitalized Firms.

	OLS CAPM*			ROBUST CAPM	
EP8403	0.1187	(0.86)	EP8403	0.3158	(2.59)
BP8403	−0.0907	(−1.56)	BP8403	−0.0923	(−1.92)
CP8403	−0.0150	(−0.15)	CP8403	−0.0867	(−1.04)
SP8403	−0.0461	(−0.55)	SP8403	0.0771	(1.05)
REP8403	−0.7583	(−3.45)	REP8403	−0.7065	(−3.94)
RBP8403	−0.0648	(−0.94)	RBP8403	0.0648	(1.12)
RCP8403	0.0269	(0.34)	RCP8403	0.0122	(0.18)
RSP8403	0.0836	(1.24)	RSP8403	−0.0720	(1.26)
PRGR8403	0.2724	(4.10)	PRGR8403	0.3055	(5.48)
F Value	4.4320		F Value	7.1010	
R-square	0.0421		R-square	0.0724	

	OLS CAPM			ROBUST CAPM	
EP8406	0.4243	(4.55)	EP8406	0.4537	(5.59)
BP8406	0.1098	(2.10)	BP8406	0.1100	(2.41)
CP8406	−0.2949	(−3.18)	CP8406	−0.3034	(−3.77)
SP8406	0.1412	(1.91)	SP8406	0.1326	(2.05)
REP8406	0.1579	(1.15)	REP8406	0.1561	(1.29)
RBP8406	−0.0401	(−0.83)	RBP8406	−0.0505	(−1.17)
RCP8406	0.2464	(1.90)	RCP8406	0.3038	(2.67)
RSP8406	0.0800	(1.52)	RSP8406	0.0831	(1.79)
PRGR8406	0.1309	(1.86)	PRGR8406	0.1298	(2.11)
F Value	4.4220		F Value	6.4870	
R-square	0.0385		R-square	0.0603	

	OLS CAPM			ROBUST CAPM	
EP8409	0.5041	(5.37)	EP8409	0.6517	(7.60)
BP8409	0.1020	(1.95)	BP8409	0.1580	(3.43)
CP8409	−0.2649	(−2.89)	CP8409	−0.3602	(−4.29)
SP8409	0.1436	(1.97)	SP8409	0.1210	(1.91)
REP8409	0.0122	(0.09)	REP8409	0.0716	(0.59)
RBP8409	−0.0773	(−1.47)	RBP8409	−0.1065	(−2.36)
RCP8409	0.2338	(1.77)	RCP8409	0.2226	(1.95)
RSP8409	0.0764	(1.34)	RSP8409	0.1161	(2.36)
PRGR8409	0.1953	(3.41)	PRGR8409	0.2465	(4.94)
F Value	7.2380		F Value	14.3240	
R-square	0.0682		R-square	0.1354	

Continued on following page

TABLE 7.5 (Continued)

R38412

	OLS CAPL			ROBUST CAPL	
EP8412	−0.0173	(−0.09)	EP8412	0.1302	(0.84)
BP8412	−0.1259	(−1.74)	BP8412	−0.1592	(−2.69)
CP8412	0.0427	(0.29)	CP8412	−0.1001	(−0.84)
SP8412	0.0960	(0.90)	SP8412	0.0701	(0.81)
REP8412	−0.3092	(−1.90)	REP8412	−0.2609	(−2.04)
RBP8412	−0.2616	(−2.82)	RBP8412	−0.2793	(−3.72)
RCP8412	0.2674	(1.89)	RCP8412	0.2682	(2.36)
RSP8412	0.2483	(3.11)	RSP8412	0.2960	(4.53)
PRGR8412	0.1229	(2.33)	PRGR8412	0.1152	(2.73)
F Value	3.7700		F Value	7.1200	
R-square	0.0288		R-square	0.0615	

R38503

	OLS CAPL			ROBUST CAPL	
EP8503	0.6910	(6.57)	EP8503	0.7928	(8.53)
BP8503	−0.0089	(−0.16)	BP8503	0.0236	(0.49)
CP8503	−0.2614	(−3.28)	CP8503	−0.3361	(−4.76)
SP8503	0.1392	(2.06)	SP8503	0.1064	(1.80)
REP8503	−0.1993	(−1.05)	REP8503	−0.1851	(−1.14)
RBP8503	−0.5722	(−5.11)	RBP8503	−0.6073	(−6.11)
RCP8503	−0.1028	(−0.54)	RCP8503	−0.1923	(−0.88)
RSP8503	0.1370	(2.11)	RSP8503	0.1478	(2.54)
PRGR8503	0.2426	(4.16)	PRGR8503	0.2384	(4.78)
F Value	15.4090		F Value	22.7160	
R-square	0.1331		R-square	0.1878	

R38506

	OLS CAPL			ROBUST CAPL	
EP8506	−0.5116	(−4.32)	EP8506	−0.5018	(−5.27)
BP8506	−0.0696	(−1.50)	BP8506	−0.0680	(−1.83)
CP8506	0.3587	(3.97)	CP8506	0.3502	(4.77)
SP8506	0.0471	(0.74)	SP8506	0.0102	(0.20)
REP8506	0.2863	(1.35)	REP8506	0.2744	(1.65)
RBP8506	0.1425	(0.88)	RBP8506	0.1551	(1.17)
RCP8506	−0.0137	(−0.12)	RCP8506	0.0077	(0.09)
RSP8506	−0.0548	(−0.86)	RSP8506	−0.0858	(−1.62)
PRGR8506	−0.0325	(−0.49)	PRGR8506	−0.0824	(−1.54)
F Value	3.1740		F Value	4.5600	
R-square	0.0224		R-square	0.0361	

	OLS CAPM			ROBUST CAPM	
EP8412	−0.2642	(−1.64)	EP8412	−0.2659	(−1.90)
BP8412	−0.1674	(−2.00)	BP8412	−0.1793	(−2.55)
CP8412	0.0330	(0.24)	CP8412	0.0526	(0.45)
SP8412	0.2289	(2.47)	SP8412	0.2092	(2.69)
REP8412	0.1610	(1.07)	REP8412	0.2168	(1.62)
RBP8412	0.1121	(1.30)	RBP8412	0.1121	(1.54)
RCP8412	0.0486	(0.30)	RCP8412	0.0071	(0.05)
RSP8412	0.0585	(0.77)	RSP8412	0.0565	(0.89)
PRGR8412	0.2102	(3.55)	PRGR8412	0.2179	(4.38)
F Value	3.4750		F Value	4.7320	
R-square	0.0291		R-square	0.0433	

	OLS CAPM			ROBUST CAPM	
EP8503	0.5435	(4.96)	EP8503	0.5742	(5.96)
BP8503	0.0147	(0.20)	BP8503	0.0864	(1.35)
CP8503	−0.2950	(−3.02)	CP8503	−0.2824	(−3.34)
SP8503	0.1650	(2.16)	SP8503	0.1367	(2.10)
REP8503	0.0816	(0.68)	REP8503	0.0548	(0.51)
RBP8503	−0.3514	(−3.18)	RBP8503	−0.3786	(−3.89)
RCP8503	0.0624	(0.20)	RCP8503	0.0492	(0.18)
RSP8503	0.2232	(3.26)	RSP8503	0.2186	(3.60)
PRGR8503	0.2864	(4.12)	PRGR8503	0.3138	(5.19)
F Value	6.3010		F Value	9.4370	
R-square	0.0610		R-square	0.0936	

	OLS CAPM			ROBUST CAPM	
EP8506	0.2068	(1.76)	EP8506	0.1977	(1.93)
BP8506	−0.0544	(−0.90)	BP8506	−0.0493	(−0.93)
CP8506	0.0135	(0.13)	CP8506	0.0096	(0.10)
SP8506	0.1838	(2.84)	SP8506	0.1561	(2.79)
REP8506	0.0268	(0.17)	REP8506	0.0566	(0.39)
RBP8506	−0.0701	(−0.47)	RBP8506	−0.0908	(−0.69)
RCP8506	−0.3012	(−2.77)	RCP8506	−0.3267	(−3.46)
RSP8506	0.0943	(1.63)	RSP8506	0.1049	(2.07)
PRGR8506	0.1957	(2.90)	PRGR8506	0.1854	(3.17)
F Value	3.4620		F Value	4.1870	
R-square	0.0281		R-square	0.0361	

Continued on following page

TABLE 7.5 (Continued)

R38509

	OLS CAPL			ROBUST CAPL	
EP8509	0.7763	(7.15)	EP8509	0.8694	(8.94)
BP8509	−0.0209	(−0.51)	BP8509	−0.0179	(−0.50)
CP8509	−0.4856	(−5.63)	CP8509	−0.5733	(−7.35)
SP8509	0.1619	(3.16)	SP8509	0.1762	(3.95)
REP8509	−0.0736	(−0.52)	REP8509	−0.0293	(−0.24)
RBP8509	−0.5360	(−3.80)	RBP8509	−0.5240	(−4.19)
RCP8509	0.0838	(0.93)	RCP8509	0.0542	(0.68)
RSP8509	0.3282	(5.34)	RSP8509	0.3096	(5.66)
PRGR8509	0.2327	(4.44)	PRGR8509	0.1994	(4.35)
F Value	11.4770		F Value	15.2180	
R-square	0.1006		R-square	0.1318	

R38512

	OLS CAPL			ROBUST CAPL	
EP8512	0.5993	(4.19)	EP8512	0.5763	(4.63)
BP8512	0.2017	(3.34)	BP8512	0.1794	(3.46)
CP8512	−0.6432	(−5.69)	CP8512	−0.6861	(−7.05)
SP8512	0.0409	(0.66)	SP8512	0.0491	(0.94)
REP8512	0.3354	(1.76)	REP8512	0.3039	(1.82)
RBP8512	−0.6221	(−4.10)	RBP8512	−0.6261	(−4.64)
RCP8512	0.2056	(2.17)	RCP8512	0.1861	(2.27)
RSP8512	−0.0301	(−0.35)	RSP8512	−0.0548	(−0.71)
PRGR8512	0.0896	(1.56)	PRGR8512	0.0812	(1.68)
F Value	12.5360		F Value	19.0920	
R-square	0.1094		R-square	0.1617	

R38603

	OLS CAPL			ROBUST CAPL	
EP8603	0.4693	(4.27)	EP8603	0.4277	(4.34)
BP8603	−0.2881	(−4.13)	BP8603	−0.2767	(−4.55)
CP8603	−0.1928	(−1.78)	CP8603	−0.1541	(−1.58)
SP8603	−0.0676	(−1.20)	SP8603	−0.0866	(−1.75)
REP8603	0.0676	(0.28)	REP8603	−0.0780	(−0.28)
RBP8603	−0.0844	(−0.72)	RBP8603	−0.2035	(−1.84)
RCP8603	0.2011	(0.30)	RCP8603	0.4427	(0.74)
RSP8603	−0.0067	(−0.08)	RSP8603	0.0918	(1.23)
PRGR8603	0.1155	(2.11)	PRGR8603	0.1269	(2.67)
F Value	11.3250		F Value	14.2180	
R-square	0.0998		R-square	0.1243	

	OLS CAPM			ROBUST CAPM	
EP8509	0.2450	(2.05)	EP8509	0.2498	(2.41)
BP8509	−0.0791	(−1.27)	BP8509	−0.0577	(−1.06)
CP8509	−0.4464	(−4.22)	CP8509	−0.4430	(−4.82)
SP8509	0.1311	(1.99)	SP8509	0.1331	(2.36)
REP8509	−0.1286	(−0.55)	REP8509	−0.2720	(−1.23)
RBP8509	0.0742	(0.54)	RBP8509	0.0788	(0.66)
RCP8509	0.2505	(2.33)	RCP8509	0.2736	(2.79)
RSP8509	0.0892	(1.42)	RSP8509	0.0798	(1.47)
PRGR8509	0.2518	(3.80)	PRGR8509	0.2852	(4.95)
F Value	5.8600		F Value	7.8000	
R-square	0.0557		R-square	0.0763	

	OLS CAPM			ROBUST CAPM	
EP8512	0.6161	(4.60)	EP8512	0.6683	(5.73)
BP8512	−0.1677	(−2.13)	BP8512	−0.1489	(−2.16)
CP8512	−0.3475	(−3.17)	CP8512	−0.3826	(−4.08)
SP8512	0.3198	(5.05)	SP8512	0.2962	(5.36)
REP8512	−0.1436	(−0.88)	REP8512	−0.1857	(−1.30)
RBP8512	−0.2570	(−1.90)	RBP8512	−0.2489	(−2.15)
RCP8512	−0.0769	(−0.68)	RCP8512	−0.1248	(−1.29)
RSP8512	−0.0661	(−0.87)	RSP8512	−0.0435	(−0.67)
PRGR8512	0.0727	(1.14)	PRGR8512	0.1016	(1.85)
F Value	12.0550		F Value	15.9780	
R-square	0.1202		R-square	0.1562	

	OLS CAPM			ROBUST CAPM	
EP8603	−0.0412	(−0.34)	EP8603	−0.0191	(−0.18)
BP8603	−0.3680	(−3.75)	BP8603	−0.3563	(−4.16)
CP8603	−0.0117	(−0.09)	CP8603	−0.0159	(−0.14)
SP8603	0.0561	(0.83)	SP8603	0.0551	(0.93)
REP8603	−0.2264	(−0.78)	REP8603	−0.2399	(−0.94)
RBP8603	0.2710	(2.03)	RBP8603	0.1646	(1.38)
RCP8603	0.6857	(3.50)	RCP8603	0.6493	(3.89)
RSP8603	−0.0321	(−0.36)	RSP8603	0.0073	(0.09)
PRGR8603	0.2699	(3.93)	PRGR8603	0.2606	(4.37)
F Value	5.7950		F Value	7.4440	
R-square	0.0560		R-square	0.0739	

Continued on following page

TABLE 7.5 (Continued)

R38606

	OLS CAPL			ROBUST CAPL	
EP8606	−0.2592	(−1.80)	EP8606	−0.4424	(−3.56)
BP8606	0.3124	(5.64)	BP8606	0.3086	(6.56)
CP8606	0.5747	(5.23)	CP8606	0.7127	(7.61)
SP8606	−0.1508	(−2.79)	SP8606	−0.1876	(−4.07)
REP8606	−0.1864	(−1.79)	REP8606	−0.1436	(−1.61)
RBP8606	−0.1013	(−0.88)	RBP8606	−0.1587	(−1.64)
RCP8606	−0.2293	(−0.62)	RCP8606	−0.1780	(−0.58)
RSP8606	0.1416	(2.41)	RSP8606	0.1491	(3.00)
PRGR8606	0.0394	(0.62)	PRGR8606	0.0184	(0.35)
F Value	16.8430		F Value	26.4770	
R-square	0.1432		R-square	0.2121	

R38609

	OLS CAPL			ROBUST CAPL	
EP8609	−0.2906	(−2.08)	EP8609	−0.3710	(−3.21)
BP8609	−0.2483	(−4.36)	BP8609	−0.2534	(−5.37)
CP8609	0.2438	(2.37)	CP8609	0.3239	(3.78)
SP8609	0.0625	(1.05)	SP8609	0.0382	(0.79)
REP8609	−0.0961	(−0.98)	REP8609	−0.1358	(−1.69)
RBP8609	0.1333	(1.00)	RBP8609	0.1263	(1.14)
RCP8609	0.6289	(2.18)	RCP8609	0.5623	(2.47)
RSP8609	−0.0408	(−0.60)	RSP8609	−0.0499	(−0.87)
PRGR8609	0.1271	(2.25)	PRGR8609	0.1260	(2.76)
F Value	3.7920		F Value	6.2370	
R-square	0.0289		R-square	0.0530	

R38612

	OLS CAPL			ROBUST CAPL	
EP8612	−1.2206	(−7.59)	EP8612	−1.1918	(−8.86)
BP8612	−0.4855	(−6.59)	BP8612	−0.5049	(−8.20)
CP8612	0.5986	(4.97)	CP8612	0.5278	(5.25)
SP8612	0.0834	(1.06)	SP8612	0.0618	(0.95)
REP8612	0.2073	(1.96)	REP8612	0.2204	(2.31)
RBP8612	0.9848	(4.19)	RBP8612	0.9545	(4.75)
RCP8612	−2.3343	(−1.13)	RCP8612	2.0348	(0.86)
RSP8612	−0.0154	(−0.17)	RSP8612	−0.0583	(−0.75)
PRGR8612	0.1851	(3.07)	PRGR8612	0.1530	(3.03)
F Value	15.4690		F Value	21.4520	
R-square	0.1334		R-square	0.1787	

	OLS CAPM			ROBUST CAPM	
EP8606	0.3075	(2.47)	EP8606	0.3851	(3.53)
BP8606	0.3346	(4.86)	BP8606	0.3955	(6.67)
CP8606	0.0374	(0.38)	CP8606	0.0047	(0.05)
SP8606	0.0320	(0.49)	SP8606	0.0106	(0.19)
REP8606	−0.0439	(−0.38)	REP8606	−0.0224	(−0.22)
RBP8606	−0.0044	(−0.04)	RBP8606	−0.0629	(−0.74)
RCP8606	0.8218	(2.70)	RCP8606	0.8723	(3.47)
RSP8606	0.0630	(1.27)	RSP8606	0.0641	(1.50)
PRGR8606	0.2728	(4.23)	PRGR8606	0.2494	(4.51)
F Value	8.1760		F Value	12.0750	
R-square	0.0810		R-square	0.1197	

	OLS CAPM			ROBUST CAPM	
EP8609	0.0482	(0.35)	EP8609	0.0325	(0.27)
BP8609	−0.3272	(−4.63)	BP8609	−0.2795	(−4.56)
CP8609	−0.1342	(−1.30)	CP8609	−0.1072	(−1.20)
SP8609	0.1877	(2.82)	SP8609	0.1339	(2.34)
REP8609	0.1058	(0.82)	REP8609	0.1489	(1.35)
RBP8609	0.2951	(2.75)	RBP8609	0.2794	(3.05)
RCP8609	−0.3339	(−1.14)	RCP8609	−0.3352	(−1.38)
RSP8609	0.0300	(0.57)	RSP8609	0.0124	(0.28)
PRGR8609	0.1951	(3.13)	PRGR8609	0.2046	(3.87)
F Value	5.2260		F Value	6.0060	
R-square	0.0509		R-square	0.0597	

	OLS CAPM			ROBUST CAPM	
EP8612	−0.4268	(−3.35)	EP8612	−0.4903	(−4.38)
BP8612	−0.1731	(−2.11)	BP8612	−0.1650	(−2.34)
CP8612	0.1087	(1.15)	CP8612	0.1473	(1.77)
SP8612	−0.0331	(−0.40)	SP8612	−0.0348	(−0.48)
REP8612	0.2415	(1.84)	REP8612	0.2261	(1.99)
RBP8612	−0.0487	(−0.29)	RBP8612	0.0783	(0.54)
RCP8612	0.7956	(3.81)	RCP8612	0.7846	(4.53)
RSP8612	0.1179	(1.74)	RSP8612	0.0476	(0.79)
PRGR8612	0.2088	(3.29)	PRGR8612	0.1634	(3.01)
F Value	6.3250		F Value	7.6500	
R-square	0.0657		R-square	0.0807	

Continued on following page

TABLE 7.5 (Continued)

R38703

	OLS CAPL			ROBUST CAPL	
EP8703	−0.5424	(−4.34)	EP8703	−0.5192	(−4.79)
BP8703	−0.2457	(−3.79)	BP8703	−0.2472	(−4.59)
CP8703	0.4062	(3.79)	CP8703	0.3371	(3.71)
SP8703	0.1686	(2.79)	SP8703	0.1582	(3.13)
REP8703	−0.3781	(−2.56)	REP8703	−0.4142	(−3.36)
RBP8703	0.6172	(1.74)	RBP8703	1.0476	(3.35)
RCP8703	−0.0362	(−0.10)	RCP8703	−0.0505	(−0.16)
RSP8703	−0.0178	(−0.24)	RSP8703	−0.0931	(−1.45)
PRGR8703	0.1757	(2.76)	PRGR8703	0.1953	(3.63)
F Value	6.4310		F Value	8.9350	
R-square	0.0548		R-square	0.0783	

R38706

	OLS CAPL			ROBUST CAPL	
EP8706	−0.2860	(−1.58)	EP8706	−0.3013	(−2.01)
BP8706	−0.0239	(−0.44)	BP8706	−0.0655	(−1.45)
CP8706	0.0393	(0.24)	CP8706	0.0113	(0.08)
SP8706	0.0325	(0.61)	SP8706	0.0301	(0.69)
REP8706	0.3130	(1.25)	REP8706	0.3117	(1.57)
RBP8706	0.0018	(0.03)	RBP8706	−0.0167	(−0.30)
RCP8706	0.0236	(0.12)	RCP8706	0.0832	(0.52)
RSP8706	0.1372	(2.42)	RSP8706	0.1574	(3.38)
PRGR8706	0.1723	(2.77)	PRGR8706	0.1245	(2.45)
F Value	2.7750		F Value	4.7340	
R-square	0.0180		R-square	0.0372	

R38709

	OLS CAPL			ROBUST CAPL	
EP8709	0.3218	(1.63)	EP8709	0.3783	(2.13)
BP8709	0.1609	(2.71)	BP8709	0.1929	(3.59)
CP8709	0.2688	(1.56)	CP8709	0.2322	(1.48)
SP8709	−0.3480	(−6.41)	SP8709	−0.3791	(−7.90)
REP8709	−0.4368	1.17)	REP8709	−0.4709	(−1.48)
RBP8709	−0.2228	(−3.31)	RBP8709	−0.2321	(−3.88)
RCP8709	−0.1023	(−0.61)	RCP8709	−0.1261	(−0.87)
RSP8709	0.0867	(1.48)	RSP8709	0.0918	(1.76)
PRGR8709	0.3007	5.08)	PRGR8709	0.2917	(5.59)
F Value	12.8790		F Value	17.7970	
R-square	0.1098		R-square	0.1485	

	OLS CAPM			ROBUST CAPM	
EP8703	−0.3822	(−2.86)	EP8703	−0.3667	(−2.97)
BP8703	−0.0701	(−0.81)	BP8703	−0.0737	(−0.98)
CP8703	0.2035	(2.01)	CP8703	0.1982	(2.14)
SP8703	0.1120	(1.41)	SP8703	0.1121	(1.64)
REP8703	0.8982	(1.84)	REP8703	0.7141	(1.73)
RBP8703	0.5514	(2.69)	RBP8703	0.4495	(2.45)
RCP8703	0.3445	(1.66)	RCP8703	0.2397	(1.36)
RSP8703	0.1430	(2.40)	RSP8703	0.1129	(2.15)
PRGR8703	0.0391	(0.58)	PRGR8703	0.0529	(0.90)
F Value	5.0970		F Value	4.4920	
R-square	0.0517		R-square	0.0444	

	OLS CAPM			ROBUST CAPM	
EP8706	−0.2518	(−1.52)	EP8706	−0.2328	(−1.47)
BP8706	−0.0629	(−1.02)	BP8706	−0.0747	(−1.36)
CP8706	0.2098	(1.45)	CP8706	0.1484	(1.03)
SP8706	0.0555	(0.96)	SP8706	0.0762	(1.50)
REP8706	0.0615	(0.22)	REP8706	0.0081	(0.03)
RBP8706	0.0963	(1.50)	RBP8706	0.0751	(1.33)
RCP8706	0.0667	(0.46)	RCP8706	0.0454	(0.31)
RSP8706	0.1132	(2.08)	RSP8706	0.1066	(2.21)
PRGR8706	0.2492	(3.60)	PRGR8706	0.2239	(3.70)
F Value	3.5210		F Value	3.6540	
R-square	0.0303		R-square	0.0319	

	OLS CAPM			ROBUST CAPM	
EP8709	0.3265	(1.37)	EP8709	0.4275	(1.91)
BP8709	0.2772	(3.51)	BP8709	0.3328	(4.55)
CP8709	0.2620	(1.18)	CP8709	0.1936	(0.92)
SP8709	−0.2973	(−4.59)	SP8709	−0.3090	5.19)
REP8709	0.2686	(0.58)	REP8709	0.2646	(0.63)
RBP8709	−0.0533	(−0.75)	RBP8709	−0.0396	(−0.61)
RCP8709	0.2767	(1.57)	RCP8709	0.2809	(1.72)
RSP8709	−0.1337	2.15)	RSP8709	−0.1661	(−2.91)
PRGR8709	0.2225	(3.30)	PRGR8709	0.2162	(3.51)
F Value	8.3730		F Value	11.2970	
R-square	0.0844		R-square	0.1140	

Continued on following page

182 HISTORICAL DATA AND ANALYSTS' FORECASTS

TABLE 7.5 (Continued)

R38712

	OLS CAPL			ROBUST CAPL	
EP8712	−0.4125	(−2.50)	EP8712	−0.4524	(−3.37)
BP8712	0.0444	(0.84)	BP8712	0.0588	(1.36)
CP8712	0.1782	(1.11)	CP8712	0.2040	(1.54)
SP8712	0.2703	(4.58)	SP8712	0.2041	(4.12)
REP8712	0.8000	(2.09)	REP8712	−1.0973	(−2.06)
RBP8712	0.0956	(1.17)	RBP8712	0.1361	(2.05)
RCP8712	0.1005	(0.67)	RCP8712	−0.2290	(−1.53)
RSP8712	0.0524	(0.68)	RSP8712	0.0455	(0.73)
PRGR8712	−0.0109	(−0.22)	PRGR8712	−0.0119	(−0.30)
F Value	8.1190		F Value	11.5360	
R-square	0.0689		R-square	0.0992	

R38803

	OLS CAPL			ROBUST CAPL	
EP8803	0.3756	(2.09)	EP8803	0.4230	(2.93)
BP8803	0.2171	(3.52)	BP8803	0.1944	(3.76)
CP8803	−0.3384	(−1.92)	CP8803	−0.3465	(−2.48)
SP8803	0.0853	(1.46)	SP8803	0.0941	(1.97)
REP8803	−0.0748	(−1.04)	REP8803	−0.0622	(−1.05)
RBP8803	0.1871	(2.20)	RBP8803	0.1656	(2.39)
RCP8803	0.0130	(0.22)	RCP8803	0.0319	(0.61)
RSP8803	0.0148	(0.28)	RSP8803	0.0244	(0.56)
PRGR8803	0.2194	(5.48)	PRGR8803	0.2307	(7.14)
F Value	8.3880		F Value	12.2640	
R-square	0.0714		R-square	0.1052	

R38806

	OLS CAPL			ROBUST CAPL	
EP8806	0.2456	(2.42)	EP8806	0.2855	(3.46)
BP8806	0.1192	(2.09)	BP8806	0.1207	(2.58)
CP8806	−0.1312	(−1.17)	CP8806	−0.1599	(−1.75)
SP8806	0.1459	(2.59)	SP8806	0.1116	(2.37)
REP8806	−0.0920	(−0.73)	REP8806	−0.0895	(−0.84)
RBP8806	−0.0521	(−0.85)	RBP8806	−0.0268	(−0.53)
RCP8806	0.1671	(1.43)	RCP8806	0.0695	(0.69)
RSP8806	0.0378	(0.72)	RSP8806	0.0460	(1.03)
PRGR8806	−0.0788	(−1.29)	PRGR8806	−0.0940	(−1.89)
F Value	3.5440		F Value	4.8540	
R-square	0.0252		R-square	0.0377	

	OLS CAPM			ROBUST CAPM	
EP8712	−0.6777	(−3.21)	EP8712	−0.5168	(−2.83)
BP8712	−0.1404	(−2.01)	BP8712	−0.0822	(−1.39)
CP8712	0.5746	(2.73)	CP8712	0.4864	(2.64)
SP8712	0.2692	(3.96)	SP8712	0.2974	(5.10)
REP8712	−1.0582	(−1.96)	REP8712	−0.6954	(−1.48)
RBP8712	0.0487	(0.62)	RBP8712	0.0320	(0.48)
RCP8712	−0.1923	(−0.97)	RCP8712	−0.1743	(−1.06)
RSP8712	0.0228	(0.32)	RSP8712	0.0732	(1.22)
PRGR8712	0.1852	(3.00)	PRGR8712	0.1668	(3.26)
F Value	5.7790		F Value	8.0190	
R-square	0.0575		R-square	0.0822	

	OLS CAPM			ROBUST CAPM	
EP8803	0.2654	(1.24)	EP8803	0.2029	(1.15)
BP8803	−0.2407	(−3.03)	BP8803	−0.2560	(−3.90)
CP8803	−0.1268	(−0.60)	CP8803	−0.1953	(−1.12)
SP8803	0.2577	(3.81)	SP8803	0.2475	(4.39)
REP8803	−0.1116	(−1.35)	REP8803	−0.0837	(−1.22)
RBP8803	0.0895	(1.01)	RBP8803	0.1000	(1.35)
RCP8803	0.0987	(1.34)	RCP8803	0.0874	(1.35)
RSP8803	0.0446	(0.76)	RSP8803	0.0181	(0.37)
PRGR8803	0.1829	(3.63)	PRGR8803	0.1595	(3.90)
F Value	4.5640		F Value	5.3130	
R-square	0.0438		R-square	0.0525	

	OLS CAPM			ROBUST CAPM	
EP8806	−0.1227	(−0.92)	EP8806	−0.0949	(−0.83)
BP8806	−0.1522	(−2.15)	BP8806	−0.0873	(−1.42)
CP8806	0.1920	(1.09)	CP8806	0.1711	(1.07)
SP8806	0.2461	(3.99)	SP8806	0.2413	(4.55)
REP8806	0.0154	(0.14)	REP8806	0.0187	(0.21)
RBP8806	0.1608	(2.14)	RBP8806	0.1457	(2.30)
RCP8806	−0.1801	(−1.65)	RCP8806	−0.2104	(−2.20)
RSP8806	−0.0131	(−0.23)	RSP8806	−0.0154	(−0.32)
PRGR8806	0.0806	(1.30)	PRGR8806	0.0881	(1.68)
F Value	3.2970		F Value	4.5000	
R-square	0.0264		R-square	0.0396	

Continued on following page

TABLE 7.5 (Continued)

R38809

	OLS CAPL			ROBUST CAPL	
EP8809	−0.3066	(−2.00)	EP8809	−0.3349	(−2.66)
BP8809	−0.1239	(−1.66)	BP8809	−0.1098	(−1.80)
CP8809	0.5367	(3.82)	CP8809	0.5311	(4.59)
SP8809	0.0774	(1.23)	SP8809	0.0446	(0.87)
REP8809	−0.0345	(−0.32)	REP8809	−0.0398	(−0.46)
RBP8809	0.0274	(0.46)	RBP8809	0.0615	(1.24)
RCP8809	−0.1023	(−0.97)	RCP8809	0.0028	(0.03)
RSP8809	−0.0333	(−0.62)	RSP8809	−0.0528	(−1.20)
PRGR8809	0.1913	(3.31)	PRGR8809	0.1704	(3.62)
F Value	3.6890		F Value	4.5950	
R-square	0.0267		R-square	0.0354	

R38903

	OLS CAPL			ROBUST CAPL	
EP8903	−0.2156	(−0.86)	EP8903	−0.1258	(−0.62)
BP8903	−0.0384	(−0.37)	BP8903	−0.0013	(−0.01)
CP8903	0.1173	(0.48)	CP8903	0.0061	(0.03)
SP8903	−0.0589	(−0.67)	SP8903	−0.1326	(−1.73)
REP8903	0.1139	(0.59)	REP8903	0.1171	(0.77)
RBP8903	0.1347	(2.26)	RBP8903	0.1147	(2.35)
RCP8903	0.2155	(0.68)	RCP8903	0.6611	(1.05)
RSP8903	−0.0786	(−1.40)	RSP8903	−0.0887	(−1.93)
PRGR8903	0.0321	(0.64)	PRGR8903	0.0640	(1.55)
F Value	0.8140		F Value	1.6640	
R-square	−0.0019		R-square	0.0067	

R38906

	OLS CAPL			ROBUST CAPL	
EP8906	0.2834	(2.76)	EP8906	0.2785	(3.34)
BP8906	−0.2319	(−2.99)	BP8906	−0.1935	(−3.10)
CP8906	−0.1892	(−2.29)	CP8906	−0.1746	(−2.62)
SP8906	−0.0047	(−0.06)	SP8906	−0.0093	(−0.16)
REP8906	0.3992	(1.91)	REP8906	0.4155	(2.48)
RBP8906	0.0366	(0.79)	RBP8906	0.0399	(1.07)
RCP8906	−9.0321	(−2.84)	RCP8906	−8.9696	(−3.62)
RSP8906	−0.0071	(−0.17)	RSP8906	−0.0237	(−0.70)
PRGR8906	0.3195	(5.36)	PRGR8906	0.2763	(5.73)
F Value	7.2840		F Value	9.2170	
R-square	0.0583		R-square	0.0752	

	OLS CAPM			ROBUST CAPM	
EP8809	−0.4151	(−1.80)	EP8809	−0.3951	(−1.99)
BP8809	−0.0066	(−0.07)	BP8809	−0.0237	(−0.28)
CP8809	0.3566	(1.44)	CP8809	0.3187	(1.49)
SP8809	0.0112	(0.14)	SP8809	0.0179	(0.26)
REP8809	0.0649	(0.77)	REP8809	0.0381	(0.52)
RBP8809	0.1377	(2.02)	RBP8809	0.1422	(2.41)
RCP8809	0.0702	(0.66)	RCP8809	0.0654	(0.73)
RSP8809	−0.1200	(−2.21)	RSP8809	−0.1163	(−2.50)
PRGR8809	0.0988	(1.51)	PRGR8809	0.1190	(2.14)
F Value	1.2790		F Value	1.7540	
R-square	0.0033		R-square	0.0089	

	OLS CAPM			ROBUST CAPM	
EP8903	−0.2653	(−0.85)	EP8903	−0.3048	(−1.05)
BP8903	−0.4557	(−3.36)	BP8903	−0.4621	(−3.92)
CP8903	0.5814	(1.71)	CP8903	0.6565	(2.08)
SP8903	−0.2008	(−1.73)	SP8903	−0.1912	(−1.90)
REP8903	−1.0837	(−0.61)	REP8903	−0.4694	(−0.98)
RBP8903	0.0913	(1.31)	RBP8903	0.0779	(1.27)
RCP8903	−0.0255	(−0.09)	RCP8903	−0.0308	(−0.11)
RSP8903	0.1674	(2.76)	RSP8903	0.1525	(2.84)
PRGR8903	0.2009	(3.26)	PRGR8903	0.1904	(3.54)
F Value	5.7590		F Value	6.6280	
R-square	0.0538		R-square	0.0630	

	OLS CAPM			ROBUST CAPM	
EP8906	0.3963	(3.13)	EP8906	0.3851	(3.38)
BP8906	−0.3183	(−3.48)	BP8906	−0.3084	(−3.87)
CP8906	−0.2589	(−2.57)	CP8906	−0.2345	(−2.57)
SP8906	0.0067	(0.08)	SP8906	0.0236	(0.31)
REP8906	−4.4730	(−0.98)	REP8906	−4.3947	(−1.14)
RBP8906	0.1716	(3.07)	RBP8906	0.1202	(2.43)
RCP8906	−0.1560	(−0.56)	RCP8906	−0.1759	(−0.69)
RSP8906	0.0734	(1.72)	RSP8906	0.0745	(2.00)
PRGR8906	0.1816	(2.96)	PRGR8906	0.1815	(3.43)
F Value	7.7490		F Value	7.8790	
R-square	0.0687		R-square	0.0700	

Continued on following page

TABLE 7.5 (Continued)

R38909

	OLS CAPL			ROBUST CAPL	
EP8909	−0.7120	(−4.04)	EP8909	−0.8415	(−4.99)
BP8909	−0.4266	(−2.66)	BP8909	−0.4094	(−2.84)
CP8909	0.4466	(3.57)	CP8909	0.5618	(5.69)
SP8909	−0.8123	(−5.11)	SP8909	−0.8413	(−5.86)
REP8909	0.4002	(1.75)	REP8909	0.3921	(2.00)
RBP8909	0.0705	(1.15)	RBP8909	0.0643	(1.16)
RCP8909	12.7265	(3.22)	RCP8909	13.1151	(3.82)
RSP8909	−0.1073	(−1.89)	RSP8909	−0.1154	(−2.27)
PRGR8909	0.2009	(3.18)	PRGR8909	0.2146	(3.85)
F Value	10.4690		F Value	13.8350	
R-square	0.0853		R-square	0.1122	

R38912

	OLS CAPL			ROBUST CAPL	
EP8912	−1.0076	(−2.60)	EP8912	−0.8754	(−2.67)
BP8912	−2.2954	(−4.00)	BP8912	−2.1673	(−4.47)
CP8912	0.7790	(2.17)	CP8912	0.6903	(2.28)
SP8912	0.3318	(0.65)	SP8912	0.4176	(0.98)
REP8912	0.1049	(0.49)	REP8912	0.1089	(0.53)
RBP8912	0.1662	(1.86)	RBP8912	0.1796	(2.35)
RCP8912	−21.4049	(−1.73)	RCP8912	−21.8883	(−2.07)
RSP8912	0.0310	(0.35)	RSP8912	0.0060	(0.08)
PRGR8912	0.1340	(2.21)	PRGR8912	0.0964	(1.89)
F Value	3.8150		F Value	4.3480	
R-square	0.0271		R-square	0.0321	

R39003

	OLS CAPL			ROBUST CAPL	
EP9003	0.3321	(1.05)	EP9003	0.5009	(1.78)
BP9003	−2.1816	(−7.95)	BP9003	−2.2237	(−9.57)
CP9003	−0.5379	(−3.11)	CP9003	−0.6453	(−4.14)
SP9003	0.5286	(1.71)	SP9003	0.7208	(2.83)
REP9003	0.0065	(0.09)	REP9003	0.0086	(0.14)
RBP9003	0.8034	(2.35)	RBP9003	0.8987	(3.13)
RCP9003	0.2855	(2.36)	RCP9003	0.2708	(2.67)
RSP9003	0.1435	(2.31)	RSP9003	0.1025	(1.98)
PRGR9003	0.2388	(3.92)	PRGR9003	0.1723	(3.43)
F Value	15.2170		F Value	21.1110	
R-square	0.1226		R-square	0.1653	

	OLS CAPM			ROBUST CAPM	
EP8909	−0.5146	(−2.14)	EP8909	−0.2062	(−0.94)
BP8909	−0.5773	(−3.51)	BP8909	−0.6249	(−4.26)
CP8909	0.5450	(2.99)	CP8909	0.4800	(3.00)
SP8909	−0.5363	(−2.80)	SP8909	−0.5730	(−3.40)
REP8909	−0.9296	(−0.23)	REP8909	0.1360	(0.04)
RBP8909	−0.0955	(−1.34)	RBP8909	−0.1192	(−1.92)
RCP8909	0.4637	(1.53)	RCP8909	0.4681	(1.85)
RSP8909	0.0590	(1.00)	RSP8909	0.0639	(1.23)
PRGR8909	0.0901	(1.31)	PRGR8909	0.1167	(1.96)
F Value	4.6190		F Value	7.1270	
R-square	0.0390		R-square	0.0643	

	OLS CAPM			ROBUST CAPM	
EP8912	−0.1828	(−0.40)	EP8912	−0.3256	(−0.78)
BP8912	−2.8148	(−3.95)	BP8912	−2.6414	(−4.30)
CP8912	0.1103	(0.21)	CP8912	0.1036	(0.23)
SP8912	0.6785	(1.14)	SP8912	0.4631	(0.90)
REP8912	−6.2630	(−2.21)	REP8912	−5.2213	(−2.03)
RBP8912	−0.0819	(−0.88)	RBP8912	−0.0695	(−0.86)
RCP8912	0.3040	(1.08)	RCP8912	0.3086	(1.23)
RSP8912	0.1490	(1.75)	RSP8912	0.1518	(2.06)
PRGR8912	0.1780	(2.97)	PRGR8912	0.1660	(3.22)
F Value	4.3030		F Value	5.2240	
R-square	0.0362		R-square	0.0458	

	OLS CAPM			ROBUST CAPM	
EP9003	0.2933	(0.61)	EP9003	−0.0472	(−0.11)
BP9003	−1.3490	(−4.30)	BP9003	−1.3854	(−5.07)
CP9003	−0.5825	(−2.00)	CP9003	−0.3769	(−1.50)
SP9003	−0.0772	(−0.20)	SP9003	−0.0676	(−0.20)
REP9003	−0.1194	(−1.13)	REP9003	−0.1424	(−1.54)
RBP9003	0.2543	(0.95)	RBP9003	0.4561	(1.53)
RCP9003	0.0451	(0.51)	RCP9003	0.0951	(1.17)
RSP9003	0.2639	(4.10)	RSP9003	0.1944	(3.19)
PRGR9003	0.2399	(3.84)	PRGR9003	0.2453	(4.55)
F Value	7.9880		F Value	9.6840	
R-square	0.0738		R-square	0.0901	

Continued on following page

TABLE 7.5 (Continued)

R39006

	OLS CAPL			ROBUST CAPL	
EP9006	−0.8071	(−3.45)	EP9006	−1.2333	(−5.43)
BP9006	−0.0655	(−1.07)	BP9006	−0.0690	(−1.28)
CP9006	1.0100	(5.60)	CP9006	1.4097	(7.57)
SP9006	−0.3080	(−3.87)	SP9006	−0.4132	(−5.72)
REP9006	−0.2790	(−2.24)	REP9006	−0.2907	(−2.76)
RBP9006	0.0294	(0.28)	RBP9006	−0.0146	(−0.16)
RCP9006	−0.1213	(−0.85)	RCP9006	−0.0849	(−0.61)
RSP9006	−0.1261	(−2.33)	RSP9006	−0.1486	(−3.14)
PRGR9006	0.0252	(0.32)	PRGR9006	0.0552	(0.82)
F Value	6.6460		F Value	12.6210	
R-square	0.0517		R-square	0.1010	

R39009

	OLS CAPL			ROBUST CAPL	
EP9009	0.3710	(1.67)	EP9009	0.4172	(2.08)
BP9009	−0.4086	(−5.95)	BP9009	−0.3524	(−5.79)
CP9009	−0.1986	(−1.08)	CP9009	−0.2294	(−1.40)
SP9009	−0.0311	(−0.34)	SP9009	−0.1057	(−1.35)
REP9009	0.0458	(0.45)	REP9009	0.0459	(0.51)
RBP9009	0.2833	(3.35)	RBP9009	0.1885	(2.53)
RCP9009	0.1288	(−1.37)	RCP9009	0.0851	(1.00)
RSP9009	0.1608	(3.06)	RSP9009	0.2279	(4.84)
PRGR9009	0.2597	(4.29)	PRGR9009	0.2423	(4.62)
F Value	10.7600		F Value	13.6260	
R-square	0.0860		R-square	0.1085	

R39012

	OLS CAPL			ROBUST CAPL	
EP9012	1.0409	(3.03)	EP9012	1.7025	(5.30)
BP9012	−0.4279	(−4.62)	BP9012	−0.3924	(−5.09)
CP9012	−1.3150	(−4.54)	CP9012	−1.7368	(−6.61)
SP9012	1.3436	(5.54)	SP9012	1.4384	(6.94)
REP9012	−0.0510	(−0.62)	REP9012	−0.0347	(−0.49)
RBP9012	0.1084	(1.17)	RBP9012	0.0545	(0.70)
RCP9012	−0.0179	(−0.22)	RCP9012	−0.0280	(−0.43)
RSP9012	0.4584	(6.52)	RSP9012	0.4644	(7.82)
PRGR9012	0.2150	(4.69)	PRGR9012	0.1783	(4.69)
F Value	19.9080		F Value	27.7670	
R-square	0.1538		R-square	0.2052	

	OLS CAPM			ROBUST CAPM	
EP9006	−0.4646	(−2.60)	EP9006	−0.4385	(−2.58)
BP9006	0.1821	(3.01)	BP9006	0.2033	(3.73)
CP9006	0.4803	(2.93)	CP9006	0.4537	(2.89)
SP9006	−0.2629	(−3.55)	SP9006	−0.2863	(−4.25)
REP9006	−0.0838	(−0.63)	REP9006	−0.0716	(−0.57)
RBP9006	0.1106	(1.28)	RBP9006	0.1014	(1.28)
RCP9006	−0.0208	(−0.26)	RCP9006	0.0114	(0.15)
RSP9006	−0.0226	(−0.51)	RSP9006	−0.0336	(−0.83)
PRGR9006	0.0894	(1.42)	PRGR9006	0.0931	(1.64)
F Value	3.2600		F Value	4.0770	
R-square	0.0232		R-square	0.0313	

	OLS CAPM			ROBUST CAPM	
EP9009	0.7690	(3.33)	EP9009	0.8684	(3.84)
BP9009	−0.4210	(−5.49)	BP9009	−0.4275	(−6.29)
CP9009	−0.4406	(−2.03)	CP9009	−0.5737	(−2.68)
SP9009	0.0054	(0.06)	SP9009	−0.0039	(−0.05)
REP9009	−0.0461	(−0.30)	REP9009	−0.0279	(−0.21)
RBP9009	0.0374	(0.43)	RBP9009	0.0744	(0.93)
RCP9009	0.1562	(1.79)	RCP9009	0.1939	(2.47)
RSP9009	0.2227	(4.18)	RSP9009	0.1975	(4.14)
PRGR9009	0.2398	(3.87)	PRGR9009	0.2344	(4.31)
F Value	9.1230		F Value	11.8760	
R-square	0.0807		R-square	0.1052	

	OLS CAPM			ROBUST CAPM	
EP9012	−0.5486	(−1.50)	EP9012	−0.4470	(−1.40)
BP9012	0.1380	(1.55)	BP9012	0.0709	(0.91)
CP9012	−0.2476	(−0.73)	CP9012	−0.3402	(−1.18)
SP9012	−0.1038	(−0.45)	SP9012	0.0688	(0.34)
REP9012	0.0855	(0.55)	REP9012	0.0466	(0.35)
RBP9012	−0.1001	(−1.16)	RBP9012	−0.0819	(−1.04)
RCP9012	−0.0948	(−1.16)	RCP9012	−0.0820	(−1.20)
RSP9012	0.2876	(4.76)	RSP9012	0.2874	(5.49)
PRGR9012	−0.0031	(−0.06)	PRGR9012	−0.0187	(−0.41)
F Value	5.6500		F Value	7.6990	
R-square	0.0485		R-square	0.0684	

Continued on following page

TABLE 7.5 (Continued)

R39103

	OLS CAPL			ROBUST CAPL	
EP9103	0.3881	(2.50)	EP9103	0.3850	(2.99)
BP9103	−0.1204	(−1.55)	BP9103	−0.0962	(−1.50)
CP9103	−0.0809	(−0.62)	CP9103	−0.0643	(−0.60)
SP9103	0.0970	(0.82)	SP9103	0.1034	(1.07)
REP9103	0.1453	(1.10)	REP9103	0.1206	(1.12)
RBP9103	−0.1809	(−0.77)	RBP9103	−0.1899	(−1.01)
RCP9103	0.2016	(1.75)	RCP9103	0.2413	(2.55)
RSP9103	0.2291	(4.83)	RSP9103	0.2127	(5.45)
PRGR9103	0.0063	(0.12)	PRGR9103	−0.0000	(−0.00)
F Value	7.0000		F Value	9.6030	
R-square	0.0546		R-square	0.0765	

R39106

	OLS CAPL			ROBUST CAPL	
EP9106	0.1010	(0.81)	EP9106	0.1709	(1.48)
BP9106	0.0648	(0.93)	BP9106	0.0817	(1.35)
CP9106	−0.2214	(−2.11)	CP9106	−0.2365	(−2.51)
SP9106	−0.2587	(−4.22)	SP9106	−0.2534	(−4.79)
REP9106	1.2052	(1.32)	REP9106	1.4536	(1.84)
RBP9106	−0.2167	(−1.14)	RBP9106	−0.2836	(−1.74)
RCP9106	−0.6480	(−2.51)	RCP9106	−0.7512	(−3.43)
RSP9106	0.1391	(3.14)	RSP9106	0.1424	(3.65)
PRGR9106	0.2765	(5.34)	PRGR9106	0.2865	(6.32)
F Value	8.7580		F Value	12.2690	
R-square	0.0695		R-square	0.0979	

R39109

	OLS CAPL			ROBUST CAPL	
EP9109	−0.1842	(−1.20)	EP9109	−0.1589	(−1.17)
BP9109	−0.1710	(−2.28)	BP9109	−0.1300	(−2.04)
CP9109	−0.3558	(−2.41)	CP9109	−0.3105	(−2.41)
SP9109	−0.0546	(−0.80)	SP9109	−0.0585	(−1.00)
REP9109	3.4852	(2.84)	REP9109	3.2331	(3.06)
RBP9109	0.3466	(−1.38)	RBP9109	0.2479	(1.10)
RCP9109	−1.2811	(−3.23)	RCP9109	−1.2092	(−3.52)
RSP9109	−0.0517	(−1.08)	RSP9109	−0.0756	(−1.79)
PRGR9109	0.3137	(5.49)	PRGR9109	0.3152	(6.38)
F Value	10.7060		F Value	13.5900	
R-square	0.0860		R-square	0.1087	

	OLS CAPM			ROBUST CAPM	
EP9103	0.3688	(2.48)	EP9103	0.4342	(3.32)
BP9103	−0.2951	(−3.88)	BP9103	−0.2797	(−3.94)
CP9103	−0.2165	(−1.54)	CP9103	−0.2558	(−2.11)
SP9103	0.0317	(0.32)	SP9103	0.0639	(0.69)
REP9103	0.0984	(0.40)	REP9103	0.0839	(0.38)
RBP9103	0.0265	(0.13)	RBP9103	0.1016	(0.52)
RCP9103	−0.0053	(−0.04)	RCP9103	0.0106	(0.08)
RSP9103	0.2914	(6.17)	RSP9103	0.2752	(6.52)
PRGR9103	0.1818	(3.28)	PRGR9103	0.1880	(3.83)
F Value	7.0660		F Value	8.8780	
R-square	0.0632		R-square	0.0806	

	OLS CAPM			ROBUST CAPM	
EP9106	0.2019	(1.51)	EP9106	0.2259	(1.94)
BP9106	−0.1703	(−2.18)	BP9106	−0.1395	(−2.01)
CP9106	−0.2587	(−2.10)	CP9106	−0.2352	(−2.18)
SP9106	−0.0860	(−1.27)	SP9106	−0.0666	(−1.13)
REP9106	−0.8388	(−1.27)	REP9106	−1.0553	(−1.65)
RBP9106	0.3238	(1.59)	RBP9106	0.3088	(1.73)
RCP9106	−0.2769	(−1.81)	RCP9106	−0.2268	(−1.38)
RSP9106	0.1259	(2.86)	RSP9106	0.1365	(3.48)
PRGR9106	0.3535	(6.05)	PRGR9106	0.3675	(7.18)
F Value	7.4660		F Value	9.3660	
R-square	0.0615		R-square	0.0782	

	OLS CAPM			ROBUST CAPM	
EP9109	0.3018	(1.93)	EP9109	0.2690	(1.76)
BP9109	−0.1558	(−1.78)	BP9109	−0.1796	(−2.19)
CP9109	−0.5187	(−3.26)	CP9109	−0.5232	(−3.42)
SP9109	0.0006	(0.01)	SP9109	−0.0079	(−0.12)
REP9109	1.9101	(1.04)	REP9109	1.3182	(0.79)
RBP9109	−0.2761	(−1.07)	RBP9109	−0.2291	(−0.92)
RCP9109	−0.4269	(−1.75)	RCP9109	−0.3620	(−1.67)
RSP9109	0.0567	(1.21)	RSP9109	0.0484	(1.13)
PRGR9109	0.1184	(1.92)	PRGR9109	0.1341	(2.49)
F Value	3.6130		F Value	4.5710	
R-square	0.0277		R-square	0.0374	

Continued on following page

TABLE 7.5 (Continued)

R39112

	OLS CAPL			ROBUST CAPL	
EP9112	0.9016	(3.11)	EP9112	0.9917	(3.71)
BP9112	0.8853	(6.18)	BP9112	0.8308	(6.54)
CP9112	−0.9699	(−4.00)	CP9112	−1.1340	(−5.01)
SP9112	0.3807	(5.39)	SP9112	0.3878	(6.16)
REP9112	−0.8209	(−0.74)	REP9112	−0.7039	(−0.75)
RBP9112	−0.1381	(−0.33)	RBP9112	−0.1343	(−0.38)
RCP9112	−0.3976	(−1.33)	RCP9112	−0.4318	(−1.70)
RSP9112	0.2037	(3.34)	RSP9112	0.2516	(4.72)
PRGR9112	0.0929	(2.16)	PRGR9112	0.0974	(2.61)
F Value	25.6440		F Value	33.5350	
R-square	0.1961		R-square	0.2436	

R39203

	OLS CAPL			ROBUST CAPL	
EP9203	0.4333	(2.43)	EP9203	0.4464	(2.79)
BP9203	0.7087	(7.06)	BP9203	0.6936	(7.97)
CP9203	−0.3207	(−2.05)	CP9203	−0.1571	(−1.14)
SP9203	−0.2077	(−2.56)	SP9203	−0.2327	(−3.29)
REP9203	−0.1132	(−1.10)	REP9203	−0.0844	(−0.91)
RBP9203	0.1075	(1.24)	RBP9203	0.0868	(1.16)
RCP9203	−0.1077	(−1.21)	RCP9203	−0.1872	(−2.30)
RSP9203	0.1588	(2.91)	RSP9203	0.1344	(2.83)
PRGR9203	0.0025	(0.04)	PRGR9203	−0.0237	(−0.45)
F Value	13.9090		F Value	17.6240	
R-square	0.1141		R-square	0.1423	

R39206

	OLS CAPL			ROBUST CAPL	
EP9206	0.1509	(1.31)	EP9206	0.1407	(1.45)
BP9206	−0.1278	(−3.25)	BP9206	−0.1415	(−4.31)
CP9206	0.1188	(1.34)	CP9206	0.1198	(1.60)
SP9206	−0.0671	(−1.50)	SP9206	−0.0818	(−2.19)
REP9206	0.3303	(2.84)	REP9206	0.3122	(3.21)
RBP9206	0.0588	(0.80)	RBP9206	0.0488	(0.78)
RCP9206	0.1406	(1.85)	RCP9206	0.1382	(2.21)
RSP9206	0.0343	(0.82)	RSP9206	0.0659	(1.84)
PRGR9206	0.0706	(1.29)	PRGR9206	0.0565	(1.22)
F Value	6.5040		F Value	10.1370	
R-square	0.0501		R-square	0.0805	

	OLS CAPM			ROBUST CAPM	
EP9112	0.0754	(0.26)	EP9112	0.2637	(1.06)
BP9112	0.5747	(3.37)	BP9112	0.6355	(4.32)
CP9112	−0.0788	(−0.27)	CP9112	−0.1548	(−0.62)
SP9112	0.2725	(3.67)	SP9112	0.2654	(4.09)
REP9112	1.1680	(0.65)	REP9112	1.0434	(0.63)
RBP9112	0.2309	(0.67)	RBP9112	0.1618	(0.54)
RCP9112	−0.5653	(−3.07)	RCP9112	−0.5184	(−3.32)
RSP9112	0.0805	(1.62)	RSP9112	0.0773	(1.75)
PRGR9112	0.1607	(3.30)	PRGR9112	0.1553	(3.68)
F Value	10.9360		F Value	14.2150	
R-square	0.1007		R-square	0.1296	

	OLS CAPM			ROBUST CAPM	
EP9203	0.4219	(2.25)	EP9203	0.4874	(2.77)
BP9203	0.5701	(5.16)	BP9203	0.5996	(5.97)
CP9203	−0.1743	(−0.84)	CP9203	−0.2752	(−1.42)
SP9203	−0.0889	(−1.00)	SP9203	−0.0987	(−1.27)
REP9203	−0.0551	(−0.38)	REP9203	−0.0572	(−0.45)
RBP9203	−0.0086	(−0.09)	RBP9203	0.1235	(1.39)
RCP9203	0.0304	(0.38)	RCP9203	0.0033	(0.05)
RSP9203	0.1950	(3.47)	RSP9203	0.1294	(2.44)
PRGR9203	0.0252	(0.40)	PRGR9203	0.0016	(0.03)
F Value	9.5150		F Value	12.8710	
R-square	0.0890		R-square	0.1199	

	OLS CAPM			ROBUST CAPM	
EP9206	−0.1672	(−1.01)	EP9206	0.0039	(0.03)
BP9206	−0.0750	(−1.37)	BP9206	−0.0614	(−1.28)
CP9206	0.2765	(2.13)	CP9206	0.1552	(1.33)
SP9206	−0.0362	(−0.64)	SP9206	−0.0286	(−0.57)
REP9206	−0.0493	(−0.28)	REP9206	0.0063	(0.04)
RBP9206	−0.0276	(−0.31)	RBP9206	−0.0130	(−0.17)
RCP9206	0.1260	(1.47)	RCP9206	0.1751	(2.22)
RSP9206	0.1608	(3.34)	RSP9206	0.1338	(3.14)
PRGR9206	0.3900	(5.68)	PRGR9206	0.3626	(6.02)
F Value	5.6320		F Value	6.3750	
R-square	0.0453		R-square	0.0522	

Continued on following page

194 HISTORICAL DATA AND ANALYSTS' FORECASTS

TABLE 7.5 (Continued)

R39209

	OLS CAPL			ROBUST CAPL	
EP9209	0.5772	(4.02)	EP9209	0.6968	(5.77)
BP9209	−0.0095	(−0.20)	BP9209	0.0269	(0.69)
CP9209	−0.6308	(−6.25)	CP9209	−0.7191	(−8.36)
SP9209	0.2644	(4.75)	SP9209	0.2754	(5.91)
REP9209	−0.2191	(−1.84)	REP9209	−0.1880	(−1.77)
RBP9209	−0.0253	(−0.29)	RBP9209	0.0062	(0.09)
RCP9209	0.0976	(1.11)	RCP9209	0.0587	(0.77)
RSP9209	0.0395	(0.75)	RSP9209	0.0351	(0.80)
PRGR9209	0.2594	(4.41)	PRGR9209	0.2143	(4.34)
F Value	8.5210		F Value	12.1870	
R-square	0.0679		R-square	0.0978	

R39212

	OLS CAPL			ROBUST CAPL	
EP9212	−0.6283	(−4.34)	EP9212	−0.7302	(−5.81)
BP9212	0.3794	(7.43)	BP9212	0.3590	(8.22)
CP9212	0.4359	(4.30)	CP9212	0.4436	(4.93)
SP9212	−0.1050	(−1.88)	SP9212	−0.1314	(−2.76)
REP9212	−0.0500	(−0.42)	REP9212	−0.0536	(−0.54)
RBP9212	−0.0555	(−0.57)	RBP9212	−0.0860	(−1.01)
RCP9212	−0.0882	(−1.00)	RCP9212	−0.0514	(−0.68)
RSP9212	−0.1251	(−2.32)	RSP9212	−0.1351	(−2.84)
PRGR9212	0.2518	(4.10)	PRGR9212	0.2340	(4.48)
F Value	14.8340		F Value	19.1020	
R-square	0.1197		R-square	0.1510	

R39303

	OLS CAPL			ROBUST CAPL	
EP9303	−0.1743	(−1.35)	EP9303	−0.1372	(−1.32)
BP9303	0.0249	(0.49)	BP9303	0.0173	(0.42)
CP9303	0.0875	(0.92)	CP9303	0.0831	(1.08)
SP9303	−0.0438	(−0.80)	SP9303	−0.0039	(−0.09)
REP9303	0.0906	(0.74)	REP9303	0.1628	(1.48)
RBP9303	0.0535	(0.56)	RBP9303	0.0438	(0.57)
RCP9303	−0.0707	(−0.64)	RCP9303	−0.1006	(−1.16)
RSP9303	0.0595	(1.06)	RSP9303	0.0588	(1.30)
PRGR9303	0.2069	(3.10)	PRGR9303	0.2156	(4.00)
F Value	1.5590		F Value	2.5820	
R-square	0.0055		R-square	0.0155	

SIMULATING THE COMPOSITE MODEL STRATEGY 195

	OLS CAPM			ROBUST CAPM	
EP9209	0.3837	(1.82)	EP9209	0.6048	(3.16)
BP9209	−0.2385	(−4.00)	BP9209	−0.2539	(−4.94)
CP9209	−0.3784	(−2.58)	CP9209	−0.5546	(−4.13)
SP9209	0.1632	(2.59)	SP9209	0.1756	(3.20)
REP9209	−0.0862	(−0.56)	REP9209	0.0011	(0.01)
RBP9209	0.0087	(0.10)	RBP9209	−0.0022	(−0.03)
RCP9209	0.2054	(2.30)	RCP9209	0.2181	(2.72)
RSP9209	0.0836	(1.69)	RSP9209	0.1046	(2.41)
PRGR9209	0.0566	(0.94)	PRGR9209	0.0782	(1.50)
F Value	3.8260		F Value	6.8000	
R-square	0.0291		R-square	0.0579	

	OLS CAPM			ROBUST CAPM	
EP9212	−0.1791	(−0.94)	EP9212	−0.0451	(−0.26)
BP9212	0.3799	(5.55)	BP9212	0.4054	(6.67)
CP9212	0.2935	(2.12)	CP9212	0.2130	(1.74)
SP9212	0.0096	(0.14)	SP9212	0.0124	(0.20)
REP9212	−0.1049	(−0.63)	REP9212	−0.0277	(−0.18)
RBP9212	−0.0724	(−0.70)	RBP9212	−0.0839	(−0.87)
RCP9212	−0.0451	(−0.48)	RCP9212	−0.0528	(−0.63)
RSP9212	−0.1301	(−2.38)	RSP9212	−0.1266	(−2.57)
PRGR9212	0.1827	(2.63)	PRGR9212	0.1675	(2.79)
F Value	8.6510		F Value	11.1420	
R-square	0.0772		R-square	0.0998	

	OLS CAPM			ROBUST CAPM	
EP9303	−0.8762	(−5.21)	EP9303	−0.9166	(−6.37)
BP9303	−0.0588	(−0.99)	BP9303	−0.0314	(−0.62)
CP9303	0.3871	(3.07)	CP9303	0.3833	(3.58)
SP9303	−0.2004	(−3.35)	SP9303	−0.1897	(−3.76)
REP9303	−0.0180	(−0.18)	REP9303	0.0206	(0.24)
RBP9303	0.0801	(1.07)	RBP9303	0.0981	(1.50)
RCP9303	−0.0051	(−0.04)	RCP9303	−0.0128	(−0.11)
RSP9303	0.0646	(1.27)	RSP9303	0.0514	(1.17)
PRGR9303	0.1187	(1.84)	PRGR9303	0.1066	(1.95)
F Value	5.4610		F Value	7.7700	
R-square	0.0472		R-square	0.0700	

Continued on following page

TABLE 7.5 (Continued)

R39306

	OLS CAPL			ROBUST CAPL	
EP9306	0.0096	(0.10)	EP9306	0.0994	(1.20)
BP9306	−0.0579	(−1.40)	BP9306	−0.0491	(−1.43)
CP9306	0.0988	(1.13)	CP9306	0.0417	(0.57)
SP9306	−0.0236	(−0.44)	SP9306	−0.0046	(−0.10)
REP9306	−0.0523	(−0.62)	REP9306	0.0124	(0.17)
RBP9306	−0.0999	(−1.09)	RBP9306	−0.1494	(−1.85)
RCP9306	−0.1603	(−1.65)	RCP9306	−0.1569	(−1.96)
RSP9306	−0.1303	(−3.17)	RSP9306	−0.1157	(−3.28)
PRGR9306	0.1968	(3.57)	PRGR9306	0.1815	(3.98)
F Value	7.3250		F Value	10.2940	
R-square	0.0574		R-square	0.0822	

R39309

	OLS CAPL			ROBUST CAPL	
EP9309	−0.0352	(−1.81)	EP9309	−0.0304	(−1.79)
BP9309	−0.0296	(−4.47)	BP9309	−0.0364	(−6.31)
CP9309	0.0043	(0.28)	CP9309	0.0015	(0.11)
SP9309	0.0292	(3.29)	SP9309	0.0365	(4.63)
REP9309	0.0246	(2.01)	REP9309	0.0196	(1.81)
RBP9309	0.0165	(1.18)	RBP9309	0.0217	(1.73)
RCP9309	−0.0088	(−0.59)	RCP9309	−0.0081	(−0.62)
RSP9309	0.0272	(4.18)	RSP9309	0.0241	(4.17)
PRGR9309	−0.0091	(−1.10)	PRGR9309	−0.0136	(−1.90)
F Value	12.5840		F Value	18.4050	
R-square	0.1018		R-square	0.1456	

R39312

	OLS CAPL			ROBUST CAPL	
EP9312	0.0077	(0.07)	EP9312	0.0739	(0.83)
BP9312	−0.0036	(−0.07)	BP9312	−0.0469	(−1.15)
CP9312	−0.0311	(−0.33)	CP9312	−0.0864	(−1.09)
SP9312	0.1675	(2.54)	SP9312	0.2336	(4.14)
REP9312	0.0244	(0.30)	REP9312	0.0249	(0.36)
RBP9312	0.1880	(1.80)	RBP9312	0.1824	(2.08)
RCP9312	−0.0469	(−0.39)	RCP9312	0.0177	(0.18)
RSP9312	−0.0893	(−1.45)	RSP9312	−0.0843	(−1.63)
PRGR9312	0.0910	(1.53)	PRGR9312	0.0395	(0.79)
F Value	1.4970		F Value	2.6030	
R-square	0.0048		R-square	0.0155	

	OLS CAPM			ROBUST CAPM	
EP9306	−0.2106	(−1.68)	EP9306	−0.1509	(−1.37)
BP9306	−0.1126	(−2.14)	BP9306	−0.0933	(−2.04)
CP9306	0.1270	(1.17)	CP9306	0.1152	(1.20)
SP9306	−0.0503	(−0.80)	SP9306	−0.0484	(−0.89)
REP9306	0.0962	(0.90)	REP9306	0.0472	(0.50)
RBP9306	−0.0527	(−0.53)	RBP9306	−0.0366	(−0.42)
RCP9306	−0.1848	(−1.26)	RCP9306	−0.1612	(−1.26)
RSP9306	0.0490	(1.07)	RSP9306	0.0490	(1.22)
PRGR9306	0.2431	(3.84)	PRGR9306	0.2633	(4.72)
F Value	3.1680		F Value	3.9110	
R-square	0.0216		R-square	0.0287	

	OLS CAPM			ROBUST CAPM	
EP9309	−0.0009	(−0.04)	EP9309	−0.0012	(−0.06)
BP9309	−0.0160	(−1.90)	BP9309	−0.0195	(−2.68)
CP9309	0.0071	(0.38)	CP9309	0.0084	(0.52)
SP9309	0.0346	(3.52)	SP9309	0.0377	(4.40)
REP9309	0.0059	(0.36)	REP9309	0.0055	(0.38)
RBP9309	0.0188	(1.30)	RBP9309	0.0257	(2.04)
RCP9309	−0.0068	(−0.29)	RCP9309	0.0011	(0.05)
RSP9309	0.0258	(3.58)	RSP9309	0.0216	(3.45)
PRGR9309	0.0094	(1.07)	PRGR9309	0.0041	(0.54)
F Value	5.7170		F Value	7.5460	
R-square	0.0468		R-square	0.0638	

	OLS CAPM			ROBUST CAPM	
EP9312	−0.1823	(−1.31)	EP9312	−0.0922	(−0.78)
BP9312	−0.0261	(−0.45)	BP9312	−0.0091	(−0.18)
CP9312	0.2235	(1.64)	CP9312	0.1591	(1.37)
SP9312	0.2307	(3.32)	SP9312	0.2030	(3.39)
REP9312	0.0976	(0.84)	REP9312	0.0690	(0.68)
RBP9312	−0.0266	(−0.28)	RBP9312	0.0297	(0.35)
RCP9312	−0.3037	(−1.71)	RCP9312	−0.2753	(−1.82)
RSP9312	0.0347	(0.56)	RSP9312	0.0297	(0.56)
PRGR9312	0.0837	(1.26)	PRGR9312	0.0778	(1.38)
F Value	3.4070		F Value	3.5380	
R-square	0.0259		R-square	0.0272	

Continued on following page

TABLE 7.5 (Continued)

R39403

	OLS CAPL			ROBUST CAPL	
EP9403	0.0150	(0.15)	EP9403	0.0391	(0.46)
BP9403	0.0406	(0.79)	BP9403	0.0737	(1.61)
CP9403	−0.0030	(−0.03)	CP9403	−0.0529	(−0.65)
SP9403	0.0473	(0.73)	SP9403	0.0174	(0.31)
REP9403	−0.0724	(−0.72)	REP9403	−0.0796	(−0.91)
RBP9403	0.0879	(0.45)	RBP9403	0.1038	(0.64)
RCP9403	−0.2806	(−2.19)	RCP9403	−0.2533	(−2.17)
RSP9403	0.1361	(2.98)	RSP9403	0.1046	(2.66)
PRGR9403	0.0032	(0.05)	PRGR9403	−0.0027	(−0.05)
F Value	3.2950		F Value	3.0310	
R-square	0.0222		R-square	0.0197	

CONCLUSIONS

The goal of this chapter was to present composite models, using fundamental and earnings forecasting data that can be input to a multifactor model, to create portfolios producing low-risk, positive excess returns. The specific models analyzed were the robust-regression-weighted composite model and the three-factor APT model. The principal conclusion is that regression-weighted composite models used with multifactor models with two to five factors are capable of producing significant excess returns.

NOTES

1. The real-time U.S. results are shown in Miller, Guerard, and Takano (1992), and the real-time Japanese results are found in Guerard, Takano, and Yamane (1993).
2. There are many formulations that can be made from equation (1), and Guerard, Takano, and Yamane (1993), Guerard, Blin, and Bender (1995), and Guerard, Blin, and Bender (1996b) have tested many of them. Guerard initially created FY1 and FY2 forecast variables (CIF), a revisions variable (CIR), and an FY1 and FY2 breadth (CIB) variable before creating a composite FY1 and FY2 variable (CIBF). The dividend yield (DY) and net current asset value (NCAV), the current assets less all liabilities of the firms, were also examined. Equation (1) in this study was as consistent with other equally weighted variable models in its IC, but was the best on a regression-weighted basis. In terms of equally weighted models, the following ICs were found using the largest 3,000 securities during the 1982–1994 period.

	OLS CAPM			ROBUST CAPM	
EP9403	−0.0910	(−0.73)	EP9403	−0.0144	(−0.13)
BP9403	0.1976	(3.32)	BP9403	0.2069	(3.88)
CP9403	0.1500	(1.30)	CP9403	0.0812	(0.81)
SP9403	−0.1583	(−2.50)	SP9403	−0.1646	(−2.93)
REP9403	0.1776	(0.95)	REP9403	0.1852	(1.15)
RBP9403	0.1197	(0.75)	RBP9403	0.0396	(0.25)
RCP9403	−0.0397	(−0.24)	RCP9403	−0.0160	(−0.11)
RSP9403	0.0813	(1.69)	RSP9403	0.0915	(2.11)
PRGR9403	0.1069	(1.69)	PRGR9403	0.0955	(1.72)
F Value	3.8800		F Value	4.5910	
R-square	0.0309		R-square	0.0382	

Model Variables	IC(t)
BP	.012(0.68)
EP	.038(2.10)
CIF	.046(2.50)
CIB	.064(3.46)
CIR	.039(2.11)
CIBF	.075(4.05)
EP,BP,CP,SP,REP,RBP,RCP,RSP	.045(2.46)
EP,BP,REP,RBP,CIR,CIB	.056(3.52)
EP,BP,CP,SP,CIR,CIB,REP,RBP,RCP,RSP	.081(4.23)
EP,BP,CP,SP,DY,NCAV,CIBF	.058(3.07)
EP,BP,CP,SP,REP,RBP,RCP,RSP,CIR,CIB	.057(3.05)
EP,BP,CP,SP,DY,NCAV,REP,RBP,RCP,RSP,RDY,RCAV,CIBF	.075(3.96)

3. BGG found the following scaled mean-squared forecasting errors (MSFE) for size-ranked portfolios for the 1931–1987 forecast periods:

	Factors			
Size	1	2	3	5
Smallest, 1	0.389	0.205	0.200	0.202
2	0.242	0.113	0.111	0.121
10	0.076	0.075	0.074	0.079
19	0.178	0.076	0.077	0.089
Largest, 20	0.288	0.139	0.131	0.132

4. This result is consistent with Blin, Bender, and Guerard (1996), who found that a composite proprietary I/B/E/S variable used as a quadratic variable significantly dominated the regression-weighting of fundamental and I/B/E/S variables.

REFERENCES

Arnott, R. "The Use and Misuse of Concensus Earnings." *Journal of Portfolio Management.* (1985):18–27.

Banz, R. "The Relationship Between Return and Market Value of Common Stocks." *Journal of Financial Economics* 9 (1981):3–18.

Beaton, A. E., and J. W. Tukey. "The Fitting of Power Series, Meaning Polynomials, Illustrated on Bank-Spectroscopic Data." *Technometrics* 16 (1974): 147–185.

Belsley, D. A., A. E. Kuh, and R. E. Welsch. *Regression Diagnostics—Identifying Influential Data and Sources of Collinearity.* New York: John Wiley & Sons, 1980.

Blin, J., and G. Douglas. "Stock Returns vs. Factors." *Investment Management Review* (1987).

Blin, J., S. Bender, and J. B. Guerard, Jr. "Earnings and Value Investing—How to Make the Return Worth the Risk." Presented at the 1996 I/B/E/S Research Conference, Tokyo, June 1996, forthcoming in *Research in Finance,* edited by A. Chen.

Blin, J., S. Bender, and J. B. Guerard. "Why Your Next Japanese Portfolio Should Be Market-Neutral—and How." *Journal of Investing* (1995).

Blume, M. E., M. N. Gultekin, and N. B. Gultekin. "Validating Return-Generating Models." (1990) (under journal review).

Brown, L. D. "Earnings Forecasting Research: Its Implications for Capital Markets Research." *International Journal of Forecasting* 9 (1993):295–320.

Chen, N-f., R. Roll, and S. A. Ross. "Economic Forces and the Stock Market." *Journal of Business* 59 (1986):383–403.

Chan, L., Y. Hamao, and J. Lakonishok. "Fundamentals and Stock Returns in Japan." *Journal of Finance* 46 (1991):1739–1764.

Chopra, V. K., and W. T. Ziemba. "The Effects of Errors in Means, Variances, and Covariances on Optimal Choice." *The Journal of Portfolio Management* (1993):6–11.

Clemen, R. T. "Combining Forecasts: A Review and Annotated Bibliography." *International Journal of Forecasting* 5 (1989):559–584.

Conner, G., and R. A. Korajczyk. "The Arbitrage Pricing Theory and Multifactor Models of Asset Returns." In R. Jarrow, V. Makisimovic, and W. Ziemba, *Handbooks in OR & MS,* North-Holland, 1995.

Conner, G., and R. A. Korajczyk. "Risk and Return in an Equilibrium APT: Application of a New Test Methodology." *Journal of Financial Economics* 21 (1988):255–289.

Cottle, S., R. F. Murray, and F. E. Block. *Graham and Dodd's Security Analysis.* 5th ed. New York, McGraw-Hill.

REFERENCES

Dimson, E., ed. *Stock Market Anomalies*. Cambridge, Cambridge University Press, 1988.

Dhrymes, P. J., I. Friend, N. B. Gultekin, and M. N. Gultekin. "An Empirical Examination of the Implications of Arbitrage Pricing Theory." *Journal of Banking and Finance* 9 (1985):73–99.

Dhrymes, P. J., I. Friend, and N. B. Gultekin. "A Critical Reexamination of the Empirical Evidence on the Arbitrage Pricing Theory." *Journal of Finance* 39 (1984):323–346.

Elton, E. J., and M. J. Gruber. "Analysts' Expectations and Japanese Stock Prices." *Japan and the World Economy* 2 (1989).

Elton, E. J., and M. J. Gruber. *Modern Portfolio Theory and Investment Analysis*. New York: John Wiley & Sons, 1987.

Elton, E. J., M. J. Gruber, and M. Gultekin. "Expectations and Share Prices" *Management Science* 27 (1981):975–987.

Fama, E. F., and K. R. French. "Size and Book-to-Market Factors in Earnings and Returns." *Journal of Finance* 50 (1995):131–155.

Fama, E. F., and K. R. French. "Cross-Sectional Variation in Expected Stock Returns." *Journal of Finance* 47 (1992):427–465.

Farrell, J. *Guide to Portfolio Management*. New York: McGraw-Hill, 1983.

Givoly, D., and J. Lakonishok. "The Information Content of Financial Analysts' Forecasts of Earnings: Some Evidence on Semi-Strong Inefficiency." *Journal of Accounting and Economics* 3 (1979):165–185.

Graham, B. *The Intelligent Investor*. New York: Harper & Row, 1940.

Graham, B., and Dodd, D. *Security Analysis*. 4th ed. New York: McGraw-Hill, 1962.

Grundy, K., and B. G. Malkiel. "Reports of Beta's Death Have Been Greatly Exaggerated." *Journal of Portfolio Management* (Spring 1996).

Guerard, J. B., Jr. "Linear Constraints, Robust-Weighting, and Efficient Composite Modeling." *Journal of Forecasting* 6 (1987):193–199.

Guerard, J. B., Jr., and B. K. Stone. "Composite Forecasting of Annual Corporate Earnings." In *Research in Finance* 10, edited by A. Chen. Greenwich, Conn.: JAI Press, 1992 205–230.

Guerard, J. B., Jr., and M. Takano. "Stock Selection and Composite Modeling in Japan." *Security Analysts Journal* 29 (1991):1–13.

Guerard, J. B., Jr., A. S. Bean, and B. K. Stone. "Goal Setting for Effective Corporate Planning." *Management Science* 36 (1990):359–367.

Guerard, J. B., Jr., J. Blin, and S. Bender. "Earnings Revisions in the Estimation of Efficient Market-Neutral Japanese Portfolios." *Journal of Investing* (spring 1996a).

Guerard, J. B. Jr., J. Blin, and S. Bender. "The Riskiness of Global Portfolios." Presented at the Society of Quantitative Analysts, Annual Fuzzy Day Seminar, New York, May 1996b.

Guerard, J. B., Jr., M. Gultekin, and B. K. Stone. "Time Decay of Earnings Estimates." Presented at the Seminar on Corporate Earnings Analysis, New York, April 1996, forthcoming in *Handbook of Corporate Earnings Analysis*, Volume 2, edited by B. Bruce.

Guerard, J. B., Jr., M. Takano, and Y. Yamane. "The Development of Efficient Portfolios in Japan with Particular Emphasis on Sales and Earnings Forecasting." *Annals of Operations Research* 45 (1993):91–108.

Gultekin, M. N., and N. B. Gultekin. "Stock Return Anomalies and the Tests of the Arbitrage Pricing Theory." *Journal of Finance* 42 (1987):1213–1224.

Gunst, R. F., and R. L. Mason. *Regression Analysis and Its Application.* New York: Marcel Dekker, 1980.

Hawkins, E., S. C. Chamberlain, and W. E. Daniel. "Earnings Expectations and Security Prices." *Financial Analysts Journal* 24–29 (1984):30–38.

Jacobs, B., and K. Levy. "Disentangling Equity Return Regularities: New Insights and Investment Opportunities." *Financial Analysts Journal* (1988):18–43.

Keon, E. "Earnings Expectations in Financial Theory and Investment Practice." *I/B/E/S Research Bibliography,* edited by L. D. Brown. 5th ed.

King, B. F. "Market and Industry Factors in Stock Price Behavior." *Journal of Business* 39 (1966):139–190.

Kothari, S. P., J. Shanken, and R. G. Sloan. "Another Look at the Cross-Section of Expected Stock Returns." *Journal of Finance* 50 (185–224).

Latane, H., D. Tuttle, and C. Jones. *Security Analysis and Portfolio Management.* New York: Ronald, 1975.

Markowitz, H. M. *Portfolio Selection: Efficient Diversification of Investments.* New York: John Wiley & Sons, 1959.

Miller, J. D., Guerard, J. B., Jr., and M. Takano. "Bridging the Gap Between Theory and Practice in Equity Selection Modeling: Case Studies of U.S. and Japanese Models" (under journal review).

Morgan, A., and I. Morgan. "Measurement of Abnormal Returns from Small Stocks." *Journal of Business & Economic Statistics* 5 (1987):121–129.

Roll, R., and S. A. Ross. "An Empirical Investigation of the Arbitrage Pricing Theory." *Journal of Finance* 35 (1980):1073–1103.

Rudd, A., and B. Rosenberg. "Realistic Portfolio Optimization." In *Portfolio Theory: 25 Years Later,* edited by E. Elton and M. Gruber. Amsterdam: North-Holland, 1979.

Tinic, S. M., and R. R. West. "Risk and Return: January vs. the Rest of the Year." *Journal of Financial Economics* (1984):561–574.

Vu, J. D. "An Anomalous Evidence Regarding Market Efficiency: The Net Current Asset Rule." In A. Chen, *Research in Finance* 8 (1990):241–254.

Webster, J. T., R. F. Gunst, and R. L. Mason. "Latent Root Regression Analysis." *Technometrics* 16 (1974):513 522.

Wheeler, L. "Changes in Consensus Earnings Estimates and Their Impact on Stock Returns." Presented at the Institute for Quantitative Research in Finance. In *The Handbook of Corporate Earnings Analysis,* edited by B. Bruce and C. Epstein. Chicago: Probus Publishing Co., 1990.

Ziemba, W. T. *Invest Japan.* Chicago: Probus Publishing Co., 1992.

Ziemba, W. T. "Fundamental Factors in U.S. and Japanese Stock Returns." Presented at the Berkeley Program in Finance, Santa Barbara, Calif., 1990.

8
ESTIMATION OF EFFICIENT MARKET-NEUTRAL JAPANESE AND U.S. PORTFOLIOS

This chapter examines the creation and estimation of market-neutral portfolios in Japan and the United States, with particular emphasis on earnings forecasting, revisions, and momentum. Market-neutral portfolios, created when one purchases undervalued securities and sells short overvalued securities, are shown to produce much higher returns for a given level of risk than those achieved by merely creating efficient (long) portfolios. A multifactor risk model is useful for creating market-neutral portfolios. We find that the inclusion of consensus I/B/E/S revisions increases the market-neutral portfolio average annual return by more than 600 basis points, relative to using only historical data, in Japan and by more than 1,200 basis points in the United States.

BACKGROUND

In this study the universe comprises the first-section securities of the Tokyo Stock Exchange (TSE), approximately 1,200 firms, and the U.S. securities included in the WorldScope database, approximately 2,800 securities in 1995. The WorldScope financial database is the source of earnings and book value data for the 1987–1996 period. The underlying composite model describing total security returns is econometrically estimated using fundamental variables: the price-to-earnings ratio (PE), price-to-book ratio, (PB), and how these current fundamental ratios compare with their five-year means (the "relative" variables, denoted by "R" preceding the fundamental ratio; i.e., RPE is the relative PE). The composite model uses the "traditional" formulation of the PE multiple, price divided by 12-month trailing earnings, which ignores stocks with zero or negative earnings. We refer to the purchase of the lowest PE stocks as the "low PE" strategy that has been traditionally analyzed on Wall Street. A value-based strategy prefers to purchase stocks with the lowest PE, RB, RPE, and RPB values (see Latane, Tuttle, and Jones (1975), Guerard and Takano (1992), and Guerard, Takano, and Yamane (1993), as we discussed in the preceding chapter). The following valuation model is used to identify the most undervalued Japanese and U.S. securities:

$$TR_T = a_0 + a_1 PE_t + a_2 PB_t + a_3 RPE_t + a_4 RPB_t + e_t \qquad (8.1)$$

where

TR_T = total returns (for three months for quarterly data) following the calculation of fundamental financial variables

PE, PB, RPE, RPB = previously defined financial variables

$a_0, a_1, \ldots a_8$ = regression parameters

e_t = randomly distributed error terms

The benefits of the underlying fundamentally based model is that it is in keeping with the recent anomalies literature, particularly with the PE and PB variables. See Dimson (1989), Jacobs and Levy (1988), and Ziemba (1990, 1992). One can use ordinary least squares (OLS) and robust (ROB) regression to address the problem of outliers in the data. The Beaton-Tukey (1974) biweight outlier-adjustment procedure is used in this study; see Guerard (1987), Guerard and Stone (1992), and Guerard, Takano, and Yamane (1993). The underlying composite model is created by equally weighting the four factors. We showed in Chapter 7 how a regression weighted composite model outperformed an equally weighted model in the United States. The regression-weighted model was created in three steps: (1) identifying the independent variables with positive coefficients that are statistically significant at the 10% level, (2) using the regression coefficients in step (1) as the one-period variable weights, and (3) averaging the weights over the past four quarters to produce the coming quarterly weights. The valuation approach in Japan produced statistically significant information coefficients (ICs) for the 1974–1990 period. The ICs are the regression coefficients of ranked subsequent total returns on ranked financial variables.

We find that the ICs are statistically significant and positive in 85% of the quarters in this study, as was the case in Blin, Bender and Guerard (1995).

We used the WorldScope fundamental data and created an equally weighted composite model using the consolidated Japanese PE, PB, RPE, and RPB variables described in Guerard and Takano (1992). We used the traditional approach to equity modeling in which one buys low PE and low PB stocks, particularly when they are low relative to their five-year means. We did not invest in any negative PE stocks; neither did Guerard and Takano (1991, 1992) in the initial Japanese modeling approaches. The statistical significance of the Japanese fundamental model is interesting, given the traditionally high PE multiples in Japan. In June 1987 the 12-month trailing PE was 98.40, whereas the corresponding U.S. PE was 25.99. One has been able to use the value strategy to outperform the Japanese equity market. The Japanese and U.S. WorldScope PE, PB, RPE, and RPB variable means for June 1987–1996 are shown in Table 8.1. Equation (8.1) uses four variables and is a subset of the eight-variable model used with parent company data reported in Guerard, Takano, and Yamane (1993). The cash flow and interim sales and their respective relative variables are not available on a consolidated basis for the entire first-section securities (generally only about 700 of the 1,200 firms).

A robust regression-based strategy outperformed the equally weighted strategy by more than 600 basis points annually in the Japanese market-neutral test described in Guerard, Blin, and Bender (1996). We will restrict this study to a comparison of equally weighted Japanese and U.S. WorldScope composite models, because our purpose is to show the effectiveness of the integration of value and growth (I/B/E/S forecasts) strategies. The

TABLE 8.1 Japan and U.S. Valuation Means

Variable	87	88	89	90	91	92	93	94	95
				Japan					
PE	98.4	114.4	90.71	88.65	68.31	65.04	118.5	165.5	92
PB	5.15	5.78	5.7	5.53	3.84	2.14	3.42	3.82	2
RPE	2.05	1.92	1.63	1.87	9.03	0.59	0.99	1.01	0.62
RPB	1.65	1.68	1.27	1.46	0.99	0.64	0.82	0.91	0.61
				United States					
PE	25.99	23.4	21.25	28.9	27.42	37.36	35.05	25.7	25.33
PB	2.92	2.37	2.4	2.92	2.85	2.61	6.82	4.01	6.09
RPE	1.41	1.26	1.31	1.26	1.21	1.2	1.39	1.22	1.49
RPB	1.34	1.12	1.15	1.09	1.08	1.13	1.27	1.16	1.26

underlying composite model in Japan and the United States produces statistically significant ICs and a very stable ranking approach, such that a strategy of purchasing the upper quintile of the value-based Japanese and U.S. securities was extremely effective during the 1974–1990 period, as reported in Guerard, Takano, and Yamane (1993). The reader is referred to the Guerard and Takano (1991, 1992) studies for a comparison of regression techniques in Japanese stock selection models, that is, the more sophisticated models (the weighted latent root regression model, WLRR) outperformed the OLS estimated model. Once we have collected value and earnings data for all stocks in each country at the close of each quarter and created the stock ranking score, we must also create efficient portfolios.

PORTFOLIO CONSTRUCTION: THE APT MODEL

An efficient portfolio has minimal risk for any given level of return. To compute efficient portfolios requires a risk model as well as the stock ranking list or expected return. The most powerful risk model available stems from Ross's Arbitrage Pricing Theory (APT, 1976). The goal is to measure the volatility of any asset and trace its lineage; to determine how much risk is shared and how much is company-specific. To do so, APT builds on a simple observation: asset prices tend to move together, or at least systematically. The APT model formalizes this fact.

Assume that the actual returns of an asset universe break down into three basic components: the expected return (E(r)), the return due to the joint impact of shared "shocks" (i.e., macroeconomic events), and the individual asset return due to company-specific shocks, notably earnings). Formally,

$$r = E(r) + Bf + e \qquad (8.2)$$

where

$r = n$-vector of asset returns
$E(. . .) = $ expected value
$f = k$-vector of factor returns
$B = (n \times k)$ matrix of factor sensitivities
$e = n$-vector of asset-specific random return

Expected asset returns will "tend" to display the following relationship:

$$E(r) = \sim 1r^* + Bp \qquad (8.3)$$

where

- r^* = zero-risk asset return (the risk-free rate),
- p = equilibrium "risk price" vector (factor risk premia)

Arbitrage pricing creates the linkage across assets in equilibrium such that returns are consistent with factor risk premia. As investors shift their portfolios, they enforce a very strong structure into the covariance matrix. In an arbitraged market, Ross (1976) shows, the covariance matrix will not be just any matrix. Arbitrage will force consistent risk prices across all assets. Each risk carries a risk premia that investors impute into the price of every asset as they see it.

Mathematically, Ross's remarkable insight implies a simple structure behind the covariance matrix. It breaks up into the sum of two square matrices, one measuring each stock's sensitivity to each latent risk factor, and one measuring what is unique about the stock's risk. In effect, the original covariance matrix is but a single estimator of the risk of stock investment one stock at a time. But if we in fact know that the matrix can be broken down into a much simpler structure, each stock's risk profile being simply its sensitivity to a set of common factors in addition to its own idiosyncratic risk, then we can treat the individual entries of the matrix as many different samples, taken across different strata and different shock scenarios. Thus one can get pooled estimates of risk. The latent matrix structure allows the specification to find more efficient and robust estimates of risk.

Rather than using the full covariance matrix (c), one can extract more information by estimating its inner structure, its eigenvectors and eigenvalues. The eigenvectors are simply the factor returns generating the observed asset returns given the asset factor profiles. Formally, since

$$[r - E(r)] = Bf + e \tag{8.4}$$

and $E(e\,f) = 0$

Setting $E(f\!f') = I$

$$C = E\{[r - E(r)][r - E(r)]'\} = BB' + D \tag{8.5}$$

where

$$D = E(e\,e') \tag{8.6}$$

In summary, to estimate B and e, one estimates the eigenvalue decomposition of C, first extracting the factor returns (f) vectors, which form an orthonormal basis for the observed asset returns. In this basis, one maps the assets by estimating their factor sensitivities (shared-risk exposures and k − vector b for each of the assets, creating a n x K matrix, B); and their asset-specific risk, e. An Arbitrage Pricing Theory model was estimated by John Blin and Steve Bender in the 1980s, which assumed that U.S. stock returns were generated by 20 factors that could be estimated using 2.5 years of weekly price data. See Blin and Douglas (1988) for a more complete description of the APT model. The factors are orthogonal, and the factors and error terms are independent. Blin and Bender use maximum-likelihood techniques to estimate the betas and specific asset standard deviations. The advantage of the Blin and Bender multifactor approach is that one does not start with a prespecified set of macroeconomic, industry, or homemade variables, but allows the researcher to estimate factor realizations through least squares. Blin and Bender estimated their APT model in Japan for the 1987–1996 period.

The results of the APT model estimation are very powerful. The APT model captures much of the historical returns variance. In Japan, for instance, the APT model explains more than 66% of the total weekly return variance, implying that 33% of the performance of the average Japanese stock is wholly company-specific. Unless we can accurately decompose returns, we cannot build portfolios devoid of implicit risk exposures.

SINGLE OR MULTIPLE SCORES, LINEAR OR QUADRATIC MAPS?

To pick a portfolio, we need an additional mapping of the global scores, showing each stock's overall attractiveness, *and* the stocks' risk profiles, into a single number, the "utility" of the portfolio. For the moment, note that the APT portfolio optimizer maps each stock's attractiveness and risk score into a global portfolio predicted volatility (risk) number. As one alters the portfolio weights, the predicted volatility changes. In short, one can pick the portfolio in two steps: by first mapping across return-generating attributes, then mapping across risk and return given a level of risk aversion (a "tilt" level).

A traditional way of mapping return-generating attributes into an overall score is through a simple linear map, arriving at a weighted sum of the individual attributes. For instance, one can take the four variables PE, PB, RPE, and RPB and weight them equally. Alternatively, one can assign 100%

weighting to whichever variable was the best predictor of subsequent performance in the last quarter. One can estimate which weights would have best predicted subsequent performance, as did Guerard and Takano (1991, 1992) and Guerard, Takano, and Yamane (1993), using regression techniques.

Whether one uses regression or equal weights, one can produce a stock global score first and *then* produce the optimal portfolio through the APT system. But there is, in fact, another way to proceed. Rather than aggregating across attributes all at once, through statistical estimation of the function mapping attributes into expected returns, one can also let the optimizer do the "aggregation" directly, in effect letting the interaction (if any) "carry through" to the optimization stage. In this case the optimizer becomes the final arbiter between these diverse attributes.

To use this approach, one must specify a functional form to effect the appropriate tradeoffs. A classical formulation in optimization theory is to set a quadratic penalty function, whereby one picks a preferred level of an attribute for the portfolio (e.g., a portfolio P/E equal to 10) and penalizes deviations from this level. As one increases the penalty weight, the value of the penalty function increases, making it less and less desirable to the optimizer.

Formally, let a denote the attribute vector for all the stocks in the universe, x denote the stock proportions in the portfolio, A denote the desired level of the portfolio attribute (e.g., P/E = 10, a scalar), and p denote the (scalar) penalty weight. Then, a quadratic penalty function is:

$$p * (a \cdot x - A)^{\wedge}2$$

where

* * denotes the scalar product
* · the dot denotes product of the a and x vectors
* $^{\wedge}2$ denotes the square of the expression inside the parentheses

For now, let us assume that one can add such quadratic terms to the quadratic optimizer objective function. And as one varies the value of the penalty weight (the "shadow cost" of not meeting the preferred portfolio attribute level exactly—the Lagrange multipler), one can make it more or less "expensive" to deviate from the "soft" constraint expressed by such a term. Here the cost of such deviation is the incremental predicted volatility of the portfolio—the added risk. In this way the optimizer becomes the arbiter of the tradeoffs. To the extent that one estimates the tradeoffs in the tilt term directly without accounting for the systematic (factor) risk, one may find that carrying each attribute separately and weighting them through the optimizer yields better results.

FINDINGS: BUILDING EFFICIENT PORTFOLIOS

An efficient portfolio x minimizes the following quadratic function:

$$x' [(w) \cdot B B' + (2 - w) <e>] x - h \cdot c \cdot x \qquad (8.7)$$

where x represents the optimal portfolio vector, $<e>$ is the diagonal matrix of stock-specific volatilities, c is the vector of stock scores (for instance, the one obtained through regression-based weights of the value variables), w is a number between 0 and 2, to vary the emphasis on systematic versus specific risk, and h is the scalar portfolio tilt parameter. As we increase the tilt, the portfolio risk increases along with the expected return. The tilt term is linear and subsumes in a single number all the components we may use to predict expected returns. As mentioned earlier, one can also add quadratic terms, with different penalty weights, to let the optimizer itself effect the tradeoffs between the returns components. Combining tilt and quadratic terms offers the maximum flexibility.

A value investor who let the model pick stocks would end up with high *expected* returns stocks, not necessarily high *actual* returns. Even if value investing "worked"; that is, if value stocks outperformed growth stocks, the actual portfolio returns depend on the common factors driving all stocks. Witness, for instance, the drastic drop in Japanese equities when the Japanese market collapsed in 1990. The market factor swept all performance away. But what if the investor had hedged his long stock position with Topix futures? The hedge may or may not have mimicked the investor's portfolio. It may have been a portfolio of other stocks to track his long portfolio, a "market-neutral" portfolio. More precisely "factor-neutral."

Whether we use earnings or value-investing or a combination, we can find both long and short stock candidates. Thus we can build long-only efficient portfolios or long/short (market-neutral) portfolios. Clearly, if the stock picking model is valid, it is best to use it on both sides—long and short—barring, of course, institutional constraints on short selling.

The market-neutral (or long/short) portfolios substantially reduce standard deviations of returns relative to long-only portfolios. The shares sold short in Japan increased tremendously during the 1987–1996 period such that one could implement the strategy by 1987. Guerard, Blin, and Bender (1996) reported that the value-based market-neutral portfolios returned a 12-month rolling average of 23.6% during the 1987–1994 period, whereas the three-month Gensaki rate averaged only 5.1% and the TOPIX less financial firms index averaged only 1.88% during the corresponding period (a do-nothing market). An I/B/E/S earnings revisions quadratic function increased the average annual return by more than 200 basis points during the 1987–1994 period.

BASIC RESULTS

Before considering our results, imagine that we wanted to take account of the fact that earnings are fundamentally stock-specific. One way to emphasize that the optimizer is not to concern itself with the residual (stock-specific) part of the risk is to choose a weight w in equation (8.7) near 2, rather than the default setting of 1. Conversely, if we felt that the strategy had a more systematic bias, clearly not the case in this instance, one could pick a weight nearer 0. As we vary the weight from 1 to 2, the number of stocks in the portfolio (Nb stocks in Portf) declines, since we are in effect instructing the system to take the scores at their face value, even at the risk of some sizeable adverse shocks for any one stock. Conversely, as we reduce the weight from 1 to 0, the number of stocks increases. We run the tests for w values ranging from 1 to 1.5 and 1.9, and the market-neutral portfolio return increases (Performance). Clearly I/B/E/S-driven earnings-only strategies in Japan add substantial performance when the portfolio is truly market-neutral (see Figure 8.1).

Japanese market-neutral portfolios dominate the long-only portfolio (which beat the market by a substantial margin in backtest during the 1992–1996 period). If one constructs a long-only portfolio employing the composite model as described in Guerard, Takano, and Yamane (1993) using equation (8.1), and one assumes a 3% upper bound on security weights and 200 basis points of transactions costs on all purchases and sales, one produces a set of portfolios that average 67 stocks, generating a 3.72% rolling 12-month average return with a 16.75% annualized standard deviation. The TOPIX produced an average rolling 12-month return of 0.32%, and thus the annual excess return on this portfolio is 340 basis points. The use of I/B/E/S forecasts, revisions, and momentum as a quadratic penalty variable increases the return to 6.22% annually, or an excess return of 591

TABLE 8.2 Market-Neutral Strategies

Japan/IBES/$w = 1$	Mean	An Mu/S
Nb Stocks in Portf	184	
Predict Fact Rsk/yr	4.56	
Predict Resid Rsk/yr	0.54	
Predict Total Rsk/yr	4.60	
Long Perform/month	4.20	0.63
Short Perform/month	1.92	0.30
Mkt Neutral Perform/mo	2.28	0.79
Turnover	85.30	
Performance	2.28	0.79
Rolling 12-Month Performance		
Long	11%	0.46
Short	0%	0.01
Net	11%	1.56

Japan/IBES/$w = 1.5$	Mean	An Mu/S
Nb Stocks in Portf	202.77	
Predict Fact Rsk/yr	5.43	
Predict Resid Rsk/yr	0.88	
Predict Total Rsk/yr	5.51	
Long Perform/month	5.93	0.77
Short Perform/month	2.63	0.36
Mkt Neutral Perform/mo	3.30	1.21
Turnover	85.33	
Performance	3.30	1.21
Rolling 12-Month Performance		
Long	15.25%	0.52
Short	1.44%	0.06
Net	13.81%	1.84

Japan/IBES/$w = 1.9$	Mean	An Mu/S
Nb Stocks in Portf	187.00	
Predict Fact Rsk/yr	5.84	
Predict Resid Rsk/yr	0.56	
Predict Total Rsk/yr	5.87	
Long Perform/month	6.31	0.81
Short Perform/month	2.85	0.39
Mkt Neutral Perform/mo	3.45	1.17
Turnover	84.78	
Performance	3.45	1.17
Rolling 12-Month Performance		
Long	18.23%	0.58
Short	1.39%	0.06
Net	16.84%	1.42

TABLE 8.2 (Continued)

Japan/IBES + Value	Mean		An Mu/S
Nb Stocks in Portf	143.92		
Predict Fact Rsk/yr	4.29		
Predict Resid Rsk/yr	1.83		
Predict Total Rsk/yr	4.68		
Long Perform/month	4.64		0.72
Short Perform/month	0.05		0.01
Mkt Neutral Perform/mo	4.60		2.74
Turnover	23.72		
Performance	4.60		2.74
Rolling 12-Month Performance			
Long	6.72%		0.38
Short	−10.17%		−0.59
Net	16.89%		4.53

Japan/Value Only	Mean	StdDevn	An Mu/S
Nb Stocks in Portf	169.31		
Predict Fact Rsk/yr	4.37		
Predict Resid Rsk/yr	1.82		
Predict Total Rsk/yr	4.77		
Long Perform/month	4.76		0.67
Short Perform/month	1.60		0.20
Mkt Neutral Perform/mo	3.16		0.74
Turnover	17.27		
Performance	3.16		0.74
Rolling 12-Month Performance			
Long	5.59%		0.31
Short	−4.42%		−0.21
Net	10.01%		0.93

US/Value + IBES	Mean		An Mu/S
Nb Stocks in Portf	199.91		
Predict Fact Risk/yr	4.92		
Predict Resid Rsk/yr	3.05		
Predict Total Rsk/yr	5.83		
Long Perform/month	5.19		0.99
Short Perform/month	2.13		0.82
Mkt Neutral Perform/mo	3.05		0.96
Turnover	19.34		
Performance	3.05		0.96
Rolling 12-Month Performance			
Long	20.85%		0.90
Short	7.96%		0.71
Net	12.89%		0.89

Continued on following page

TABLE 8.2 (Continued)

US/Value Only	Mean	An Mu/S
Nb Stocks in Portf	264.30	
Predict Fact Rsk/yr	4.64	
Predict Resid Rsk/yr	1.88	
Predict Total Rsk/yr	5.03	
Long Perform/month	4.03	0.84
Short Perform/month	4.60	1.18
Mkt Neutral Perform/mo	−0.57	−0.16
Turnover	11.90	
Performance	−0.57	−0.16
Rolling 12-Month Performance		
Long	14.93%	0.86
Short	18.79%	1.18
Net	−3.85%	−0.42

Long-Only Strategies

Japan/IBES + Value	Mean	StdDevn	An Mu/S
Nb Stocks in Portf	72.46		
Predict Fact Rsk/yr	2.79		
Predict Resid Rsk/yr	3.77		
Predict Total Rsk/yr	4.70		
Strategy Perform/month	4.29		0.81
Index Perform/month	2.23		0.45
Net Perform/mo	2.06		1.46
Turnover	26.20		
Performance	3.77		0.70
Rolling 12-Month Performance			
Strategy	6.22%		0.41
Index	0.32%		0.02
Net	5.91%		1.03

Japan/Value Only	Mean	StdDevn	An Mu/S
Nb Stocks in Portf	67.46		
Predict Fact Rsk/yr	2.89		
Predict Resid Rsk/yr	3.26		
Predict Total Rsk/yr	4.37		
Strategy Perform/month	3.74		0.62
Index Perform/month	2.23		0.45
Net Perform/mo	1.50		0.94
Turnover	7.62		
Performance	3.58		0.60

BASIC RESULTS 217

TABLE 8.2 (Continued)

Japan/Value Only	Mean		An Mu/S
Rolling 12-Month Performance			
Strategy	3.72%		0.22
Index	0.32%		0.02
Net	3.40%		0.78
US/Value + IBES	**Mean**		**An Mu/S**
Nb Stocks in Portf	92.73		
Predict Fact Rsk/yr	2.94		
Predict Resid Rsk/yr	2.83		
Predict Total Rsk/yr	4.09		
Strategy Perform/month	4.79		1.29
Index Perform/month	3.09		1.20
Net Perform/mo	1.70		0.91
Turnover	18.00		
Performance	4.62		1.25
Rolling 12-Month Performance			
Strategy	19.46%		1.25
Index	12.13%		1.07
Net	7.32%		0.79
US/Value Only	**Mean**	**StdDevn**	**An Mu/S**
Nb Stocks in Portf	101.58		
Predict Fact Rsk/yr	2.76		
Predict Resid Rsk/yr	2.74		
Predict Total Rsk/yr	3.90		
Strategy Perform/month	3.91		1.18
Index Perform/month	3.09		1.20
Net Perform/mo	0.83		0.54
Turnover	3.65		
Performance	3.88		1.17
Rolling 12-Month Performance			
Strategy	16%		1.22
Index	12%		1.07
Net	4%		0.54

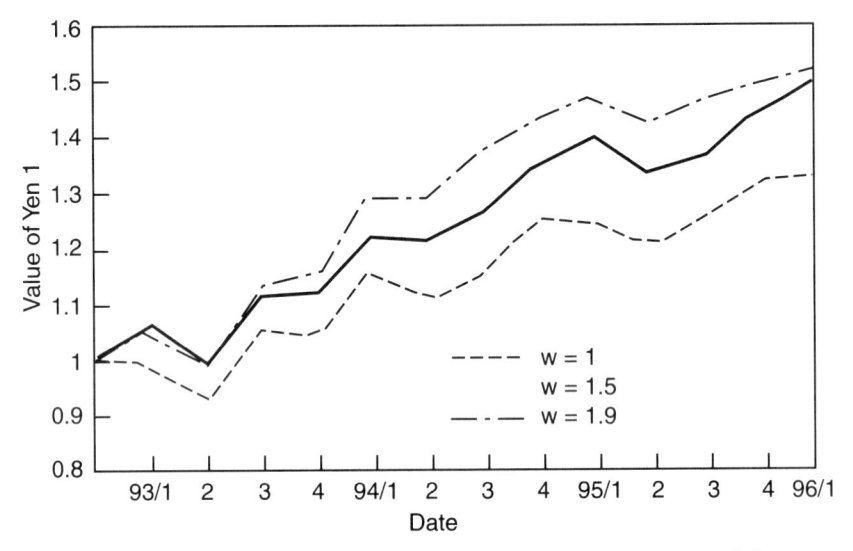

Figure 8.1 Growth of IBES-driven market-neutral Japanese portfolios.

basis points on the long-only portfolios. The use of the proprietary Japanese I/B/E/S variable increased the long-only return by more than 250 basis points annually and reduced the annualized portfolio standard deviation to 15.34%. The use of earnings revisions in the United States has been substantiated in the works of Givoly and Lakonishok (1979), Hawkins, Chamberlain, and Daniel (1984), Arnott (1985), and others summarized in Brown (1993). We find support for the creation of composite I/B/E/S variables as was put forth by Wheeler (1990). The long-only Japanese portfolios with the I/B/E/S-quadratic penalty variable produced R-squared values between the optimal portfolio and target index of approximately 95% and predicted tracking errors of approximately 4.70%. For approximately the same level of risk as the target, one can outperform by more than 500 basis points on the long-only portfolios. The portfolio turnover rates are approximately 26.2% per quarter for the historical and I/B/E/S revisions models.

The long-only side of the value-only market-neutral portfolio outperformed the TOPIX by more than 550 basis points on a 12-month rolling average basis during the 1992–1996 period. The short securities underperformed the TOPIX by more than 450 basis points per quarter (-442 vs. 0.32 basis points) during the corresponding period. Thus, the market-neutral portfolio produces a 3.16% quarterly return after subtracting 2.0% transactions costs each way, or 10.01% on a rolling 12-month basis with a tilt of 12 (with a Sharpe ratio, the ratio of portfolio excess returns divided by the portfolio standard deviation, of 0.93). If a manager uses the quarterly I/B/E/S variable as a penalty quadratic variable with the four-variable equally weighted composite model, the annualized portfolio return rises to 16.89% and annualized portfolio standard deviation falls to 3.73% (with a Sharpe ratio of 4.53). The annualized market-neutral tracking error is approximately 4.6%. Thus, for the same tracking error as a long-only portfolio, one can effectively dominate the long-only portfolio by shorting securities that are predicted to be less favored by the market. An equally weighted historical data market-neutral portfolio produced an average annual return of 19.50% during the 1987–1994 period, providing support for the initial analysis of the model. Had one created a market-neutral composite model using the conventional formulation (i.e., buying high EP, BP, REP, and RBP stocks in Japan), the market-neutral model would have earned a $1.68 cumulative wealth ratio for every dollar invested during the December 1987–September 1994 period, whereas the traditional value-only model produced a $2.95 cumulative wealth ratio over the corresponding period. One should not short negative PE stocks in Japan. For a comparison of the traditional and conventional approaches to composite modeling, see Guerard and Takano (1992). The conclusion of Japanese concensus forecasts and revisions is consistent with the Elton, Gruber, and Gultekin (1981) U.S. analysis and the earlier Guerard, Blin, and Bender (1996a) Japanese results, in which they found no statistical evidence that concensus forecasts produce excess returns, but that revisions produce significantly positive excess returns, and, most important, firms with the highest actual earnings-per-share (eps) growth for the 1973–1975 period produced the highest excess returns.

In the United States we find a similar result for the use of the same form of the I/B/E/S proprietary variable. A value-only long-only U.S. composite model for the December 1987–March 1996 period using equation (8.1), 3% upper bounds on security weights, and 1% (each-way) transactions costs produced a 12-month rolling average return of 16.99%, whereas the corresponding Standard & Poors (S&P) 500 return was 12.13%. The value-only, long-only portfolios outperform the S&P 500 by more than 375 basis points annually, a figure quite consistent with the Guerard and Takano (1992) studies. The Sharpe ratio of the value-only, long-only portfolio was 1.22, reflecting its 13.07% standard deviation, whereas the S&P 500 had a Sharpe ratio of 1.07. The use of the I/B/E/S variable increases the 12-month average return to 19.46% and its annualized standard deviation rises to 15.53%, producing a Sharpe ratio of 1.25. The use of I/B/E/S produced an excess return of 732 basis points, a figure quite consistent with the Guerard, Gultekin, and Stone (1996) results using Gultekin's prespecified APT model, as discussed in Chapter 7. Had one created a market-neutral portfolio using only the value component, one would have lost 3.86% annually (an average return of 14.93% on longs and 18.79 percent on the shorts) during the 1987–1996 period in the United States, whereas the market-neutral I/B/E/S quadratic model would have produced a 17.11% return on the longs and a 8.59% return on the short side of the portfolio.

Long-only investing in the United States with value only is better, averaging about 3.8% per year with 7% volatility. Adding earnings doubles the returns to 7.3% with a slightly higher volatility, about 9%. The contribution of earnings variables in the United States is as compelling as it is in Japan. In fact, it makes the difference between perverse results derived from strict value investing and reasonable results from a combo strategy.

One gains more than 1,200 basis points on the U.S. market-neutral portfolio using the I/B/E/S variable, relative to the use of the value-only strategy. It appears to be the case that the I/B/E/S variable added about 250 basis points in the European nations during the past four years.[1] More important, the Japanese market-neutral model outperforms the Gensaki (the Japanese government bond) rate and is enhanced by the I/B/E/S analysis, whereas the I/B/E/S variable is necessary in our formulation of the U.S. market-neutral model.[2]

UPDATED ANALYSIS

Guerard, Blin, and Bender (1997) updated the Japanese and U.S. market-neutral portfolio analysis using an updated IBES international database CD-ROM. The Japanese and the U. S. portfolios were constructed for the December 1987–June 1997 period. Similar results to those presented in this chapter were found (see Table 8.3). IBES is adding more value in Japan and is necessary for a U.S. market-neutral portfolio.

TABLE 8.3 Japanese and U. S. Portfolios for the December 1987–June 1997 Period

	ExR%		Mu/S
		Japan	
Value-only	14.20		1.05
Value+IBES	20.80		1.93
		United States	
Value-only	−3.25		NA
Value+IBES	12.54		1.03

NOTES

1. In our initial modeling of the G5 nations, the United Kingdom, Germany, France, Italy, and Canada, we find that the I/B/E/S quadratic variable adds over 250 basis points to the market-neutral historical data-only model excess returns. In the G5 nations for the December 1992–March 1996 period, we find that the WorldScope four-value factor market-neutral model produces excess returns of 7.41% annually, whereas the market-neutral excess returns including the I/B/E/S quadratic variable are 10.22% annually.
2. In a recent presentation, Guerard, Blin, and Bender (1996b) reported results using an expanded U.S. model that was formulated as:

$$TR = f(EP, BP, CP, SP, DY, \text{ and } NCAV, I/B/E/S) \tag{8.8}$$

where

　　CP = cashflow-to-price; i.e., net income plus depreciation
　　SP = sales-to-price
　　DY = dividend yield
　　NCAV = net current asset value-to-price; i.e., current assets minus all liabilities

Guerard, Blin, and Bender (1996b) found that the long-only, value-only model using the non-I/B/E/S components underperformed the S&P 500 by 134 basis points annually as compared to the 201 basis-point outperformance by adding the I/B/E/S quadratic variable. The excess returns rise to 475 basis points when one regression-weights equation (8.8) as discussed in Guerard, Takano, and Yamane (1993), with the I/B/E/S variable having an average quarterly weight of 0.35 during the 1982–1995 period. The expanded model in equation (8.8) produced an average F-statistic of 15.82, whereas equation (8.1) produced an average F-statistic of 20.10. Guerard, Takao, and Yamane (1993) estimated a slightly different equation:

$$TR = f(FEP, BP, CP, SP, REP, RBP, RCP, \text{ and } RSP) \tag{8.9}$$

where

FEP = I/B/E/S forecast of the current year's eps (FY1)/price
REP, RBP, RCP, RSP are the respective five-year relative variables in which the current EP, . . . , SP are divided by their respective year-year means

The Guerard, Takano, and Yamane (1993) formulation produced an average F-statistic of 18.10. The alternative model formulation are highly statistically significant and the ICs are almost identical (0.13–0.153); however, we find that the

I/B/E/S variable is more effective when used as a quadratic variable than as a variable in a multiple regression model. Turnover increases from 11.5% quarterly with the regression-weighted model to 28.5% quarterly using the quadratic variable approach.

REFERENCES

Arnott, R. "The Use and Misuse of Concensus Earnings." *Journal of Portfolio Management* (1985):18–27.

Banz, R. "The Relationship Between Return and Market Value of Common Stocks." *Journal of Financial Economics* 9 (1981):3–18.

Beaton, A. E., and J. W. Tukey. "The Fitting of Power Series, Meaning Polynomials, Illustrated on Bank-Spectroscopic Data." *Technometrics* 16 (1974):147–185.

Belsley, D. A., A. E. Kuh, and R. E. Welsch. *Regression Diagnostics—Identifying Influential Data and Sources of Collinearity.* New York: John Wiley & Sons, 1980.

Blin, J., and G. Douglas. "Stock Returns vs. Factors." *Investment Management Review* (1987).

Blin, J., S. Bender, and J. B. Guerard, Jr. "Earnings and Value Investing—How to Make the Return Worth the Risk." Presented at the 1996 I/B/E/S Research Conference, Tokyo, June 1996, forthcoming in *Handbook of Corporate Earnings Analysis,* Volume 2, edited by B. Bruce.

Blin, J., S. Bender, and J. B. Guerard. "Why Your Next Japanese Portfolio Should Be Market-Neutral—and How." *Journal of Investing* (1995).

Blume, M. E., M. N. Gultekin, and N. B. Gultekin. "Validating Return-Generating Models." (1990) (under journal review).

Brown, L. D. "Earnings Forecasting Research: Its Implications for Capital Markets Research." *International Journal of Forecasting* 9 (1993):295–320.

Chen, N-f., R. Roll, and S. A. Ross. "Economic Forces and the Stock Market." *Journal of Business* 59 (1986):383–403.

Chan, L., Y. Hamao, and J. Lakonishok. "Fundamentals and Stock Returns in Japan." *Journal of Finance* 46 (1991):1739–1764.

Chopra, V. K., and W. T. Ziemba. "The Effects of Errors in Means, Variances, and Covariances on Optimal Choice." *The Journal of Portfolio Management* (1993):6–11.

Clemen, R. T. "Combining Forecasts: A Review and Annotated Bibliography." *International Journal of Forecasting* 5 (1989):559–584.

Conner, G., and R. A. Korajczyk. "The Arbitrage Pricing Theory and Multifactor Models of Asset Returns." In R. Jarrow, V. Makisimovic, and W. Ziemba, *Handbooks in OR & MS,* North-Holland, 1995.

Conner, G., and R. A. Korajczyk. "Risk and Return in an Equilibrium APT: Application of a New Test Methodology." *Journal of Financial Economics* 21 (1988):255–289.

Cottle, S., R. F. Murray, and F. E. Block. *Graham and Dodd's Security Analysis.* 5th ed. New York: McGraw-Hill.

Dimson, E., ed. *Stock Market Anomalies.* Cambridge, Cambridge University Press, 1988.

Dhrymes, P. J., I. Friend, N. B. Gultekin, and M. N. Gultekin. "An Empirical Examination of the Implications of Arbitrage Pricing Theory." *Journal of Banking and Finance* 9 (1985):73–99.

Dhrymes, P. J., I. Friend, and N. B. Gultekin. "A Critical Reexamination of the Empirical Evidence on the Arbitrage Pricing Theory." *Journal of Finance* 39 (1984):323–346.

Elton, E. J., and M. J. Gruber. "Analysts' Expectations and Japanese Stock Prices." *Japan and the World Economy* 2 (1989).

Elton, E. J., and M. J. Gruber. *Modern Portfolio Theory and Investment Analysis.* New York: John Wiley & Sons, 1987.

Elton, E. J., M. J. Gruber, and M. Gultekin. "Expectations and Share Prices" *Management Science* 27 (1981):975–987.

Fama, E. F., and K. R. French. "Size and Book-to-Market Factors in Earnings and Returns." *Journal of Finance* 50 (1995):131–155.

Fama, E. F., and K. R. French. "Cross-Sectional Variation in Expected Stock Returns." *Journal of Finance* 47 (1992):427–465.

Farrell, J. *Guide to Portfolio Management.* New York: McGraw-Hill, 1983.

Givoly, D., and J. Lakonishok. "The Information Content of Financial Analysts' Forecasts of Earnings: Some Evidence on Semi-Strong Inefficiency." *Journal of Accounting and Economics* 3 (1979):165–185.

Graham, B. *The Intelligent Investor.* New York: Harper & Row, 1940.

Graham, B., and Dodd, D. *Security Analysis.* 4th ed. New York: McGraw-Hill, 1962.

Grundy, K., and B. G. Malkiel. "Reports of Beta's Death Have Been Greatly Exaggerated." *Journal of Portfolio Management* (Spring 1996).

Guerard, J. B., Jr. "Linear Constraints, Robust-Weighting, and Efficient Composite Modeling." *Journal of Forecasting* 6 (1987):193–199.

Guerard, J. B., Jr., and B. K. Stone. "Composite Forecasting of Annual Corporate Earnings." In *Research in Finance* 10, edited by A. Chen. Greenwich, Conn.: JAI Press, 1992 205–230.

Guerard, J. B., Jr., and M. Takano. "Stock Selection and Composite Modeling in Japan." *Security Analysts Journal* 29 (1991):1–13.

Guerard, J. B., Jr., A. S. Bean, and B. K. Stone. "Goal Setting for Effective Corporate Planning." *Management Science* 36 (1990):359–367.

Guerard, J. B., Jr., J. Blin, and S. Bender. "Earnings Revisions in the Estimation of

Efficient Market-Neutral Japanese Portfolios." *Journal of Investing* (spring 1996a).

Guerard, J. B. Jr., J. Blin, and S. Bender. "The Riskiness of Global Portfolios." Presented at the Society of Quantitative Analysts, Annual Fuzzy Day Seminar, New York, May 1996b.

Guerard, J. B., Jr., J. Blin, and S. Bender. "Forecasting Earnings Composite Variables, Financial Anomalies, and Efficient Japanese and U. S. Portfolios." *International Journal of Forecasting* (in press).

Guerard, J. B., Jr., M. Takano, and Y. Yamane. "The Development of Efficient Portfolios in Japan with Particular Emphasis on Sales and Earnings Forecasting." *Annals of Operations Research* 45 (1993):91–108.

Gultekin, M. N., and N. B. Gultekin. "Stock Return Anomalies and the Tests of the Arbitrage Pricing Theory." *Journal of Finance* 42 (1987):1213–1224.

Gunst, R. F., and R. L. Mason. *Regression Analysis and Its Application.* New York: Marcel Dekker, 1980.

Hawkins, E., S. C. Chamberlain, and W. E. Daniel. "Earnings Expectations and Security Prices." *Financial Analysts Journal* 24–29 (1984):30–38.

Jacobs, B., and K. Levy. "Disentangling Equity Return Regularities: New Insights and Investment Opportunities." *Financial Analysts Journal* (1988):18–43.

Keon, E. "Earnings Expectations in Financial Theory and Investment Practice." In *I/B/E/S Research Bibliography,* edited by L. D. Brown. 5th ed., 1996.

King, B. F. "Market and Industry Factors in Stock Price Behavior." *Journal of Business* 39 (1966):139–190.

Kothari, S. P., J. Shanken, and R. G. Sloan. "Another Look at the Cross-Section of Expected Stock Returns." *Journal of Finance* 50 (185–224).

Latane, H., D. Tuttle, and C. Jones. *Security Analysis and Portfolio Management.* New York: Ronald, 1975.

Markowitz, H. M. *Portfolio Selection: Efficient Diversification of Investments.* New York: John Wiley & Sons, 1959.

Miller, J. D., Guerard, J. B., Jr., and M. Takano. "Bridging the Gap Between Theory and Practice in Equity Selection Modeling: Case Studies of U.S. and Japanese Models" (under journal review).

Morgan, A., and I. Morgan. "Measurement of Abnormal Returns from Small Stocks." *Journal of Business & Economic Statistics* 5 (1987):121–129.

Roll, R., and S. A. Ross. "An Empirical Investigation of the Arbitrage Pricing Theory." *Journal of Finance* 35 (1980):1073–1103.

Rudd, A., and B. Rosenberg. "Realistic Portfolio Optimization." In *Portfolio Theory: 25 Years Later,* edited by Elton and Gruber. Amsterdam: North-Holland, 1979.

Tinic, S. M., and R. R. West. "Risk and Return: January vs. the Rest of the Year." *Journal of Financial Economics* (1984):561–574.

Vu, J. D. "An Anomalous Evidence Regarding Market Efficiency: The Net Current Asset Rule." In A. Chen, *Research in Finance* 8 (1990):241–254.

Webster, J. T., R. F. Gunst, and R. L. Mason. "Latent Root Regression Analysis." *Technometrics* 16 (1974):513–522.

Wheeler, L. "Changes in Consensus Earnings Estimates and Their Impact on Stock Returns." Presented at the Institute for Quantitative Research in Finance. In *The Handbook of Corporate Earnings Analysis,* edited by B. Bruce and C. Epstein. Chicago: Probus Publishing Co., 1990.

Ziemba, W. T. *Invest Japan.* Chicago: Probus Publishing Co., 1992.

Ziemba, W. T. "Fundamental Factors in U.S. and Japanese Stock Returns." Presented at the Berkeley Program in Finance, Santa Barbara, Calif., 1990.

9
THE (NOT SO SPECIAL) CASE OF SOCIAL INVESTING

In this chapter we address two questions concerning socially responsible investing. First, is the average return of a socially screened equity universe statistically different from the average return of an unscreened universe for the 1987–1994 period? Second, can one use an expected return model incorporating both value and growth components to select stocks and create portfolios in the socially screened and unscreened equity universes such that one can outperform both universe benchmarks? We find no statistically significant differences in the mean returns of unscreened and screened equity universes for the 1987–1994 period. We find little differences in the predictive power of the composite model to select stocks in both unscreened and screened equity universes. The estimated composite model offers the potential for substantial outperformance of socially screened and unscreened equity universes.

There is a growing literature in academic and professional investment journals that suggests that socially responsible investing may produce higher risk-adjusted portfolio returns than those achieved by merely using all available stocks in the equity universe.[1] Whereas a financial screen is applied to an investment universe to reduce potential investments, a social screen is a non-financial criterion applied in the investment process that is an expression of a social, ethical, or religious concern. The application of a social screen allows the manager to apply these concerns in the investment process (Kinder 1997). An investor might expect lower returns from companies that damage the natural environment, sell liquor and other alcoholic products, produce, design, or use nuclear power, engage in gambling, or are large defense contractors, when one considers the possible corporate expenses of fines and litigation. Is socially screened investing a "dumb" idea, as has been put forth in some recent popular media?[2] It is the case that 24 socially screened mutual funds have substantially underperformed the S&P 500 during the past five and ten years.[3] However, the difference between the average return on socially screened equity mutual funds and 2,034 unscreened equity mutual funds drops from −417 basis points over the past five years to −105 basis points over the past ten years, a less meaningful differential, particularly given the very small number of socially screened equity mutual funds with long-term track records. There are only six socially screened equity mutual funds with five-year track records in the Morningstar universe, and only Dreyfus Third Century and Parnassus have ten-year records. The College Retirement Equities Fund (CREF) Social Choice Account, a balanced account containing 62% socially screened equities and 38% debt, has matched its annualized benchmark for the past five years.[4] The equity performance of the CREF Social Choice Account provides substantial evidence that social screening need not lead to the underperformance that one finds in the recent *Morningstar* socially responsible fund universe. We will show that a socially screened universe return is not significantly different from an unscreened universe return for the 1987–1994 period. We also show that a composite model integrating value and growth components, such as that developed and estimated in Chapter 7, can consistently produce positive and statistically significant correlations between a stock's expected return ranking and its subsequent performance. Significant outperformance is generated in a socially screened investment universe. It is not "dumb" to be a socially conscious investor; rather, one must look at how a manager implements the investment process. We will examine a special case of the models estimated in Chapter 7 and show how one can construct a socially responsible portfolio with financial characteristics that are virtually identical to those of an unscreened portfolio. Furthermore, we show that efficient portfolios may be constructed using the 400 socially responsible securities in the Domini Social Index (DSI).

The purposes of this chapter are (1) to examine the returns of an unscreened equity universe composed of 1,300 equity stocks and a socially screened universe of approximately 900 stocks and test as to whether there are statistically significant differences in the average returns of the two equity universes, and (2) to determine whether a composite model using both value and growth components is as effective in a screened universe as in an unscreened universe in identifying undervalued securities, and whether these can be combined into portfolios that may outperform the screened universe benchmark. Guerard (1997a) showed that there is no significant difference between the average monthly returns of the screened and unscreened universes during the 1987–1994 period. Indeed, from January 1987 to December 1994, there is less than a 15-basis-point differential in equally weighted annualized stock returns. We also show that a composite model using both value and growth (I/B/E/S) components produces statistically significant information coefficients (ICs) in the unscreened and screened stock universes. There are no significant differences in stock selection modeling between screened and unscreened universes, and significant excess returns may be realized using quantitative models in the screened universe. The screens used in this analysis, provided by Kinder, Lydenberg, and Domini (KLD), are as follows:

Military
Nuclear Power
Product (Alcohol, Tobacco, and Gambling)
Environment

The Vantage Global Advisors' unscreened 1,200-stock universe generated returns such that a $1.00 investment grew to $3.84 during the December 1987–December 1996 period. A corresponding investment in the socially screened universe would have grown to $3.57. There is no statistically significant difference in the respective return series, and more important, there is no economically meaningful difference between the return differentials. The variability of the two return series is almost equal during the 1987–1996 period. One can test for statistically significant differences in the two return series using the F-test, which examines the differences in series mean (returns) relative to the standard deviations of the series. When one applies the F-test, one finds that series are not statistically different from one another.

As an example, let us examine the financial characteristics of the stocks in the unscreened and socially screened Vantage Global Advisors' (VGA) universes as of December 1994. The unscreened VGA universe of 1,300 stocks had BARRA growth and book-to-price sensitivities of 0.185 and 0.306, whereas the socially screened VGA universe had corresponding BARRA growth and book-to-price sensitivities of 0.269 and 0.279, respectively. The unscreened universe had an average market capitalization of $3.433 billion in December 1994, whereas the socially screened universe had a mean capitalization of $2.796 billion. The average BARRA growth and book-to-price sensitivities of the excluded securities were -0.164 and 0.414, respectively, and the average market capitalization of the excluded stocks exceeded $6.1 billion. Thus, socially screened-out stocks had higher market capitalizations and were more value-oriented than the unscreened universe, a condition noted by Lloyd Kurtz (Kurtz and D. DiBartolomeo 1996). There was a statistically significant difference between the unscreened VGA universe lower price-to-book ratio and the higher price-to-book ratio of the Vantage screened universe. Professors Fama and French at the University of Chicago found that smaller stocks with lower price-to-book ratios tended to outperform larger stocks with higher price-to-book ratios in the very long run.[5] The higher price-to-book ratio of the screened universe represents a risk exposure to a socially responsible investor. The screened universe is more sensitive to the BARRA growth factor return than the Vantage unscreened universe, and this exposure should help relative performance for socially responsible investors when the BARRA growth factor return outperforms the BARRA value factor return.[6]

The higher growth sensitivity helped Luck and Pilotte (1993) find that the Domini Social Index outperformed the S&P 500 Index during the May 1990–September 1992 period. Luck and Pilotte used the BARRA Performance Analysis (PAN) package and found that the 400 securities in the DSI produced an annualized active return of 233 basis points relative to the S&P 500 and specific asset selection accounted for 199 basis points of the active return. Luck and Pilotte noted that the May 1990–September 1992 period was characterized by positive growth factor and size returns (smaller stocks outperformed larger-capitalized stocks as a rule during this period). Superior asset selection may have been achieved as Kinder, Lydenberg, Domini & Co. (KLD) created the DSI in May 1990 by including non-S&P 500 stocks with "good" records on corporate citizenship, product quality, and broad representation of women and minorities KLD developed criteria to establish the records of socially responsible firms (see Kinder, Lydenberg, and Domini 1993). For example, in March 1992, KLD produced a screen of 24 publicly traded firms that dealt in or used recycled materials. A second screen of 20 companies known for quality products was developed by KLD, although one-third of these firms failed other screens. In August 1992, 12 firms were recognized by a KLD diversity screen that identified firms with four or more (or at least one-third of the members if the firm had fewer than 12 members) board seats held by women or minorities. Additional KLD screens in August 1992 identified 10 firms with women or minority CEOs and 20 firms that possessed notable records on promoting women and minorities. KLD screens established criteria to substantiate good corporate citizenship. It is important to note that these criteria did not "cost" the investor any meaningful average return during the 1987–1994 period and may have produced positive active (relative to the S&P 500) returns during some subperiods.

STOCK SELECTION IN UNSCREENED AND SCREENED UNIVERSES

In the previous section we examined the financial characteristics of unscreened and socially screened stocks, finding, as did Kurtz and DiBartolomeo (1996), that larger, more value-oriented stocks are excluded by social screening. Can a composite stock selection model, using value and growth factors, be effective in selecting securities that outperform the market in a socially-screened universe? Let us propose to use a quantitative model for all securities publicly traded on any exchange during the 1987–1996 period. The model has seven variables, six value factors and the composite, proprietary growth variable developed in Chapter 7. The six value factors are earnings-to-price, book value-to-price, cash flow-to-price, sales-to-price, dividend yield, and net current asset value. The earnings, book value, cash flow, and sales variables are traditional fundamental variables examined in the investment literature, as discussed in Chapter 8.[7] The traditional theory of value-investing holds that securities with higher earnings, book value, cash flow, and sales are preferred to those securities with lower ratios, respectively. The net current asset value is the current assets of a firm less its total liabilities. A firm is hypothesized to be undervalued when its net current asset value is less than its stock price (Graham and Dodd 1962; Vu 1990).

Financial economists have studied the effectiveness of consensus (mean values of forecasts) for more than 30 years in the United States, producing a huge literature exceeding 400 articles, recently summarized in Keon (1996). A consensus has yet to develop as to whether analysts' forecasts add value, that is, create excess returns. It has been shown that analysts' forecasts are generally more accurate than time series models, but it has not been consistently shown that the more accurate forecasts produce statistically significant excess returns, see Brown (1993) for an excellent survey of the literature on earnings forecasting. In this study we analyze three possible sources of excess returns from analysts' forecasts: (1) the forecasts themselves, (2) the changes in the mean values of earnings forecasts relative to the stock price, and (3) the breadth of the forecasts, where breadth is defined to be the monthly net number of analysts raising the forecast divided by the total number of forecasts. It is possible that the forecasts themselves may not produce excess returns; that is, simply buying securities forecasted to have the highest growth in earnings for the current fiscal year (FY1) or next fiscal year (FY2) may not add value. Cragg and Malkiel (1968) and Niederhoffer and Regan (1972) found that analysts could not effectively forecast annual earnings relative to naive time series models, and Elton, Gruber, and

Gultekin (1981) found little excess returns associated with purchasing securities solely on the basis of predicted earnings forecasts (EP). Niederhoffer and Regan (1972) and Elton, Gruber, and Gultekin (1981) found that the securities achieving the highest earnings growth produced significant excess returns. Thus, there is a significant reward to correctly forecasting earnings, but analysts' forecasts may be not sufficient.

Guerard, Blin, and Bender (1996a) found that analysts' forecasts were not sufficient in Japan to outperform the market during the 1987–1994 period. The lack of excess returns associated with consensus forecasts should not be the end of the analysis, because changes in the mean values of the forecasts divided by the stock price have been shown to add value in the United States (Hawkins, Chamberlain, and Daniel 1984; Wheeler 1990) and Japan (Guerard, Blin, and Bender 1996a). The changes in mean forecasts are referred to as "earnings revisions" (EREV), and one purchases stocks when analysts are raising their forecasts (Keon 1996). Wheeler (1990) found substantial value to using the breadth of earnings (defined as the number of forecasts raised less the number of forecasts lowered, the result divided by the total number of forecasts) to rank stock, where one purchases stocks when a (net) increasing number of analysts are raising their forecasts (EB). The breadth measure may well be less susceptible to the undue influence of a single analyst.

A proprietary growth variable (PRGR) is created from consensus I/B/E/S forecasts, forecast revisions, and breadth of forecasts and is of the general form described in Wheeler (1990) and described in Chapter 7. The proprietary I/B/E/S variable, PRGR, greatly enhances return even after transactions costs have been included.

In this study we test several forms of an earnings forecasting (EF) variable:

1. EQ(FY1, FY2 EP)
2. EQ(FY1, FY2 EREV)
3. EQ(FY1, FY2 EB)
4. EQ(FY1 EP, EREV, EB)
5. EQ(FY2 EP, EREV, EB)
6. PRGR

The model may be summarized in equation (9.1) as follows:

$$TR_T = a_0 + a_1 EP_t + a_2 BP_t + a_3 CP_t + a_4 SP_t + a_5 DY_t + a_6 NCAV_t + a_7 EF_t + e_t \quad (9.1)$$

where

TR is total returns for the subsequent holding period (quarter)
EP is the (net income per shares) earnings-to-price ratio
BP is the book value per share-to-price ratio
CP is the cash flow per share-to-price ratio
SP is the sales-to-price ratio
DY is the dividend yield
NCAV is the net current asset value per share
EF is a particular form of the growth variable
e is the randomly distributed error term

The expected returns are created as described in Guerard (1990), Guerard and Takano (1992), and Guerard, Takano, and Yamane (1993). That is, quarterly cross-sectional regressions are run for each quarter during the 1982–1994 period every March, June, September, and December, as seen in Chapter 7. The dependent variable is the coming return for the subsequent three months, and the independent variables are constructed from the Compustat database in which the annual data are the fundamentals assumed to be known in June of each year and monthly prices are used to construct the valuation ratios. The quarterly weights are again calculated by (1) finding the independent variables that are positive (the hypothesized sign of the coefficients) and statistically significant at the 10% level, (2) normalizing the regression coefficients to be weights that sum to one, and (3) averaging the coefficients over the past four quarters.[8] The cross-sectional regressions employ the Beaton-Tukey (1974) biweight technique in which the regressions weigh observations inversely with their ordinary least squares errors; that is, the larger the residual, the lower the observation weight in the regression.[9] The Beaton-Tukey outlier-adjustment procedure, also referred to as robust regression (ROB) has been shown to produce more efficient composite models for creating a statistically based expected return ranking model than the use of ordinary least squares (OLS) (Guerard 1990; Guerard and Stone 1992).

We tested the effectiveness of the various forms of the earnings forecasting variable by creating portfolios using an equally weighted seven-factor model for all securities with annual sales and monthly stock prices on Compustat during the 1987–1996 period, using the several forms of equation 9.1, and quarterly stock rankings were created. The advantage to using equally weighted portfolios is that one can examine the excess returns

(ExR) and portfolio turnover (Turn) of the various forms of earnings forecasting relative to the use of a equally weighted value-only model. If one runs a simulation in which one purchases securities with the highest expected return ranking, one finds that the breadth of earnings dominates earnings forecasts and revisions in the current forecast year analysis (FY1). The use of two-year-ahead forecasts, revisions, and breadth does not outperform the one-year-ahead forecasts; a result consistent with Guerard, Gultekin, and Stone (1996). Earnings forecasts themselves do not add value; a result consistent with Elton, Gruber, and Gultekin (1996) and Guerard, Blin, and Bender (1997b). Earnings breadth produces higher turnover than analysts' forecasts or revisions, but generally enhances excess returns relative to the other forms of earnings forecast variable—a conclusion supported in the regression results of Guerard, Gultekin, and Stone (1996) and in Chapter 7. The proprietary, composite earnings forecast variable produces relatively higher turnover than the individual forecast variable, but much higher excess returns. In the simulations, we assume that a portfolio manager tightly constrains portfolio security weighting and industry weights to be very similar to the S&P 500 Index. A composite earnings forecasting framework similar to that in Wheeler (1990) substantially dominates the use of individual forecast variables. The equally-weighted proprietary growth model produces 362 basis points of excess returns.

The results of this study are more consistent with those of Wheeler (1990), Guerard and Stone (1992), and Chapter 7, in that analysts added significant value, than the earlier studies of Cragg and Malkiel (1968) and Elton, Gruber, and Gultekin (1981). Perhaps the value of analysts was significant because we used a broader definition of earnings forecasting and because earnings rose substantially during the 1982–1994 period. An annual

TABLE 9.1 Excess Returns (ExR) and Turnover (Turn) of Earnings Forecast, 1987–1996

	Fiscal Year ExR(%)	Turnover(%)
Value-only	−.42	46.20
EQ(FY1, FY2 EP)	−.91	46.50
EQ(FY1, FY2 EREV)	0.03	78.28
EQ(FY1, FY2 EB)	6.37	195.5
EQ(FY1, EP, EREV, EB)	5.96	161.6
EQ(FY2, EP, EREV, EB)	4.94	211.1
EQ(PRGR)	3.62	123.1
RG(PRGR)	6.35	171.2

regression of ranked achieved earnings growth on ranked total returns produced positive and statistically significant coefficients on the achieved earnings variable in 12 of the 13 years. Earnings are certainly a major determinant of stock prices—a result consistent with Graham and Dodd (1962), Niederhoffer and Regan (1972), and Elton, Gruber, and Gultekin (1981).

We have shown that earnings forecasts breadth and revisions enhance returns relative to using only historical, value-oriented data. Now we address the question of using equally weighted or regression-weighted composite models. The application of the Beaton-Tukey outlier-adjustment procedure to equation (9.1) during the 1987–1996 period in estimating equation (9.1), using the proprietary growth variable, produced the regression coefficients scaled to become the weights shown in Figures 9.1 through 9.7, where the value variable weights average approximately 65% during the period. The proprietary growth variable weight approaches 0.50 during the 1990–1994 period and averages 0.35, quite consistent with the Guerard (1990) and Miller, Guerard, and Takano (1992) estimations. The excess returns of where robust-weighted composite model are 635 basis points, exceeding the equally weighted model despite turnover of over 170%. The regression-weighted composite model has an average F-statistic of 28 and is statistically significant at the 5% level. The composite model expected return ranking procedure described earlier produces an average information coefficient of 0.093 for the 1982–1996 period (t=value of 6.5) and an average t-value of 4.14 for the 1987–1996 period. The composite model ICs are shown in

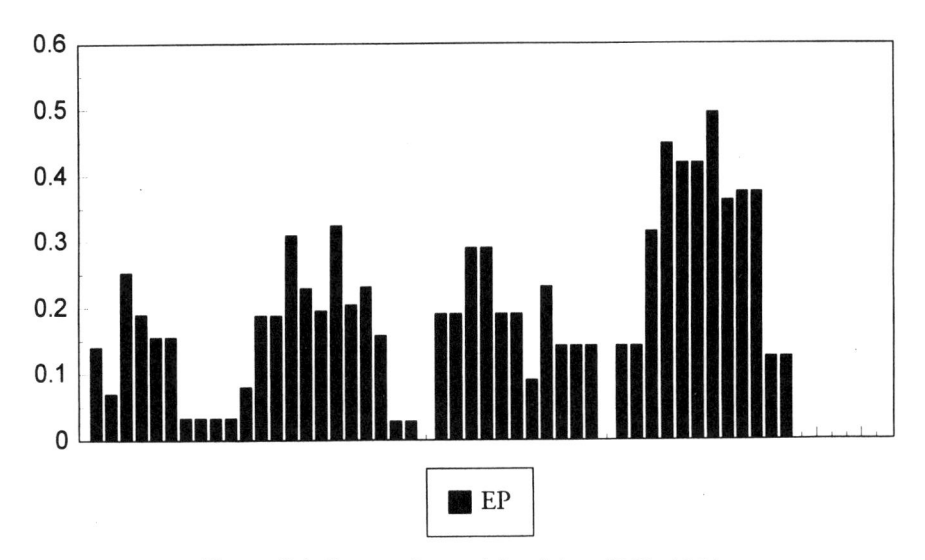

Figure 9.1 Composite model weights, 1982–1994.

236 THE (NOT SO SPECIAL) CASE OF SOCIAL INVESTING

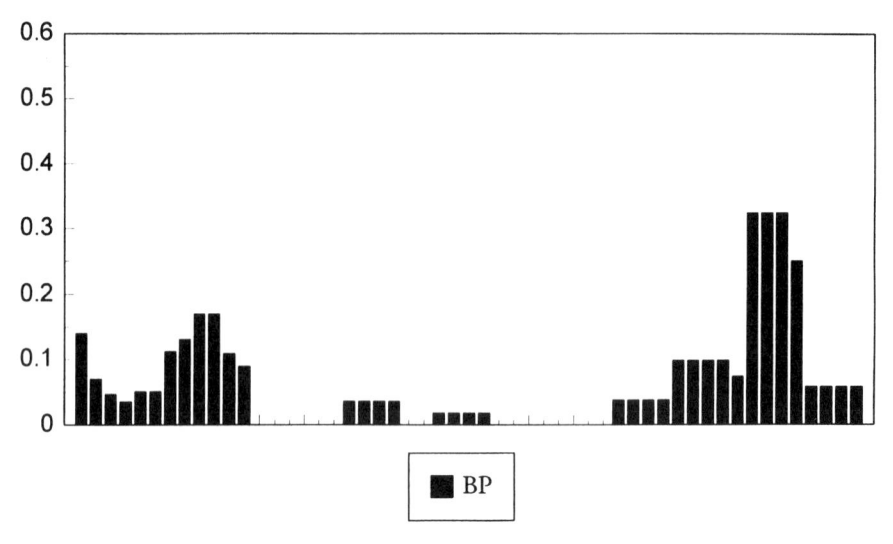

Figure 9.2 Composite model weights, 1982–1994.

Figure 9.8, as well as the upper and lower quintile returns, relative to the average universe stock return. The lower quintile (least preferred) securities consistently underperform the average stock return, and the upper quintile (most preferred) securities produce positive excess returns such that the quintile spread is positive and statistically significant. The information coef-

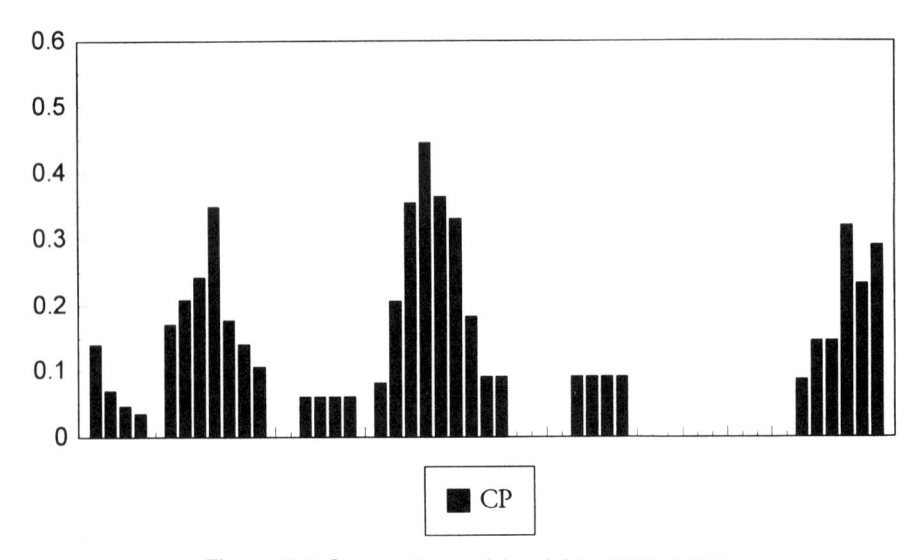

Figure 9.3 Composite model weights, 1982–1994.

THE (NOT SO SPECIAL) CASE OF SOCIAL INVESTING   **237**

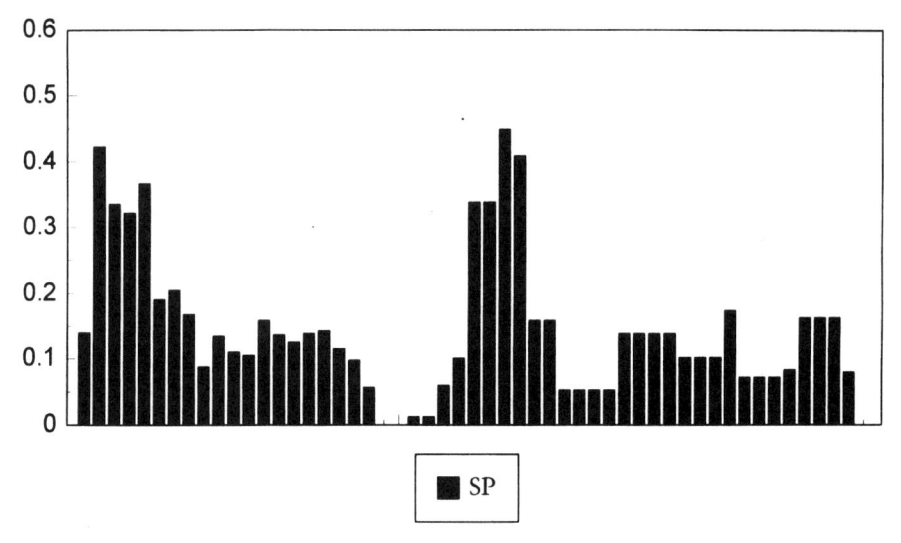

Figure 9.4 Composite model weights, 1982–1994.

ficient, measuring the association between the ranked composite model score and subsequent ranked total returns, indicates that the quantitative model is statistically significant in its ranking of securities. The IC is a standard tool used in accessing the predictive power of financial information (Farrell 1983). In a recent study, Guerard, Blin, and Bender (1996b) found

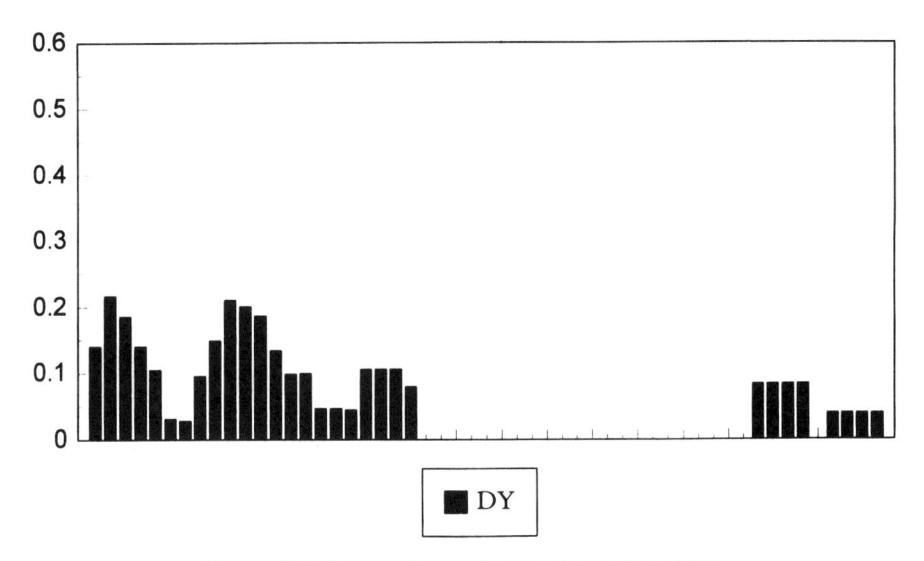

Figure 9.5 Composite model weights, 1982–1994.

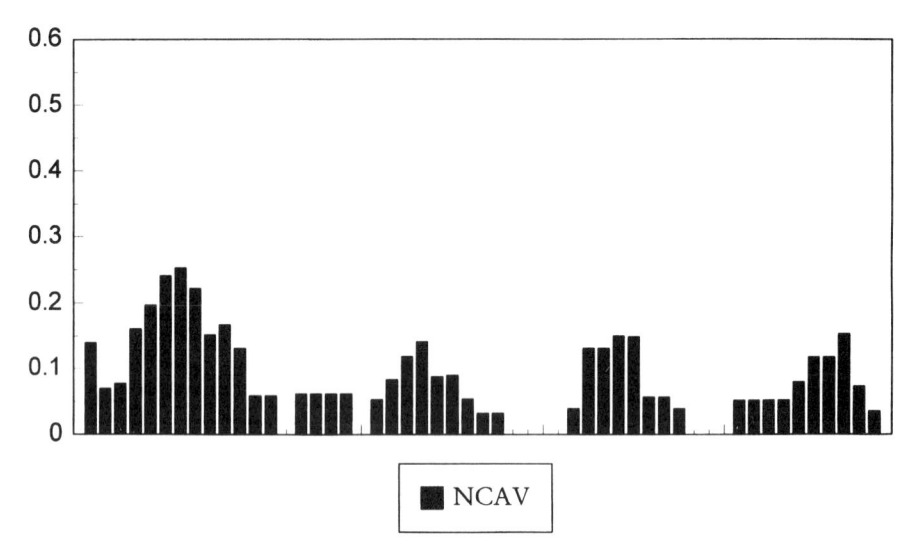

Figure 9.6 Composite model weights, 1982–1994.

that the estimated model from equation (9.1) outperformed the S&P 500 Index by 420 basis points annually during the 1987–1994 period, assuming a 3% upper bound on security weights, transactions costs of 80 basis points each way, and quarterly reoptimization. Guerard, Gultekin, and Stone (1996) found excess returns of approximately 412 basis points annually dur-

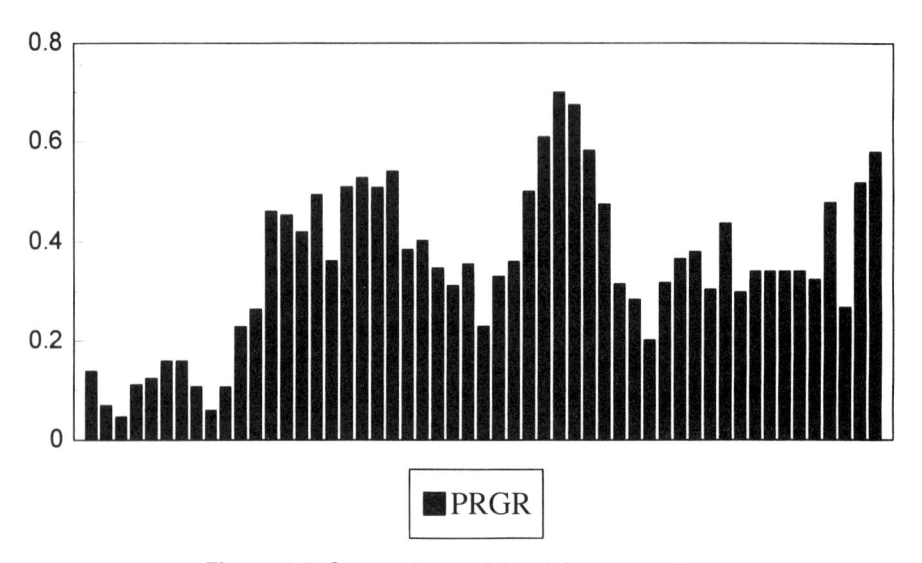

Figure 9.7 Composite model weights, 1982–1994.

Top 3000 Securities, 1982-94

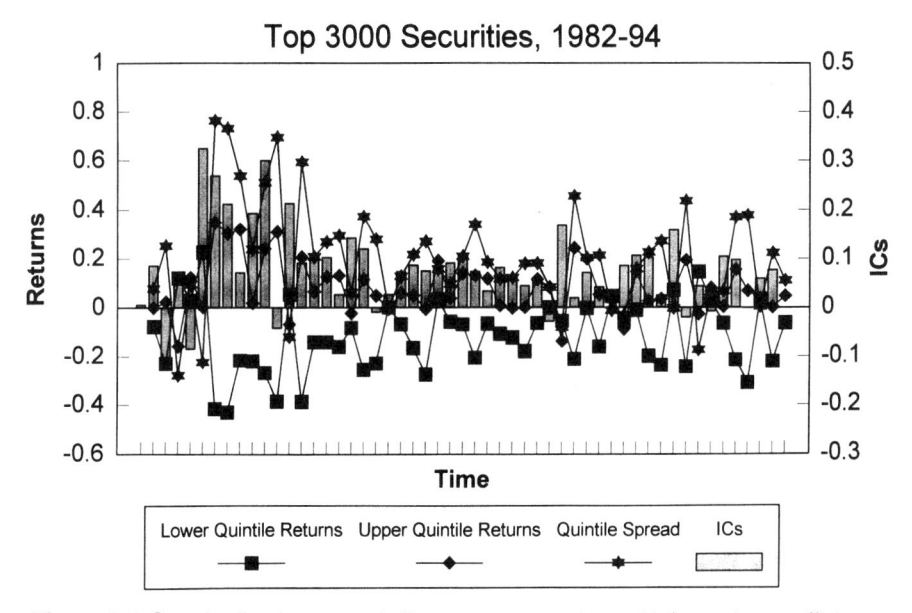

Figure 9.8 Standardized upper quintile returns, spreads, and information coefficients.

ing the 1982–1994 period using a variation on equation (9.1) that was discussed in Chapter 7.[10] The Guerard, Blin, and Bender (1996a,b) and Guerard, Gultekin, and Stone (1996) studies used unscreened investment universes.

The estimated expected return ranking model is used to create portfolios during the 1987–1995 period using a socially screened universe. The socially screened universe is created by subtracting the current KLD exclusions from a 1,200 large stock universe, resulting in a screened universe of approximately 950 stocks. A simulation is run for the January 1987–December 1995 period on the socially screened universe in which one tightly constrains industry and capitalization—weighting and a 100-basis-point transactions cost (round-trip) is accessed in the simulation. We find that the estimate composite model produces an average excess return of 743 basis points. Socially screened portfolios can outperform unscreened portfolio during the 1987–1995 period. One can invest in a socially screened portfolio and still outperform the S&P 500 socially screened benchmark.[12] It is interesting to see how the use of a socially screened universe creates a higher average weight of the proprietary growth variable in equation (9.1). The ICs of the composite model may be enhanced as one shifts from a more value-oriented weighting to a more growth-oriented weighting as one forecasts relative factor returns.[13]

STOCK SELECTION AND THE DOMINI SOCIAL INDEX SECURITIES

In this section of the study we specifically address the issue of stock with the 400 stocks of the Domini Social Index (DSI) during the June 1990–December 1994 period. If one applied the seven-factor robust regression-weighted composite model ranking to the 1,200 stock universe for the June 1990–December 1994 period and used the simulation conditions discussed in the preceding section, one would have outperformed the S&P 500 by 297 basis points instead of the 400 points of outperformance previously found. The portfolios turned over approximately 228% percent annually during the June 1990–December 1994 period. If one used an equally weighted portfolio rule in which one sold securities when the expected return ranking fell into the bottom half of the distribution, one could substantially slow down turnover and enhance performance. If one used a selling criteria of selling when the alpha fell below $-.7$, the seven-factor model would earn an excess return of 439 basis points with turnover of only 107%. If one used the same $-.7$ selling criteria and used the seven-factor model including only the 400 stocks of the DSI for the June 1990–December 1994 period, the excess returns would be 299 basis points and annualized portfolio turnover would be 79.3%. One can effectively pick stocks within the socially responsible DSI universe, and the excess returns of the 1,300 stock universe and DSI 400 will be virtually identical. Furthermore, if one wanted to be even more socially responsible and not invest in DSI stocks in which KLD has noted (with minor concerns) environmental and product concerns, one could create a portfolio strategy using the $-.7$ alpha sell-rule and outperform the S&P 500 by 323 basis points. The difference between the 323 basis points of outperformance of the environmental and products concerns portfolio, and the 299 basis outperformance of the DSI portfolios using our seven-factor model, represents the additional excess returns occurring with the implementation of two KLD screens. Clearly, more research needs to be undertaken with respect to the effective use of social screens in a socially responsible universe.

RECENT SOCIALLY RESPONSIBLE RESEARCH

Drhymes (1997) and Guerard (1997b) showed that the use of all KLD concerns and strengths should not significantly alter unscreened universe returns during the 1992–1997 period. Guerard found that only the military screen cost the investor returns during the period. A benefit to the investor is the product strength variable, which is positively associated with returns and reflects R&D leadership of the firm. Firms that engage in R&D can be recognized as being "good" socially responsible firms. If one eliminates securities that sepnd less than the median firm on R&D during the 1982–1996 period, the IC of the composite model rises from 0.093 to 0.117! R&D adds value.

SUMMARY AND CONCLUSIONS

The purpose of this study has been to show that there has been no statistically significant difference between the average returns of a socially screened and an unscreened universe during the 1987–1996 period. Socially conscious investing need not be a dumb idea, but one should be attentive when selecting a socially screened mutual fund or manager—performance can vary dramatically.

NOTES

1. The initial academic study finding that 17 socially responsible mutual funds established prior to 1985 outperformed—that is, underperformed less than, traditional mutual funds of similar risk for the 1986–1990 period—was that of Hamilton, Jo, and Statman (1993). The relative monthly outperformance of 7 basis points was not statistically different from zero. It is not obvious what criteria were used to determine the socially responsible universe in the Hamilton, Jo, and Statman study. Recent studies by J. D. Diltz, (1995a,b) found no statistically significant difference in returns for 28 stock portfolios generated from a universe of 159 securities during the 1989–1991 period. Diltz found that only the environmental and military business screens were statistically significant at the 5% level during the 1989–1991 period.
2. J. Rothchild, "Why I Invest with Sinners," *Fortune* (May 1996).
3. Morningstar, Principia for Mutual Funds, March 31, 1996.
4. The CREF Social Choice Account was a $1.174 billion account as of December 31, 1995, consisting of 61.49% socially screened equities, 37.67% bonds, and 1.72% short-term commercial paper. The CREF Social Choice Account uses environmental, weapons, nuclear power, alcohol, tobacco, and gambling products, and MacBride Principles (a code of fair employment by U.S. firms in Northern Ireland to prevent religious discrimination) screens. The CREF Social Choice Account has matched its performance benchmark for the past one- and five-year periods ending March 31, 1996, producing total returns of 23.56% and 12.70% versus its benchmarks of 24.00% and 12.32% respectively. The recent CREF Social Choice equity component is important because CREF under-

performed in its unscreened equity fund during the past one year. The CREF Bond Market account has been a market performer in its bond investment for the one- and five-year periods ending March 31, 1996, producing bond returns of 10.52% and 8.51% versus the Lehmann Aggregate Bond Yield Index of 10.79% and 8.49%, respectively. The CREF Stock Account earned one- and five-year returns of 28.81% and 13.55% versus the S&P returns of 32.10% and 14.66%, respectively. The CREF Social Choice Account has produced total returns consistent with its balanced performance benchmark and has not substantially underperformed on its equity component. The reader is referred to the College Retirement Equities Fund, *Prospectus,* Individual, Group, and Tax-Deferred Variable Annuities, April 1, 1995, for a description of the CREF Social Choice Account.

5. Fama and French (1995) actually tested whether higher book-to-price stocks outperformed the lower book-to-price stocks. It can be confusing when one thinks of the "low P/E" approach of Graham and Dodd (1962) in which an investor purchases low price-to-earnings stocks (i.e., one should not purchase a stock that has a price-earnings multiple exceeding 1.5 times the average price-earnings multiple of the market) and the higher earnings yield, or earnings-to-price (EP), approach tested in the academic literature. The two earnings formulations yield roughly the same result when applied to low-PE or high-EP decisions; see J. Guerard and M. Takano, "The Development of Mean-Variance Efficient Portfolios in Japan and the U.S.," *Journal of Investing* (fall 1992). Wall Street persons traditionally think of the low-PE and low-PB models, whereas academicians prefer the conventional EP and BP models because the conventional formulations are not plagued by small negative and positive denominators, such as with very small positive and negative earnings which can create very large positive and negative (often meaningless) PEs. See Graham and Dodd (1962) for long-run evidence supporting the low PE approach and their mixed thoughts on the price-to-book multiple.

6. The BARRA growth factor is a predictor of future growth of a company and is based on the five-year earnings-to-price ratio, historical earnings growth, recent earnings change, recent I/B/E/S change, the current earnings-to-price ratio, the I/B/E/S earnings-to-price ratio, and asset growth (BARRA, *U.S. Equity Beta Book,* January 1996).

7. Jacobs and Levy (1988) found substantial rewards for analysts' revisions and residual reversal. W. T. Ziemba (1992) found that last month's residual reversal, the one-year-ahead forecasted earnings per share growth rate, the two-year relative book value, and the low PE effect were the strongest variables in Japan and that small stocks outperformed large stocks in the United States and Japan, particularly in January.

8. The composite model-weighting scheme was advanced in Guerard (1990) and continues to produce statistically significant rankings. It is obvious that an infinite number of weighting schemes can be created; the four-period weighted re-

gression pattern produced significant real-time outperformance in the United States and Japan during the 1988–1994 period. See Miller, Guerard, and Takano (1992) and Guerard, Takano, and Yamane (1993).

9. The Beaton-Tukey biweight procedure was put forth in Beaton and Tukey (1974). The reader is referred to D. C. Montgomery and E. A. Peck, *Introduction to Linear Regression Analysis* (New York: John Wiley & Sons, 1982) for a very complete description of the outlier-adjustment process.

10. Guerard has experimented with several variations on equation (9.1) in his joint research with Blin, Bender, Gultekin, Stone, Takano, and Yamane. Let us briefly examine the average F-statistics and ICs of the various forms of equation (9.1) using the top 3,000 securities for the 1982–1994 period. In summary: (1) the BP variable has an average IC of 0.012 (*t*-value of 0.71), whereas the EP variable has an average IC of 0.039 (*t*-value of 2.10), which indicates that the low PE, or high EP, strategy worked well in identifying undervalued securities during the 1982–1994 period; (2) the use of relative variables; i.e., the relative EP (REP, the current EP divided by its five-year average of monthly ratios) increased the ICs of the four-fundamental-variable model (EP, BP, CP, SP) from 0.039 (*t*-value of 2.17) to 0.042 (*t*-value of 2.28); (3) the addition of the I/B/E/S FY1 forecast and breadth components further increased the IC in (2) to 0.072 (*t*-value of 3.82); (4) the use of equation (9.1) in this study produces an equally weighted IC of .058 (*t*-value of 3.15); and (5) the Beaton-Tukey robust regression estimation procedure increased the ICs to approximately 0.085, with little difference in the composite model ICs. Guerard initially used composite I/B/E/S revisions (CIR) and breadth (CIB) in lieu of the CIBF variable.

Model Variables	Average F-stat.	Average Reg.-wt. IC(*t*)	Average EQ.-wt. IC(*t*)
EP,BP,REP,RBP,CIR,CIB	18.87	0.066(4.82)	.067(3.52)
EP,BP,CP,SP,REP,RBP,RCP,RSP,CIBF	18.10	0.086(4.49)	.058(3.07)
EP,BP,CP,SP,DY,NCAV,CIBF	28.96	0.083(4.39)	.058(3.15)
EP,BP,CP,SP,DY,NCAV,REP,RBP,RCP, RSP,RDY,RNCAV,CIBF	15.82	0.081(4.28)	.068(3.59)

11. It is interesting to note that if one uses only the 1,300-stock universe less the socially screened stocks as the entire universe, reran the regression, and recalculated the expected returns, one finds an average F-statistic of 9.64 in the OLS analysis and 12.4 in the ROB estimations. The average IC of 0.078 is statistically significant, having an average *t*-value of 3.63. The use of a value-oriented model with the elimination of many smaller stocks does not diminish the IC; however, the weighting of the composite growth variable is approximately 0.40. If one equally weights the seven-factor model, the average IC is 0.027 with a *t*-value of 0.90; the ranking procedure is not statistically significant in the smaller, socially screened universe. One finds positive and statistically sig-

nificant ICs even using only a larger-capitalized, socially screened universe when one applies the Beaton-Tukey estimation procedure.

12. Vantage Global Advisors has been the advisor to a socially responsible fund, the Lincoln Life Social Awareness Fund in its Multi Fund Variable Annuity Family, which has produced a net return of 16.40% for the seven years ending March 13, 1996, whereas its socially responsible benchmark, the S&P 500 less its restrictions, has a generated corresponding return of 14.62%, respectively. Vantage has used a quantitative proprietary model emphasizing "growth at a reasonable price" (GARP) and will not invest in securities of firms that (1) engage in activities that damage the natural environment; (2) produce, design, or manufacture nuclear power or its equipment for the production of nuclear power; or (3) manufacture or contract for military weapons; or of (4) liquor, tobacco, and gambling industries. It is indeed possible to be a socially responsible manager and outperform the market. The seven-year returns are annualized. The performance figures include the reinvestment of dividends and other income. Past performance is not indicative of future results.

13. If one believes that BARRA value and growth factor returns can be forecast for the coming quarter using a Box-Jenkins (1976) time series model, then a random walk with drift formulation with a seasonal moving average operator can increase the CIBF weight when the BARRA growth factor return is expected to rise relative to the BARRA value factor return and can increase the predictive power of the model from an monthly IC of 0.052 (t-value of 1.66) to 0.063 (t-value of 1.99) during the 1987–1994 period.

REFERENCES

Arnott, R. "The Use and Misuse of Consensus Earnings." *Journal of Portfolio Management* (1985):18–27.

Beaton A. E., and J. W. Tukey. "The Fitting of Power Series, Meaning Polynomials, Illustrated on Bank-Spectroscopic Data." *Technometrics* 16 (1974): 147–185.

Belsley, D. A., A. E. Kuh, and R. E. Welsch. *Regression Diagnostics—Identifying Influential Data and Sources of Collinearity.* New York: John Wiley & Sons, 1980.

Blin, J., and G. Douglas. "Stock Returns vs. Factors." *Investment Management Review* (1987).

Blin, J., S. Bender, and J. B. Guerard, Jr. "Earnings and Value Investing—How to Make the Return Worth the Risk." Presented at the I/B/E/S Research Conference, Tokyo, June 1996.

Blin, J., S. Bender, and J. B. Guerard. "Why Your Next Japanese Portfolio Should Be Market-Neutral—and How." *Journal of Investing* (1995).

Bloch, M., J. Guerard, H. Markowitz, P. Todd, and G. Xu, 1993, "A Comparison of Some Aspects of the U.S. and Japanese Equity Markets." *Japan & the World Economy* 5, 3–26.

Blume, M. E., M. N. Gultekin, and N. B. Gultekin. "Validating Return-Generating Models" (under journal review).

Brown, L. D. "Earnings Forecasting Research: Its Implications for Capital Markets Research." *International Journal of Forecasting* 9 (1993):295–320.

Chopra, V. K., and W. T. Ziemba. "The Effects of Errors in Means, Variances, and Covariances on Optimal Choice." *The Journal of Portfolio Management* (1993):6–11.

Clemen, R. T. "Combining Forecasts: A Review and Annotated Bibliography." *International Journal of Forecasting* 5 (1989):559–584.

Cottle, S., R. F. Murray, and F. E. Block. *Graham and Dodd's Security Analysis*. 5th ed. New York: McGraw-Hill, 1988.

Cragg, J. G., and B. K. Malkiel. "The Consensus and Accuracy of Some Predictions of the Growth of Corporate Earnings." *Journal of Finance* 23 (1968):67–84.

Diltz, D. "Does Social Screening Affect Portfolio Performance." *Journal of Investing* spring (1995a):64–69.

Diltz, D. The Private Cost of Socially Responsible Investing." *Applied Financial Economics* (1995b):69–77.

Dhrymes, P. J. "Socially Responsible Investment: Is it Profitable?" Columbia University Working Paper, 1997.

Dhrymes, P. J., I. Friend, N. B. Gultekin, and M. N. Gultekin. "An Empirical Examination of the Implications of Arbitrage Pricing Theory." *Journal of Banking and Finance* 9 (1985):73–99.

Dimson, E., ed. *Stock Market Anomalies*. Cambridge, Cambridge University Press, 1988.

Elton, E. J., and M. J. Gruber. *Modern Portfolio Theory and Investment Analysis*. New York: John Wiley & Sons, 1987.

Elton, E. J., M. J. Gruber, and M. Gultekin. "Expectations and Share Prices." *Management Science* 27 (1981):975–987.

Fama, E. F., and K. R. French. "Size and Book-to-Market Factors in Earnings and Returns." *Journal of Finance* 50 (1995):131–155.

Fama, E. F., and K. R. French. "Cross-Sectional Variation in Expected Stock Returns." *Journal of Finance* 47 (1992):427–465.

Farrell, J. *Guide to Portfolio Management*. New York: McGraw-Hill, 1983.

Givoly, D., and J. Lakonishok. "The Information Content of Financial Analysts' Forecasts of Earnings: Some Evidence on Semi-Strong Inefficiency." *Journal of Accounting and Economics* 3 (1979):165–185.

Graham, B. *The Intelligent Investor.* New York: Harper & Row, 1940.

Graham, B., and D. Dodd. *Security Analysis*. 4th ed. New York: McGraw-Hill, 1962.

Guerard, J. B., Jr. "Linear Constraints, Robust-Weighting, and Efficient Composite Modeling." *Journal of Forecasting* 6 (1987):193–199.

Guerard, J. B., Jr. "Optimal Portfolio Analysis Using Composite Model-Tilting." Drexel Burnham Lambert, January 1990.

Guerard, John. "Is Socially Responsible Investing Too Costly?" *Pensions & Investments,* February 17, 1997a.

Guerard, J. B., Jr. "Is There a Cost to Being Socially Responsible in Investing." *Journal of Investing* (Summer 1997b).

Guerard, J. B., Jr. "Additional Evidence on the Costs of Being Socially Responsible in Investing." *Journal of Investing* (in press).

Guerard, J. B., Jr., and M. Takano. "Stock Selection and Composite Modeling in Japan." *Security Analysts Journal* 29 (1991):1–13.

Guerard, J. B., Jr., and B. K. Stone. "Composite Forecasting of Annual Corporate Earnings." In *Research in Finance* 10, edited by A. Chen. Greenwich, Conn.:" JAI Press, 1992, 205–230.

Guerard, J. B., Jr., J. Blin, and S. Bender. "Earnings Revisions in the Estimation of Efficient Market-Neutral Japanese Portfolios." *Journal of Investing* (Spring 1996a).

Guerard, J. B., Jr., J. Blin, and S. Bender. "The Riskiness of Global Portfolios." Presented at the Society of Quantitative Analysts, Annual Fuzzy Day Seminar, New York, May 1996b.

Guerard, J. B., Jr., M. Gultekin, and B. K. Stone. "Time Decay of Earnings Estimates." Presented at the Seminar on Corporate Earnings Analysis, New York, April 1996. In B. Bruce, ed., *Handbook of Corporate Earnings Analysis,* Volume 2 (in press).

Guerard, J. B., Jr., M. Takano, and Y. Yamane. "The Development of Efficient Portfolios in Japan with Particular Emphasis on Sales and Earnings Forecasting." *Annals of Operations Research* 45 (1993):91–108.

Guerard, J. B., Jr. and M. Takano, 1992, "The Development of Mean-Variance-Efficient Portfolios in Japan and the U.S., *Journal of Investing,* Fall.

Gunst, R. F., and R. L. Mason. *Regression Analysis and Its Application.* New York: Marcel Dekker, 1980.

Hamilton, S., H. Jo, and M. Statman. "Doing Well While Doing Good." *Financial Analysts Journal* (1993):62–66.

Hawkins, E., S. C. Chamberlain, and W. E. Daniel. "Earnings Expectations and Security Prices." *Financial Analysts Journal* 24–29, (1984):30–38.

Jacobs, B., and K. Levy. "Disentangling Equity Return Regularities: New Insights and Investment Opportunities." *Financial Analysts Journal* (1988):18–43.

Keon, E. "Earnings Expectations in Financial Theory and Investment Practice." In *I/B/E/S Research Bibliography,* 1996. New York 5th ed., edited by L. D. Brown.

Kinder, P. "Social Screening: Paradims Old and New." *Journal of Investing* (in press).

Kinder, P., S. D. Lydenberg, and A. L. Domini. *Making Money While Being Socially Responsible.* New York: HarperBusiness, 1993.

Kothari, S. P., J. Shanken, and R. G. Sloan. "Another Look at the Cross-Section of Expected Stock Returns." *Journal of Finance* 50 (1995):185–224.

Kurtz, L. "No Effect or No Net Effect." *Journal of Investing* (in press).

Kurtz, L., and D. DiBartolomeo. "Socially Screened Portfolios: An Attribution Analysis of Relative Performance." *Journal of Investing* (1996).

Latane, H., D. Tuttle, and C. Jones. *Security Analysis and Portfolio Management.* New York: Ronald, 1975.

Luck, C., and N. Pilotte. "Domini Social Index Performance." *Journal of Investing* (1993):60–62.

Markowitz, H. M. *Portfolio Selection: Efficient Diversification of Investments.* New York: John Wiley & Sons, 1959.

Miller, J. D., J. B. Guerard Jr., and M. Takano. "Bridging the Gap Between Theory and Practice in Equity Selection Modeling: Case Studies of U.S. and Japanese Models" (under journal review).

Niederhoffer, V., and P. J. Regan. "Earnings Changes, Analysts' Forecasts and Stock Prices." *Financial Analysts Journal* (1972):65–71.

Rudd, A., and B. Rosenberg. "Realistic Portfolio Optimization." In *Portfolio Theory: 25 Years Later,* edited by E. Elton and M. Gruber. Amsterdam: North-Holland, 1979.

Vu, J. D. "An Anomalous Evidence Regarding Market Efficiency: The Net Current Asset Rule." In *Research in Finance* 8, edited by A. Chen. JAI/Greenwich, CT 241–254.

Webster, J. T., R. F. Gunst, and R. L. Mason. "Latent Root Regression Analysis." *Technometrics* 16 (1974):513–522.

Wheeler, L. "Changes in Consensus Earnings Estimates and their Impact on Stock Returns." Presented at the Institute for Quantitative Research in Finance. In *The Handbook of Corporate Earnings Analysis,* edited by B. Bruce and A. Epstein. Chicago: Probus Publishing Co., 1990.

Ziemba, W. T. *Invest Japan.* Chicago: Probus Publishing Co., 1992.

Ziemba, W. T. "Fundamental Factors in U.S. and Japanese Stock Returns." Presented at the Berkeley Program in Finance, Santa Barbara, Calif., 1990.

10
SUMMARY AND CONCLUSIONS

In this book we have shown that corporate executives must be aware of the interdependencies of research and development, investment, dividend, and new debt expenditure decisions. One of its purposes was to analyze the determinants of corporate R&D expenditures in the United States during the 1975–1995 period and in the major industrized, G7, countries (United Kingdom, France, Germany, Canada, Italy, and Japan) during the 1982–1995 period. Our research began with a study of the interactions between the decisions concerning R&D, capital investment, dividends, and new debt financing of major industrial corporations. We found significant interdependencies in the firm's financial decision-making process. Even the presence of federal financing of a firm's R&D was insufficient to completely eliminate the potentially binding budget constraint it can experience. The initial research used a 303-firm universe of Compustat data for the 1975–1982 period. A corporate planning model was developed and estimated by the authors. We found significant correlations between stock prices and our targeted variables. In preparing the final draft of this book, we expanded our period of study in the United States for the 3,000 largest U.S. firms to include 1978 through 1995 and extended our modeling into Japan and Europe, finding interesting results. The expanded U.S. analysis supported our earlier results, and the European and Japanese results were generally supportive of the U.S. estimations; however, we are refining and expanding our European analysis.

250 SUMMARY AND CONCLUSIONS

Management attempts to manage dividends, capital expenditures, stock and debt repurchases, and R&D activities while minimizing reliance on net external funding to generate future profits. We extended the econometric model to analyze the interdependencies of the decisions on research and development, investment, dividends, and effective debt financing. The interdependence modeling results substantiated the earlier work of Dhrymes and Kurz (1964), Mueller (1967), Damon and Schramm (1972), McCabe (1979), Peterson and Benesh (1983), Jalilvand and Harris (1984), Switzer (1984), and Guerard and McCabe (1990). Higgins (1972), Fama (1974), and McDonald, Jacquillat, and Nussenbaum (1975) found little evidence of significant interdependencies of financial decisions.

Security valuation and portfolio construction constitute a major part of this book and are developed in Chapters 7, 8, and 9. In Chapter 7 we present our valuation analysis using historical fundamental data from Compustat and earnings forecasts from I/B/E/S. We found statistically significant stock selection models in the United States, Europe and Japan using both historical and earnings forecasting data. Chapter 8 extends the basic portfolio strategies introduced in Chapter 7 to include market-neutral portfolios, and we find a much greater use of earnings forecasts in the United States. Socially responsible investing is examined in Chapter 9, and we found no difference between socially screened and unscreened portfolios. One can be socially responsible and produce efficient portfolios using the models of Chapters 7 and 8. We find that valuation analysis should include both fundamental and earnings forecast data that are useful for selecting larger, smaller, and (preferably) middle-capitalized securities. The statistically based models correctly indentify securities to purchase and to sell short. Finally, we find evidence that investors should be socially responsible. It costs them nothing in return, relative to an unscreened portfolio, to invest with firms that are socially responsible. The use of the R&D variable enhances the composite model IC and adds value. The R&D variable is positively associated with the product strength criteria, and, thus, one can implement an R&D-intensive strategy and be well regarded in the socially responsible investment community, which recently passed the $1.0-trillion level. Much more research needs to be done to completely analyze the relationship between R&D and stock prices and returns.

REFERENCES

Damon, W. W., and R. Schramm. "A Simultaneous Decision Model for Production, Marketing and Finance." *Management Sci.* 18 (1972):161–172.

Dhrymes, P. J., and M. Kurz, On the Dividend Policy of Electric Utilities." *Rev. Economics and Statist.* 46 (1964):76–81.

Fama, E. F. "The Empirical Relationship Between the Dividend and Investment Decisions of Firms. " *Amer. Economic Rev.* 63 (1974):304–318.

Guerard, J. B., Jr., and G. M. McCabe. "The Integration of Research and Development Management into the Firm Decision Process." In *Management of R&D and Engineering,* edited by D. Kocaoglu. Amsterdam: North Holland, 1990.

Higgins, R. C., "The Corporate Dividend-Saving Decision." *J. Financial and Quantitative Anal.* 7(1972): 1527–1541.

Jalilvand, A., and R. S. Harris. "Corporate Behavior in Adjusting to Capital Structure and Dividend Targets: An Ecometric Study." *J. Finance 39* (1984): 127–145.

McCabe, G. M., "The Empirical Relationship Between Investment and Financing: A New Look." *J. Financial and Quantitative Anal.* 14 (1979): 119–135.

McDonald, J. G., B. Jacquillat, and M. Nussenbaum. "Dividend, Investment, and Financial Decisions: Empirical Evidence on French Firms." *J. Financial and Quantitative Anal.* 10 (1975):741–755.

Mueller, D. C. "The Firm Decision Process: An Econometric Investigation." *Quart. J. Economics* 81 (1967):58–87.

Peterson, P., and G. Benesh. "A Reexamination of the Empirical Relationship Between Investment and Financial Decisions." *J. Financial and Quantitative Anal.* 18 (1983):439–454.

Switzer, L. "The Determinants of Industrial R&D: A Funds Flow Simultaneous Equation Approach." *Rev. Economics and Statist.* 66 (1984):163–168.

Index

A

Andrews, S., 38, 97, 111
Applied research, funding patterns, 6, 9
Arbitrage pricing model theory, portfolio creation, 149–153, 208–210
Arnott, R., 218

B

Babiak, H., 40, 72, 99, 127
Basic research, funding patterns, 6, 8
Bean, A. S., 21, 38, 96, 97, 100, 111, 112, 113, 114
Beardsley, G., 96
Beaton, A. E., 141, 206, 233, 235
Belsley, D. A., 141
Bender, S., 153, 207, 210, 213, 220, 232, 236, 238, 239
Benesh, G., 2, 38, 39, 51, 99, 100, 127, 250

Ben-Zion, U., 96, 97, 112
Blin, J., 153, 207, 210, 213, 220, 232, 236, 238, 239
Blume, M. E., 149, 151
Brenner, M. S., 19
Brown, L. D., 218, 231
Burton, R. M., 114
Business segment data, ICI/CIMS study, 28–31

C

Carleton, W. T., 114
Census/NSF data *vs.* Compustat data, 95–121
    financial decision estimation results, 98–110
        expenditure data comparison, 101–104
        regression results comparison, 104–110
    future research, 116–117
    innovation and stockholder wealth, 96

Census/NSF data *vs.* Compustat data *(cont'd)*
  overview, 95–96
  prior research and current results, 111–112
  simultaneous equation approach extended, 112–115
  study model, 97–98
Chamberlain, S. C., 218, 232
Chan, L., 140, 143
Chen, N-F., 153
Composite model strategy simulation, portfolio creation, 153–198
Compustat data, *see* Census/NSF data *vs.* Compustat data
Cooper, R. G., 19, 20
Copeland, T. E., 95
Corporate laboratories, segment labs versus, ICI/CIMS study, 25, 27–28
Cragg, J. G., 231

D

Damon, W. W., 2, 38, 114, 250
Daniel, W. E., 218, 232
Development, funding patterns, 6, 10
Dhrymes, P. J., 2, 38, 39, 40, 41, 51, 97, 126, 127, 250
DiBartolomeo, D., 230, 231
Dimson, E., 206
Dividends:
  investment decisions and, 2, 38
  perfect markets hypothesis, 40, 72
Dodd, D., 140, 231, 234
Domini, A. L., 229, 230
Domini Social Index (DSI), social investing, 240
Douglas, G., 210
Downsizing, R&D contributions and, 19

E

Effective debt, 123–138
  data, 125
  model, 124–125
  overview, 123–124
  simultaneous equation results, 125–136
Elton, E. J., 219, 231, 232, 234

F

Fama, E. F., 2, 39, 40, 72, 99, 127, 140, 143, 250
Farrell, J., 236
Federal financing:
  decline in, 5, 6, 11, 17
  insufficiency of, 1
French, K. R., 140, 143

G

Germeraad, P., 19, 20
Givoly, D., 219
Grabowski, H. G., 40, 51, 99, 127
Graham, B., 140, 231, 234
Gruber, M. J., 219, 231, 232, 234
Grundy, K., 142
Guerard, J. B., Jr., 2, 38, 39, 40, 41, 50, 72, 96, 97, 98, 99, 100, 110, 111, 112, 113, 114, 124, 126, 127, 140, 141, 142, 143, 153, 206, 207, 208, 211, 213, 219, 221, 232, 233, 234, 235, 236, 237, 238, 239, 250
Gultekin, M. N., 140, 142, 149, 151, 220, 232, 234, 237, 238, 239
Gultekin, N. B., 149, 151

H

Hamao, Y., 140
Hamilton, W. F., 114

Harris, R. S., 2, 38, 39, 250
Hawkins, E., 219, 232
Higgins, R. C., 2, 39, 98, 250

I

ICI/CIMS, *see* Industrial Research Institute/Center for Innovation Management Studies (ICI/CIMS)
Imperfect markets hypothesis, statement of, 2, 38. *See also* Perfect markets hypothesis
Industrial Research Institute/Center for Innovation Management Studies (ICI/CIMS), 19–36
  data sources, 20–21
  new sales ratio declines, 19–20
  outcome data needs, 31, 36
  respondents, 21–23
  results compilation, 23–31
    business segment data, 28–31
    corporate versus segment labs, 25, 27–28
    firm data, 25, 26–27
Industrial Research Institute (IRI), 19
Interdependencies, *see* Perfect markets hypothesis
Investment decisions, dividends and, 2, 38

J

Jacobs, B., 140, 206
Jacquillat, B., 2, 39, 250
Jalivand, A., 2, 38, 39, 250
Japanese portfolios, *see* Portfolio creation
Jones, C., 206

K

Keon, E., 231, 232
Kinder, P., 229, 230

Kleinschmidt, E. J., 19
Kuh, A. E., 141
Kuh, E., 40
Kurtz, L., 230, 231, 250
Kurz, M., 2, 38, 39, 40, 41, 51, 97, 126, 127

L

Lakonishok, J., 140, 218
Latane, H., 206
Levy, K., 140, 206
Lintner, J., 40, 72, 99, 127
Luck, C., 230
Lydenberg, S. D., 229, 230

M

Malkiel, B. G., 142, 231
Mansfield, E., 40, 96, 116
Mapping, portfolio creation (market-neutral), 210–212
Market-neutral portfolio creation, *see* Portfolio creation (market-neutral)
McCabe, G. M., 2, 38, 39, 40, 41, 50, 51, 72, 96, 97, 98, 99, 100, 110, 112, 124, 127, 250
McDonald, J. G., 2, 39, 250
Meyer, J. R., 40
Miller, J. D., 235
Miller, M., 2, 38, 96
Modigliani, F., 2, 38, 40, 96
Morgan, A., 151
Morgan, I., 151
Moses, M. A., 114
Mueller, D. C., 2, 38, 39, 40, 51, 96, 99, 127, 250

N

National Science Foundation (NSF), 5, 6. *See also* Census/NSF data *vs.* Compustat data

New sales ratio, decrease in, 19–20
Niederhoffer, V., 231, 232, 234
Nussenbaum, M., 2, 39, 250

O

Obel, B., 114

P

Perfect markets hypothesis, 37–93, 95–96
  data, 41
  literature review, 38–39
  model, 39–41
  overview, 37–38
  simultaneous equation estimation results, 41–92
  statement of, 2
Performer base, R&D spending, 11, 12–16
Peterson, P., 2, 38, 39, 51, 99, 127, 250
Pilotte, N., 230
Portfolio creation (historical data and forecasts), 139–203
  arbitrage pricing model theory and estimation, 149–153
  composite model estimations, 142–149
  composite model strategy simulation, 153–198
  overview, 139–142
Portfolio creation (market-neutral), 205–225
  APT model, 208–210
  background, 205–208
  findings, 212–217
  mapping, 210–212
  results, 217–221
  value-growth decision timing, 220

R

Rapoport, J., 96
Regan, P. J., 231, 232, 234
Research & development spending, 5–18
  decline in, 5
  funding patterns, 6–11
  performer base, 11, 12–16
  sectoral differences, 11, 17
Roll, R., 153
Romeo, A., 96
Ross, S. A., 153, 208, 209

S

Scherer, F. M., 116
Schramm, R., 2, 38, 250
Screened universes, social investing, 231–239
Sectoral differences, R&D spending, 11, 17
Segment laboratories, corporate labs versus, ICI/CIMS study, 25, 27–28
Social investing, 227–247
  Domini Social Index (DSI), 240
  overview, 227–230
  screened and unscreened universes, 231–239
Stock prices, significance of, 1
Stock selection models, *see* Portfolio creation
Stone, B. K., 39, 96, 124, 126, 127, 140, 141, 142, 206, 219, 233, 234, 237, 238, 239
Switzer, L., 2, 38, 39, 40, 51, 72, 96, 112, 127, 250

T

Takano, M., 140, 141, 143, 206, 207, 208, 211, 213, 219, 221, 233, 235

Tinbergen, J., 40, 97
Tukey, J. W., 141, 206, 233, 235
Tuttle, D., 206

U

U.S. Department of Defense, 6
Universities, R&D funding by, 6
Unscreened universes, social investing, 231–239

V

Vu, J. D., 140, 231

W

Wagner, S., 96
Welsch, R. E., 141
Weston, J. F., 95
Wheeler, L., 218, 232, 234

Y

Yamane, Y., 140, 141, 206, 208, 211, 213, 221, 233

Z

Ziemba, W. T., 140, 206